Disaster and Risk Research: GADRI Book Series

Disaster and Risk Research: GADRI Book Series is published under the auspices of the Global Alliance of Disaster Research Institutes (GADRI). The global status of disaster research reflects the major strides made in the disaster sciences. These volumes present the forefront of disaster research, including key scientific findings, methodologies, policy recommendations and case studies. Whilst disaster risk needs to be managed in an integrated manner, persistently isolated applications of knowledge, practice and policy are falling short of ensuring disaster-resilient societies. Responding to this deficit calls for measurement, tools, techniques and institutional structures that can realistically support comprehensive risk assessment and management across multiple hazard landscape. As such, disaster research is now faced with a multi-disciplinary, multi-stakeholder challenge. Contributions to this series therefore address many varied and critical opportunities to advance the subject area. A cross-cutting vision shared across the Disaster and Risk Research volumes is to improve the future of scientific and technological guidance with clearly identifiable roadmaps to ensure human safety and security. The Global Alliance of Disaster Research Institutes was established in March 2015, directly after the United Nations World Conference on Disaster Risk Reduction (WCDRR 2015) in Sendai, Japan, based on the belief that a multi-institutional alliance can strengthen disaster research and its influences around the world. GADRI has a mandate to support the implementation of the Sendai Framework for Disaster Risk Reduction 2015-2030 and is a member of the Scientific and Technical Advisory Group (STAG) of the United Nations Office for Disaster Risk Reduction (UNDRR). In addition, GADRI provides a platform for scientific communities from different disciplines, backgrounds and countries, helping them share their knowledge, findings and views. This approach yields more holistic and farther-reaching insights, which can contribute to further steps in effective disaster risk management.

Hirokazu Tatano · Andrew Collins
Editors

Proceedings of the 4th Global Summit of Research Institutes for Disaster Risk Reduction

Increasing the Effectiveness and Relevance of Our Institutes

Springer

Editors
Hirokazu Tatano
Disaster Prevention Research Institute
Kyoto University
Uji, Kyoto, Japan

Andrew Collins
Department of Geography
and Environmental Sciences
Northumbria University
Newcastle upon Tyne, UK

ISSN 2524-5961 ISSN 2524-597X (electronic)
Disaster and Risk Research: GADRI Book Series
ISBN 978-981-19-5565-5 ISBN 978-981-19-5566-2 (eBook)
https://doi.org/10.1007/978-981-19-5566-2

This Springer imprint is published by the registered company Springer Nature Singapore Pte Ltd.
The registered company address is: 152 Beach Road, #21-01/04 Gateway East, Singapore 189721, Singapore

GADRI Foreword

UNISDR
United Nations Office for Disaster Risk Reduction

The 4th Global Summit of Research Institutes for Disaster Risk Reduction takes place just four years after the adoption of the Sendai Framework for Disaster Risk Reduction 2015–2030. This gathering is an important opportunity to take stock of the role that the advancement of science and technology plays in reducing disaster risk and losses. I firmly support the Summit's goal to contribute to the 2016 Science and Technology Roadmap to move forward the implementation of the Sendai Framework.

I strongly welcome the Summit's focus on how evidence-based research can be utilized to support the implementation of the Sendai Framework including the achievement of its seven targets, particularly those which seek to reduce mortality, reduce the numbers of people affected by disasters, and reduce both economic losses and damage to critical infrastructure.

Alarmingly, the four years that have passed since the adoption of the Sendai Framework have been the warmest years on record. We continue to see avoidable deaths and huge economic losses resulting from extreme weather events, as well as earthquakes, volcanic eruptions, and tsunamis.

Politicians and policy-makers need the advice and counsel of research institutes more than ever before, as we need to shift our focus from simply managing disasters to managing the underlying risk drivers which lead to those events.

It is largely thanks to the efforts of your institutes and the broader science and research community that we can assemble coherent arguments for essential prevention, mitigation, and adaptation actions: risk-informed urbanization including the enforcement of building codes, the safeguarding of protective ecosystems, and an inclusive approach to disaster risk reduction reaching out to all sections of society including the marginalized and the most vulnerable.

We are at a critical moment in the implementation of the Sendai Framework which is an important pillar of the 2030 Agenda for Sustainable Development and so vital to overall efforts to eradicate poverty and strengthen action on climate change.

It is particularly important that we see your research feeding into the national and local strategies for disaster risk reduction which need to be in place by 2020, as set out in the Sendai Framework.

I look forward to the outcome of the 4th Global Summit and to welcoming many of you to the Global Platform for Disaster Risk Reduction which takes place in Geneva, Switzerland, from 13 to 17 May. The Global Platform will provide many opportunities to discuss research findings and innovative approaches to reducing disaster risk, building resilience, and ensuring inclusivity.

I wish you a very successful Summit and look forward to strengthening our collaboration with the Global Alliance of Disaster Research Institutes in 2019 and beyond.

Mami Mizutori
Special Representative of the Secretary-General for Disaster Risk Reduction
United Nations Office for Disaster Risk Reduction (UNDRR), Geneva, Switzerland

Preface

GADRI
Global Alliance of Disaster Research Institutes

Greetings!

Welcome to the 4th Global Summit of Research Institutes for Disaster Risk Reduction (4th GSRIDRR, 2019): **Increasing the effectiveness and relevance of our institutes** held at the Disaster Prevention Research Institute (DPRI), Kyoto University, Uji Campus, Kyoto, Japan, from 13 to 15 March 2019. The 4th GSRIDRR, 2019, is sponsored by the Global Alliance of Disaster Research Institutes (GADRI) and Kyoto University. The Global Summit Series provides a platform for researchers, practitioners, policy-makers, and other stakeholders in both government and non-governmental institutes involved in disaster risk reduction and resilience to come together to discuss, share, and exchange ideas. This is also a way for the scientific community to contribute to the priority areas of the Sendai Framework for Disaster Risk Reduction Agenda 2015–2030.

The 4th Global Summit's programme comprises an impressive array of global stakeholders whose expertise and experience encompass the management of knowledge and its application for governments and industries, sharing outcomes and bridging science and decision-making. Through this process, we are confronting new scientific challenges in DRR and DRM.

The Global Summit Series through its institutional structure advocates the importance of engagement in collaborative research activities to accomplish disaster risk reduction and resilience at a global level. The Sendai Framework Agenda for 2030 calls for increased awareness of disaster risk reduction and resilience and implementation of policies to prevent new risks. GADRI commits to take a coherent and a global approach to disaster risk reduction and resilience through its network of disaster research institutes.

Disasters continue to increase in numbers and affect nations and communities by impact and intensity. Through the 4th Global Summit, GADRI will, in a challenging way, further promote and contribute to the 2030 agenda of the Sendai Framework and the 2016 Science and Technology Roadmap. This is assisted by a unique program, high-level dialogues between governmental, UN, and other leaders from the private sector and the science and technology community broadly defined. There are parallel thematic group discussion sessions, panel sessions to present synthesis reports from group discussion sessions, poster presentations, and opportunities for participants to interact and review contributions that progress disaster risk reduction and resilience activities.

Group discussions will, in particular, explore the following:

- GADRI Contributions to the 2016 Science & Technology Roadmap—How best could GADRI promote the SFDRR 2030 agenda?
- SDGs, climate change, and adaptation—What engagement mechanisms and research linkages are needed to influence research directions among policy-makers, governments, localities, media, and other groups?
- Nation's Synthesis: Concept Design and Implementation, and the SFDRR Agenda 2030—What systemic synthesis can be built?
- Research Funding—Where is the funding for disaster risk reduction activities coming from and is this invested to the right causes and the areas that are most needing support?
- Data Working Group—What are the real strengths and weaknesses of data and risk management?

Focusing on the above themes, the 4th Global Summit discussion sessions will explore the governing nature of empirical research on disaster risk reduction conducted by various research institutions nationally and globally, especially how evidence-based research results are utilized or implemented in mitigating disasters. Emphasis will be on integrated disaster research with a wider focus on collaborative research activities. Many research institutions conduct research for publication purposes and for fulfilling requirements for grants and their institutional targets. A shortcoming may be where those drivers may not be fit for purpose in terms of impacting on disaster risk reduction. Further, research is often needlessly duplicated, or at times its value may go unnoticed and simply archived.

Through our debates, we hope to achieve the following:

- Provide inputs to the 2016 Science & Technology Roadmap and submit the outcomes through the Global Platform 2019.
- Learn from the accumulated research knowledge of each research institute.
- Share information on ongoing project activities and achievements.
- Explore opportunities for collaborative and empirical research activities.

On behalf of all of us, we welcome you to the 4th Global Summit. We are excited to have all of you as part of our team and together we can reach new heights to reduce disaster risk and achieve disaster resilience.

We wish you a successful 4th Global Summit.

Prof. Hirokazu Tatano
Secretary-General, GADRI
Uji, Japan

Prof. Andrew Collins
Chair GADRI Board of Directors, Newcastle upon Tyne, UK

4th Global Summit of Research Institutes for Disaster Risk Reduction (4thGSRIDRR 2019): Increasing the Effectiveness and Relevance of our Institutes; and GADRI General Assembly, Obaku Plaza, DPRI, Kyoto University, Uji Campus, Kyoto, Japan
13 to 15 March 2019
MC: Tetsuya Takemi, Yuki Matsushi, Kazuyoshi Nishijima

Day 1 : Wednesday, 13 March 2019

9:15-10:00	Registration
10:00-11:00	Opening Ceremony Chair: Andrew Collins • Welcome Greetings, Hajime Nakagawa, Director, Disaster Prevention Research Institute (DPRI), Kyoto University, Japan • Welcome Address, Juichi Yamagiwa, President, Kyoto University, Japan • Welcome Greetings, Tadashi Yamamoto, Mayor, Uji City, Kyoto, Japan • 4th Global Summit Overview, Hirokazu Tatano, Secretary-General, GADRI; and Professor, DPRI, Kyoto University, Japan Group Photograph (A printed copy of the group photo will be available during the Welcome Reception for each participant to sign.)
11:00 - 11:30	Coffee Break - Hybrid Space
11:30 - 12:30	Plenary Session I Listening to Advice and Management of Science Knowledge by Various Stakeholders Chair: Kaoru Takara Keynote 1 Communication between Science and Technology, and Civil societies - Japan's experience regarding the Nankai Trough Earthquake, Setsuko Saya, Disaster Management Bureau, Cabinet Office, Government of Japan Keynote 2 UNESCO's Priorities on DRR Activities with Scientific Partners, Soichiro Yasukawa, Natural Sciences Sector, UNESCO, France Keynote 3 Partnerships for Mainstreaming DRM in Developing Countries, Masato Miyazaki, Special Representative Japan, The World Bank Tokyo Office, Japan Keynote 4 Nation's Synthesis - Concept Design and Implementation, Toshio Koike, Director, ICHARM, Japan
12:30 - 13:30	Lunch Break
13:30 - 14:30	Chair: Ana Maria Cruz GADRI Session - Introduction of Regional Alliances - Showcasing Collaboration and Research: Workflow and Opportunities • Science Council of Japan (SCJ) - Kaoru Takara • North American Alliance for Hazards and Disaster Research Institutes (The North American Alliance) - Jennifer Tobin on behalf of Lori Peek, Charles Scawthorn, and Paul Kovacs • African Alliance for Disaster Research Intitutes (AADRI) - Desmond Manatsa • UK Alliance for Disaster Research (UKADR) - Andrew Collins • 2nd World Bosai Forum, November 2019 - Yuichi Ono
14:30 - 15:30	Poster Session I (Hybrid Space) Coordinators: Masamitsu Onishi and Kenji Kawaike
15:30 - 16:00	Coffee break - Hybrid Space - Move to each seminar room for Group discussion sessions.
16:00 - 18:00	Group Discussion Session I : Contributions to Increase the Effectiveness and Relevance of Our Institutes in DRR (Participants are required to select a group discussion session.) I-A (SR 4&5): Water and Weather Related — Chair: Tetsuya Takemi; Co-chairs: Sameh Kantoush, Kenji Kawaike, Wei-Sen Li, Ortwin Renn, Paul Kovacs, Mahua Mukherjee I-B (S-207D): Earthquake and Volcano Disaster Related — Chair: Jim Mori; Co-chairs: Yoshiki Ikeda, Hiroshi Kawase, Khalid Mosalam, Soichi Nakajima I-C (S-143D): Geohazard Related — Chair: Yuki Matsushi; Co-chairs: Masahiro Chigira, Rodrigo Cienfuegos, Indrajit Pal I-D (S-519D): Social Dimension of Risk, and Health and DRM — Chair: Subhajyoti Samaddar, Andrew Collins; Co-chairs: Yuichi Ono; Chair: Virginia Murray (Health and DRM) I-E (SR 1&2): NaTech and its Cross-cutting Issues — Chair: Ana Maria Cruz; Co-chair: Elisabeth Krausmann
18:30 - 20:30	MC: Sameh Kantoush Welcome Reception at University Cafeteria, Uji Campus, Kyoto University Greetings: Greetings, Shuichi Yamauchi, Vice-Governor, Kyoto Prefecture, Kyoto, Japan

Day 2 : Thursday, 14 March 2019

9:15 - 10:15	Plenary Session II Fostering Uptake of Science in Governments and Industries Chair: Hirokazu Tatano and Ortwin Renn Keynote 5 The Role of Science and Technology to Enhance Risk-informed Development for a Resilient and Sustainable Future, Shoko Arakaki, Partnerships, Inter-governmental Process and Inter-agency Cooperation, UNISDR, Switzerland Keynote 6 Navigating the European Landscape of DRR – The Case of Data and Socioeconomic Factors, Jörgen Sparf, Mid Sweden University, Sweden Keynote 7 The Sendai Framework and Science Advice to Governments to Support Implementation – Suggestions on What Works, Virginia Murray, Global Disaster Risk Reduction, Public Health England, UK Keynote 8 Global Risk Assessment Framework (GRAF), UNISDR: Emphasising the Relevance of Disaster Research Institutes – Andrew Collins, DDN, University of Northumbria
10:15 - 10:45	Coffee break - Hybrid Space
10:45 - 12:30	Panel Discussion I - Discussion Summaries and Outcomes from Day 1 Group Discussion Sessions - I Chair: Srikantha Herath and Yuichi Ono Invited speakers from each group (each member to summarize group discussion session outcomes within 5-8 minutes) • Presentations by each group member • Comments by the panelists • Discussion/Comments by the audience
12:30-13:30	Lunch Break
13:30 - 14:30	Plenary Session III Bridging Science and Decision Making through Sharing of Knowledge Chair: Tomoharu Hori and Paul Kovacs Keynote 9 Perspectives on Climate Change Impact Assessment and Adaptation, Eiichi Nakakita, DPRI, Kyoto University, Japan Keynote 10 Integrated Disaster Data Management: the Experience of the DesignSafe Cyberinfrastructure, Jean-Paul Pinelli, Florida Institute of Technology (FIT), USA Keynote 11 Disaster Risk Management Knowledge Centre: A Collaborative Approach to Foster Resilience, Elisabeth Krausmann, European Commission-Joint Research Centre (EC-JRC), Italy Keynote 12 Infrastructure Resilience: A Framework for Assessment, Governance and Management, Craig Davis, American Society of Civil Engineering (ASCE), USA
14:30 - 15:30	Poster Session II (Hybrid Space) Coordinators: Masamitsu Onishi and Kenji Kawaike
15:30 - 16:00	Coffee break - Hybrid Space - Move to each seminar room for Group discussion sessions.
16:00 - 18:00	Group Discussion II: Effectiveness and Relevance of Disaster Research Institutes in DRR - Contributions to World Stakeholders (Participants are required to select a group discussion session.) II-A (S-519D): GADRI Contributions to the 2016 Science & Technology Roadmap; and how best could GADRI promote SFDRR 2030 agenda? — Chair: Hirokazu Tatano, Shoko Arakaki; Co-chairs: Ortwin Renn, Paul Kovacs, Irwan Meilano, Annisa Triyanti II-B (SR 4&5): SDGs, climate change and adaptation - what engagement mechanisms and research linkages are needed to influence research directions among policy makers, governments, localities, media and other groups? — Chair: Tetsuya Takemi, Eiichi Nakakita, Srikantha Herath; Co-chairs: Yuichi Ono, Rajib Shaw, Soichi Nakajima, Kazuyoshi Nishijima II-C (SR 1&2): Nation's Synthesis: Concept Design and Implementation, and the SFDRR Agenda 2030 – what systemic synthesis can be built? — Chair: Toshio Koike, Elisabeth Krausmann; Co-chairs: Indrajit Pal, Rodrigo Cienfuegos II-D (S-143D): Research Funding - where is the funding for disaster risk reduction activities coming from and is this invested to the right causes and the areas that are most needing support? — Chair: Kaoru Takara, Takahiro Sayama, Subhajyoti Samaddar; Co-chairs: Sameh Kantoush, Charles Scawthorn, Mahua Mukherjee, Mohsen Ashtiany II-E (S-207D): Data Working Group (DWG)-Action Data in the Implementation of Disaster Risk Reduction - what are the real strengths and weaknesses of data and risk management? — Chair: Andrew Collins; Co-chairs: Jörgen Sparf, Jean-Paul Pinelli
18:30	Side Event: DPRI Alumni Meeting and Dinner Chair: Jim Mori and Hirohiko Ishikawa

Day 3 : Friday, 15 March 2019

9:15 - 10:15	Plenary Session IV New Scientific Challenges to DRR and DRM Chair: Norio Maki and Mahua Mukherjee Keynote 13 Scientific Challenges for Disaster and Risk Governance, Ortwin Renn, Scientific Director, Institute for Advanced Sustainability Studies (IASS), Germany Keynote 14 Closing Gaps Between Scientists and Decision Makers by Creating Demand for Safety as a Basic Human Right, Mohsen Ghafory-Ashtiany, International Institute of Earthquake Engineering and Seismology (IIEES), I. R. Iran Keynote 15 Taming New Scientific Challenges to Disaster Risk Reduction (DRR) and Disaster Risk Management (DRM) in Africa, Desmond Manatsa, Associate Professor, Geography Department, Bindura University of Science, Zimbabwe Keynote 16 Advances and potentials of science and technology in implementation of Sendai Framewor, Rajib Shaw, Professor, Keio University, Japan; and Chair, UNISDR Science and Technology Advisory Group (STAG)
10:15 - 10:30	Coffee Break - Hybrid Space
10:30 - 12:00	Panel Discussion II - Discussion summaries and outcomes from Day 2 Group discussion Sessions II Chair: Shoko Arakaki and Charles Scawthorn Invited speakers from each group (each member to summarize group discussion session outcomes within 5-8 minutes) • Presentations by each group member • Comments by the panelists • Discussion/Comments by the audience
12:00 - 13:30	Lunch Break
13:30 - 14:50	Wrap-up Session Chair. Hiroshi Kawase - Resolution / report to be presented at the Global Platform for DRR, Geneva, 13 to 17 May 2019 - Jim Mori, Andrew Collins and Hirokazu Tatano - Adoption of the Final Outcomes
14:50 - 15:00	Closing Remarks • Manabu Hashimoto, Director-Elect, DPRI, Kyoto University End of the 4th Global Summit of Research Institutes for Disaster Risk Reduction We thank all of you for your participation and much valued contributions!
15:00 - 16:30	Transit to Kyoto Station - Hotel Granvia-3F (Guided by Takemi, Matsushi, Nishijima and students) (Recommending participants to take the JR Nara Line train from Obaku Station to Kyoto Station. Train schedule: 15:33, 15:42; 16:03, 16:12, and 16:33)
	GADRI Meeting of the GADRI General Assembly
16:30-17:00	Registration - Meeting of the GADRI General Assembly Hotel Granvia Kyoto Station, 3F, Kyoto (Limited to GADRI member institutes. We expect, at least, one member from each GADRI participating institutes to attend this session.)
17:00 - 18:00	Chair: Andrew Collins and Hirokazu Tatano • Introduction of the GADRI Board of Directors and the Advisory Board • Disaster and Risk Research: GADRI Book Series • Other GADRI Publications • Venue for the 5th Global Summit of Research Institutes for DRR in March 2020
18:00 - 20:30	MC: Subhajyoti Samaddar Cultural Performance & Closing Banquet - Granvia Hotel-3F, Kyoto Greetings: Kayo Inaba, Executive Vice-President, Kyoto University, Japan Daisaku Kadokawa, Mayor, Kyoto City, Japan

Acknowledgements

The Global Alliance of Disaster Research Institutes (GADRI) conveys its deep appreciation and gratitude to the Kyoto University, Japan, and the Disaster Prevention Research Institute (DPRI), Kyoto University, for hosting the 4th Global Summit of Research Institutes for Disaster Risk Reduction 2019 at the DPRI, Uji, Kyoto, Japan, and for their generous financial and logistical support rendered towards the success of the 4th Global Summit and the related events. Furthermore, a deep appreciation is expressed to the Kyoto University and the DPRI, Kyoto University, for hosting the GADRI secretariat at the DPRI, Kyoto University, without whose financial and administrative support GADRI could not have executed its targeted objectives and activities. GADRI also thank the Prefecture of Kyoto and the City of Uji for their in-kind contributions to the 4th Global Summit of GADRI.

GADRI also thank all keynote speakers and panellists for taking time off from their busy schedules to deliver keynote addresses and panel presentations during the 4th Global Summit. GADRI acknowledge, with thanks, the support and courtesy rendered by the United Nations Office for Disaster Risk Reduction (UNDRR).

GADRI also express its deep appreciation and gratitude to its members of the Board of Directors and the Advisory Board for their continued contributions and support for GADRI activities. GADRI thank all of its members for their participation and contributions towards the success of the 4th Global Summit of Research Institutes for Disaster Risk Reduction 2019 and also for continued contributions to the targets of the S&T Roadmap for the implementation of the goals and priorities of the Sendai Framework.

Wilma James
james.wilmatheonesta.6r@kyoto-u.ac.jp

Messages from the Opening Ceremony

Greetings from the President of the Kyoto University

Your Excellencies, Distinguished Guests, Colleagues, Ladies, and Gentlemen!

It is my great pleasure and honour to address all of you today at the 4th Global Summit of Research Institutes for Disaster Risk Reduction (4th GSRIDRR), and I heartily welcome all of you to Kyoto University, Disaster Prevention Research Institute (DPRI), Uji Campus!

Messages from the Opening Ceremony—(L-R—Interpreter, Mayor of Uji City, Prof. Yamagiwa, Prof. Collins, Prof. Tatano, and Prof. Nakagawa)

We are very proud to host the Global Alliance of Disaster Research Institutes (GADRI) at DPRI, Kyoto University, Uji Campus. In March 2015, GADRI was established to support the initiatives of the Third UN World Conference on Disaster Risk Reduction held in Sendai, Japan, and also to support the implementation of the Sendai Framework for Disaster Risk Reduction 2015–2030 (SFDRR) and the work of the Scientific and Technical Advisory Group of the United Nations Office for Disaster Risk Reduction (UNISDR).

Each year, we are witnesses and/or victims to an increasing number of disasters that occur in various forms and levels which hamper through losses in lives and economy. We see in statistics staggering numbers impacted by these disasters. Yet we fail to draw the attention and investments needed for disaster risk reduction within the political realm and by other key decision-makers. We have not reached the desired levels to collectively and coordinately find solutions and to bring risk-informed decision-making in our governments and communities.

GADRI is strategically moving forward as a relevant platform for bringing these players together, to launch, and find creative and innovative solutions. In accordance with its vision, GADRI are enhancing their efforts, capacities, and capabilities to share knowledge and available research resources with the ultimate goal of tackling complex global problems and contributing to disaster risk reduction and resilience.

The goals set by GADRI and the Sendai Framework Agenda for 2030 are not different from my vision for Kyoto University. We are geared towards providing not only a competitive environment, but a place that provides stimuli to foster different abilities and ideas beyond the students' own fields, enjoy dialogue, and construct cooperative and lifelong relationships. We should open windows to a world in which our students can flourish and guide them gently out into that world. To further cement this idea, I formulated the acronym "WINDOW", and it was launched in 2015. The WINDOW Concept—a vision for the future to internationalize and localize—is a set of guidelines and themes to highlight endeavours and resources of the Kyoto University. We have made significant progress under each of these six themes.

While we at Kyoto University endeavour to foster motivation, creativity, and nurture researchers capable of assuming leadership roles in society, we are committed in supporting GADRI in its firm stance to support the Sendai Framework Agenda 2030, the 2016 Science and Technology Roadmap, and the SDGs.

GADRI have taken a pivotal role to enhance their efforts to strengthen the agenda on disaster risk reduction and resilience, to find implementable solutions for existing disasters, and to come up with solutions for communities devastated by disasters to resist future occurrences.

By contribution to GADRI's resource enhancement and working together to accomplish its goals, Kyoto University will pursue the highest global standards and cultivate in its mission support and facilitation across the broad structure of education and research to pioneer in previously unexplored domains of knowledge.

I firmly believe that through the activities of the 4th Global Summit of Research Institutes for Disaster Risk Reduction (4th GSRIDRR, 2019), all of you will accomplish the goals set by 4th GSRIDRR to intensify its contributions to the global society to attain disaster risk reduction and resilience. I look forward to working together to develop and expand collaborative activities and transfer our message for implementation to policy-makers.

I wish you all the very best in your discussions and wish you a successful 4th Global Summit!

Prof. Juichi Yamagiwa
President, Kyoto University

Greetings by Mr. Shuichi Yamauchi, Vice Governor of Kyoto Prefecture

I would like to offer my heartfelt best wishes for the opening of the 4th Global Summit of Research Institutes for Disaster Risk Reduction here in Kyoto. I am delighted to see so many people, including Chair of the Board of GADRI Prof. Andrew Collins, come to Kyoto for this occasion, and I would like to extend my very warmest welcome to you all.

The Global Alliance of Disaster Research Institutes (GADRI) organizes numerous international symposiums, through which it shares information, knowledge, experiences, and ideas with institutions across the globe working on disaster research and prevention. I also understand that since its establishment in 2015, the Global Alliance has been making efforts to contribute to the promotion of disaster risk reduction and improved disaster resilience from an academic perspective.

Given the fact that large-scale disasters occur frequently around the world, I strongly believe that GADRI activities are greatly contributing to the development of the disaster countermeasures eagerly sought by all nations. As such, once again, I would like to express my deepest respect to all members of the Alliance for their great efforts.

Kyoto Prefecture also suffered from massive damage caused by earthquakes, typhoons, and floods in recent years. We are working on verifying existing and future disaster countermeasures, taking various advice from the Disaster Prevention Research Institute (DPRI), Kyoto University. We are resolved to capitalize on the achievements brought about by advanced research on disaster countermeasures.

Through this Summit, I hope that your research will greatly contribute to the further promotion of global disaster prevention measures.

In closing, I sincerely hope that this Summit will be a fruitful one. Thank you.

Shuichi Yamauchi
Vice Governor, Kyoto Prefecture

Professor Collins and Vice Mayor of Kyoto City Shuichi Yamauchi at the Opening Reception

Greetings by Mr. Tadashi Yamamoto, Mayor, Uji City

Mr. Tadashi Yamamoto, Mayor, Uji City, stressed that there is an increase in natural disasters that struck various parts of Japan and globe more than in the past, for example, the Osaka earthquake with the seismic centre in the northern Osaka regions, the Jebi typhoon no. 21 in Japan, and various other disasters that struck abroad, hurricanes hit the USA, and tsunami struck in Indonesia.

He stated that so far this year, two earthquakes have hit Kumamoto and Hokkaidao. It has been predicted that there will be a magnitude of 8–9 level earthquake along the Nankai Trough with a 70%–80% probability within the next 30 years, and it is necessary to remind oneself of the stark reality of an earthquake happening anytime in future.

The Uji City strongly promotes measures to minimize the damage and further cooperate with related organization to help once disasters occur. In this regard, in 2012 with the heavy rainfall which affected the southern part of Kyoto Prefecture, Kyoto University's Disaster Prevention Research Institute (DPRI) lent their support from the beginning and throughout the whole ordeal. They offered expert advice and cooperated with us to reconsider the regional disaster prevention plan. In addition, they sent instructors to train voluntary leaders for seminars on disaster prevention and contributed to the development of regional disaster prevention development plans.

On this Summit, I believe, sharing the research results and cross-cultural cooperation from all over the world may continue to deepen the disaster risk research. I expect the results will continue to be used to support disaster areas and propel administrative disaster prevention policies for better methodologies.

In closing, I wish GADRI prosperity and success in the 4th Global Summit of Research Institutes for Disaster Risk Reduction. Congratulations. Thank you for your attention.

Mayor of Uji City Mr. Tadashi Yamamoto (C0 is greeted by Prof. Manabu Hashimoto (R) and Prof. Hajime Nakagawa (L)

Welcome Greetings by the Director of the Disaster Prevention Research Institute (DPRI), Kyoto University

Welcome to DPRI, Kyoto University Uji Campus, and welcome to Kyoto, Japan!

Distinguished guests, ladies and gentlemen!

We are very proud to host the 4th Global Summit of Research Institutes for Disaster Risk Reduction (4th GSRIDRR, 2019) at our Disaster Prevention Research Institute (DPRI), Kyoto University Uji Campus.

As you may know, the Global Summit series was initiated by DPRI in 2011 where the participants proposed to form an international network of disaster research fostered by DPRI. This proposal was further endorsed by the participants of the 2nd Global Summit, and the Global Alliance of Disaster Research Institutes (GADRI) was established. GADRI successfully organized the 3rd Global Summit in 2017, and this year, together with the Kyoto University, GADRI organized the 4th Global Summit which all of you are attending. GADRI brings together research across a wide variety of fields that address disasters.

Today GADRI is a unique and foremost Alliance of the international community working in this area.

Each year, disasters around the world are increasing in complexity. Climate change-related disasters are drawing more attention to existing threats to humanity including universally impacting phenomena such as rising global temperatures and melting glaciers. These are adding to ongoing unresolved problems requiring our immediate attention and proposals for resolutions and action on resilience and disaster reduction. Each situation produces fresh demand for dialogue to increase the effectiveness of our research to create the crucial guidelines of the future.

Together, we can confront these challenges through our outstanding research, insightful analysis, and evidence-based results. We can draw the attention of policy-makers and other stakeholders to take necessary steps forward to combat disaster risks and climate change and advance boldly to greater disaster resilience.

Through the impressive array of keynote speeches and dialogues across diverse fields, I am positive that at the end of the 4th GSRIDRR, we will see a further uptake of world-changing messages that can be delivered to the international community, governments, and industry.

I wish you all every success in the 4th Global Summit.

Prof. Hajime Nakagawa
Director, Disaster Prevention Research Institute (DPRI)
Kyoto University

Contents

Part I
Report of Activity of the 4th Global Summit of GADRI

Chapter 1
Report on the 4th Global Summit of Research Institutes for Disaster Risk Reduction (4thGSRIDRR2019)

Increasing the Effectiveness and Relevance of our Institutes, Disaster Prevention Research Institute (DPRI), Kyoto University, Uji Campus Kyoto, Japan from 13 to 15 March 2019

Hirokazu Tatano, Andrew Collins, and Wilma James

Abstract The 4th Global Summit of Research Institutes for Disaster Risk Reduction (4thGSRIDRR2019) under the theme of Increasing the Effectiveness and Relevance of our Institutes was organized by the Global Alliance of Disaster Research Institutes (GADRI) and sponsored by the Kyoto University and the Disaster Prevention Research Institute (DPRI), Kyoto University. The summit was held at the DPRI, Uji Campus, Kyoto, Japan, from 13 to 15th of March 2019.

This is a revised version of the document published at GADRI newsletter—GADRI ACTIONS, April 2019—http://gadri.net/resources/pdf/GA_Newsletter_April2019_final_rev2.pdf.

4th Global Summit of Research Institutes for Disaster Risk Reduction (4GSRIDRR)—https://www.dpri.kyoto-u.ac.jp/news_en/11487/.

H. Tatano
Disaster Prevention Research Institute (DPRI), Kyoto University, Uji-shi, Gokasho, Kyoto 611-0011, Japan

A. Collins
Disaster and Development Network, Northumbria University, Newcastle Upon Tyne, UK

W. James (✉)
DPRI, GADRI, Kyoto University, Uji-shi, Gokasho, Kyoto 611-0011, Japan
e-mail: james.wilmatheonesta.6r@kyoto-u.ac.jp

H. Tatano and A. Collins (eds.), *Proceedings of the 4th Global Summit of Research Institutes for Disaster Risk Reduction*, Disaster and Risk Research: GADRI Book Series,
https://doi.org/10.1007/978-981-19-5566-2_1

Group photo—4th Global Summit of Research Institutes for Disaster Risk Reduction

Focusing on GADRI's vision to find implementable solutions for existing disasters and to come up with solutions for communities devastated by disasters to withstand future occurrences, GADRI have taken a pivotal role to increase awareness in DRR and strengthen its contribution to the Sendai Framework Agenda.

The summit provided a platform for researchers, practitioners, policy-makers, and other stakeholders in both government and non-governmental institutes involved in disaster risk reduction and resilience to come together to discuss, share, and exchange ideas to contribute to the priority areas of the Sendai Framework for Disaster Risk Reduction.

The 4th Global Summit of GADRI, in particular, aimed to achieve the following three targets:

1. Discussions in critical challenges facing disaster prevention and contributions to develop the Science and Technology Roadmap.
2. Sharing information on the activities and knowledge in disaster risk reduction research around the world and subsequent contributions to "Sendai Framework for Disaster Risk Reduction" to achieve its targets.
3. Sharing information on research results and achievements and explore current disaster research status in each of GADRI member institute.

Point one achieved its goal highlighted in the resolutions which summarized inputs from panel sessions and the group discussions. These results were further shared at the Global Platform 2019 held in May 2019 in Geneva, Switzerland. Point two was achieved through the sixteen keynote speeches, three panel discussions, and various suggestions received during the group discussions. In addition, these discussions aimed the target areas of the Sendai Framework for disaster risk reduction and the work of research institutes around the world. Point three specifically addressed GADRI regional alliances, i.e. North American Alliance of Hazard and Disaster Research Institutes (NAAHDRI), African Alliance of Disaster Research Institute (AADRI), and UK Alliance of Disaster Research (UKADR) and member institutes and shared information on activities and information on results through poster presentations.

The keynote speakers were chosen from various fields and areas of specializations among which were the Government of Japan, The World Bank, UNISDR, ICHARM, Public Health England, Northumbria University Newcastle, and others for their specialties and contributions to the work of the Sendai Framework Agenda and DRR. In addition, the conference included two group discussion sessions consisting of 9 smaller sessions covering various topics, two panel sessions to discuss the results of the group discussion sessions and attracted 52 poster presenters. Among the final outcomes of the conference were the contributions to the contextualization of the Science and Technology Roadmap which were presented at the Global Platform for Disaster Risk Reduction held in Geneva in May 2019. The summit was attended by over 246 participants from 33 states.

1.1 Plenary I: Listening to Advice and Management of Science Knowledge by Various Stakeholders

Chaired by Prof. Kaoru Takara, Dean, Chair: Prof. Kaoru Takara, Dean, Graduate School of Advanced Integrated Studies in Human Survivability (Shishu-kan), Kyoto University, Japan.

This session was a very significant and concentrated session by various actors on the current and existing activities on DRR and their significant contributions to society to combat disaster and prevent future disasters. Madam Saya provided a very instructive and informative explanation of the local disaster management in Japan. Mr. Yasakawa provided details introduction of UNESCO and its direct contributions to DRR activities. Mr. Miyazaki of the World Bank introduced activities, relationships with various universities and how DRR activities play a major role in their projects. The session provided constructive advice to policy-maker and practitioners.

1.1.1 Madame Setsuko Seya, Director, Disaster Management Bureau, Cabinet Office, Government of Japan, Japan

Madame Setsuko Seya, Director, Disaster Management Bureau, Cabinet Office, Government of Japan presentation was on Communication between Science and Technology, and Civil Societies—Japan's experience regarding the Nankai Trough Earthquake.[1]

Since the establishment of the basic act for disaster management in 1963, almost 60 years ago, Japan has been tirelessly on combating disasters and improving disaster risk management. With examples and lessons learned from recent experiences, Ms. Setsuko Saya stressed that the science and technology community is indispensable to the Government of Japan in its contribution to the policy-making process. The Government of Japan perceive the importance of dialogue between scientists and the local communities and the importance of scientists sharing their knowledge about local risks with the citizens. Ms. Saya highlighted the recent approaches of the Japanese government to DRR.

[1] The report is based on the presentation by Ms. Setsuko Saya at the 4th Global Summit.

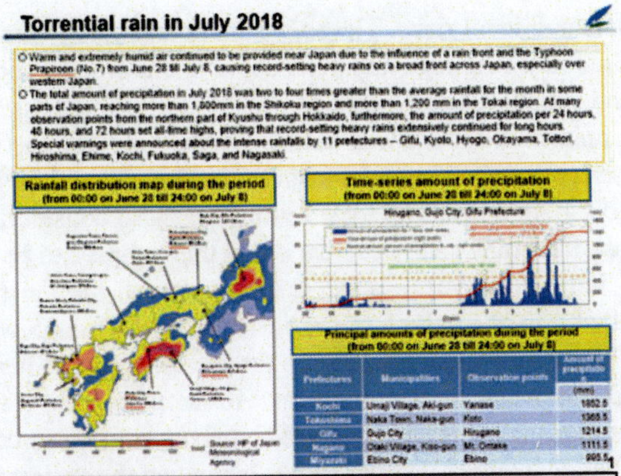

Copy of Slide from the presentation by Madam Saya.

The torrential rains in July 2019 affected approximately quarter of the Japanese mainland, and rainfall took place for about 10 days. Vast damages were experienced by the coastal area due to rivers that flow from mountainous sides. Landslides and flood inundations brought the mortality rate to more than 200 people. With more than 200 deaths, damaged houses, and many damage to infrastructure and disruption to public transportation, it was one of the worst and unusual meteorological hazards in Japan.

Preceding the event, the government called on an expert meeting consisting of many science and technology experts to find out how the situation could change policies. The deaths of 200 people were due to flooding in the area. Many people did not evacuate regardless of the warning for evacuation. There were many reasons why they did not evacuate. One can only assume that they did not hear the evacuation warning; maybe they thought the flooding will not overtake their houses; or they were not able to evacuate due to disabilities. Most of the victims were elderly people. The government expert group concluded and summarized the lessons from the threshold, and the report was published in December 2018. Main message from the expert group meeting was that citizens have to be aware that they have to save their lives by themselves. The government would provide any necessary support for their citizens to enable them to evacuate. But, the citizens are requested to understand the local risk and proper way to evacuation.

To support the recommendation, three approaches were proposed by the experts especially for children and the elderly.

- Awareness of local risks and evacuation procedures regardless of age and physical disability.
- DRR education for the local communities.
- Government to provide risk education support to enable evacuation under any circumstances by themselves using their own understanding.

This is a big mission for those ordinary people to learn about science in their local areas. Children are studying at schools. However, education at school is not enough for them to understand local risks. The science and technology must communicate with those children in a language understandable to children to know the risks in a scientific way and the evacuation method.

Another example highlighted was about the Nankai Trough (NT) earthquake which is estimated to occur at around a magnitude of 8 to 9 within the next 30 years with a probability of 70–80%. The expert group was also considering the nature of the NT earthquake, and they have concluded that there appear to be several patterns on how the NT will occur. Among the several patterns, the most critical pattern designated was that it may be along the eastern part of the trough with a magnitude of 8 which may bring immediate damage to the western part. This again is difficult to predict as it is not possible to know what will happen in the western part if the earthquake takes place in the eastern part. The expert group also concluded that at the end, there should be an action plan for all Japan and the tsunami warning of the most critical should also be issued for all Japan.

Further recommendation was that if the plan to work, all players have to be included and involved, for example, the meteorological agency, governments, local governments, citizens, and companies. In addition, a timeline was provided to assist these efforts which detailed by after several minutes, 30 min, one hour, one week, etc. and what each actor has to do. The government will play a major role in supporting or rescue activities and recovery activities. But, it is important to know that citizens and companies are all integrated in this entire programme. Without their understanding and without their action, the programme will fail. Even though it is an elaborated and a detailed programme, the important part of action is to be realized by citizens and companies. That is how science and technology knowledge must be communicated with the citizen. A lot of communication does go through the government, but unless science and technology people provide good understandable language and clear message to the government, their message will not be communicated to the citizens.

The community disaster management plan based on the Basic Act of Disaster Management Plan was revised after the 2011 earthquake. The 2011 earthquake taught a lesson to the citizens that they have to be aware of the local risks, and the basic act added one chapter on community disaster management programme. This programme is quite simple. Residents will discuss within their community about their disaster management programme and come up with a proposal for community disaster plan. And the community will submit a draught proposal to the municipality governments. The community municipality government will consider whether to integrate community disaster plan within their own disaster management plan. This action has been in implementation for approximately for the last five years. We have counted how many communities are working through this plan. It is not mandatory nor there is financial support from the government to implement the programme. Yet to the surprise of the government, they discovered that more than 3000 communities are working with the programme. Since this programme is integrated into the local disaster management plan, the local disaster management is always in the scheme of the entire national disaster management plan. The government is encouraging citizens to join in the initiative, so they are able to make up their own disaster management programme.

In this exercise, it also consists of questions from the citizens. For example, most of them want to know about the local risk in their communities as they have been living in the area for many years. As they are not scientists, their knowledge may not be the latest. They might not know the latest technology or scientific phenomena occurring in their neighbourhood.

In addition, the Government of Japan is in the process of organizing a network among the municipality governments which will support such initiatives of local citizens in the not too distant a future. She concluded by stressing that the scientists are requested to talk with citizens about their local risks as it is very important to communicate and contribute to science and technology for the community so that they can control their own disaster management plans.

1.1.2 Mr. Soichiro Yasukawa, Coordinator, Natural Sciences Sector, UNESCO, France

Presentation on the "UNESCO's Priorities on DRR Activities with Scientific Partners" was delivered by Mr. Soichiro Yasukawa, Coordinator, Natural Sciences Sector, UNESCO, France.[2]

His presentation specifically focused on the GADRI 4th Global Summit theme and tapped on to UNESCO's programmes and projects and the multi-stakeholder engagement for disaster risk reduction. Mr. Yasukawa shared how UNESCO's each mandated area—education, science, social science, culture, and communication information—has an element for disaster risk reduction. For example, Education Sector is promoting disaster risk reduction in the school curriculum, Natural Sciences Sector works on the disaster risk management in earthquakes, flood, tsunami, etc., and the Culture Sector especially promoting the resilience for the UNESCO World Heritage sites. The Communication and Information Sector is focusing on AI or information technology to prepare for natural hazards.

[2] The report is based on the presentation by Mr. Soichiro Yasukawa, UNESCO, France at the 4th Global Summit of GADRI in 2019 and it has been reviewed and edited by Mr. Yasukawa.

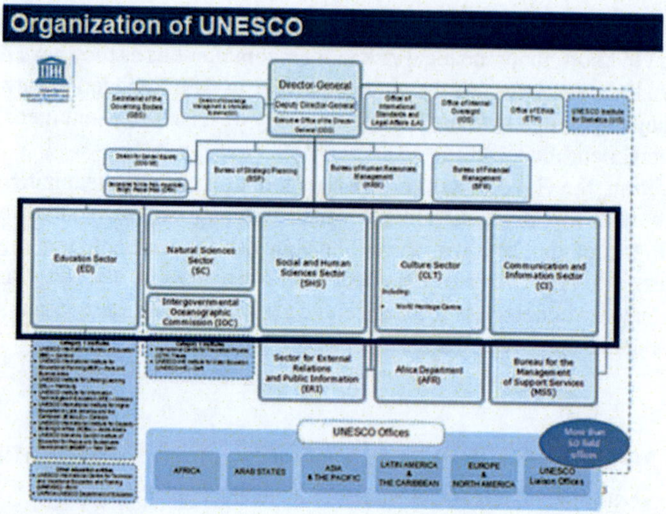

In addition to UNESCO Headquarters and Field Offices, there are UNESCO Chairs or UNITWINS programmes. Due to the human and budgetary resource constraints, UNESCO vastly relies on academic institutions' expert to complement its capacities.

In supporting Priority of Action 2 of the Sendai Framework, UNESCO operates with multi-stakeholder in disaster risk reduction—governments, international organizations or academic science, civil societies including young people, industries, etc.; he shared a few examples of UNESCO activities.

In Japan, ICHARM, conducting research on flood management, works with UNESCO on technical point or policy point of view on flood management. For example, for the projects in Pakistan, ICHARM is the main implementation partner and similar project is expected to start in the Niger River basin in Africa.

The UNESCO Chairs programme promotes universities to connect together—to share information and to learn from each other. For example, Kyoto University UNESCO Chairs programme is a flood management programme to identify gaps and fill these gaps through education, training, and contribution to the society and the UNESCO Chair for Italian universities work on risk assessment of the school facilities.

Youth is one of the UNESCO's priority together with gender and Africa. Young people need more opportunities to contribute to the society. The U-INSPIRE network is a youth network of young professional in disaster risk reduction in Indonesia. U-INSPIRE, consisting of academics, research institutes, governments, etc., engages young professional in disaster risk reduction to directly work on disaster risk reduction issues and supported by the UNESCO Regional Office in Jakarta. The U-INSPIRE encourages inter- and transdisciplinary research engaged in science-based DRR policies. U-INSPIRE is an affiliated organization of the Indonesian government on disaster risk reduction. The group is requested by the Indonesian government to share post-disaster reports on lesson to be learned and review gaps and issues for build back better.

Project in Africa on water encourages youth to come up with innovative solutions such as water quality or flood managements.

There are many challenges in small islands, developing countries, developing islands—mainly due to lack of budget or scientists to work with issues in these countries. Rise of sea levels and coastal erosions, etc. are happening, but base data is not available to think about solutions or to find out gaps in coastal erosions.

The SANDWATCH, a Science Education Scheme through Sustainable Coastal Monitoring project, on civil societies encouraging the school curriculum and kids to go to coastal areas to get data within the school curriculum thereby complementing scientific community obtains data through school curriculum even though they need further scientific validations to find out what is happening in the coastal erosions.

He highlighted an interesting example of how science, local governments, and civil societies work together to best optimize efforts prevention. During the Great East Japan Earthquake in March 2011, the marine eco-system was heavily damaged by the tsunami due to magnitude of the tsunami. UNESCO worked with Tokyo Marine University on collecting data, local knowledge, and techniques to recover the fisheries industry especially oyster farming. Through knowledge sharing and discussion among scientists, local fisherman, and municipalities, it was decided to put up a sea wall of 200 m inside the cost line to optimize disaster risk prevention system and the eco-marine systems.

Note: Since the 4th Global Summit held in March 2019, the above-mentioned UNESCO projects and DRR activities have evolved to greater progress. Visit the UNESCO link to see the updates on these activities. ENG: https://youtu.be/pE6c_oNuQmU.

1.1.3 *Mr. Masato Miyazaki, Special Representative Japan, The World Bank Tokyo Office, Japan*

Mr. Masato Miyazaki, Special Representative Japan, The World Bank Tokyo Office, Japan

Mr. Masato Miyazaki, Special Representative Japan,[3] The World Bank Tokyo Office, Japan presented on "Partnerships for Mainstreaming DRM in Developing Countries".[4]

He began by expressing sincere condolences for the devastation caused by several natural disasters that hit Japan in 2018. He expressed confidence that Japan will overcome the challenging consequences of these disasters and through lessons learned will be able to guide the world in resilience-building practices.

He highlighted the importance of strong leadership and commitment from the global leaders and experts in partnering with stakeholders around the world to strengthen resilience against natural disasters. World Bank research estimates that the impact of extreme natural disasters is equivalent to US$520 billion loss in annual consumption globally and forces some 26 million people into poverty each year.

---

[3] The report is based on the presentation by Mr. Masato Miyazaki at the 4th Global Summit. It has been reviewed and edited by the World Bank Office in Tokyo.

[4] The World Bank— https://www.worldbank.org/.

This negative impact will continue to increase due to population growth, rapid urbanization, and climate change, while the most vulnerable will continue to suffer the most.

Indeed, disasters affect the poor and vulnerable disproportionately, especially women, children, elderly, and those recovering from the impact of conflicts. Without urgent action, it is estimated that the impact of climate change could push an addition 100 million into poverty by 2030.

Recognizing these challenges, in 2014, a joint programme between Japan and the World Bank on "Mainstreaming Disaster Management in Developing Countries" was launched with a budget of US$100 million through the generous support from Japan's Ministry of Finance, with the following two overarching objectives:

1. Supporting developing countries in mainstreaming disaster risk management (DRM) in national development planning and in investment programmes through the World Bank operations.
2. Connecting Japanese and global expertise in DRM with developing countries and World Bank teams.

The programme was a success and was recently renewed to enter the second phase.

More broadly, the World Bank recognizes the importance of the integration of science and technology for better outcomes in DRM and strengthening resilience. Therefore, one of the investment priorities in DRM is furthering analytics and research to make disaster risk information open, accessible, understandable, and usable by all interested parties as called for by the Sendai Framework, including coordination within the scientific community in support of the decision-making processes. The following examples were highlighted as to how the World Bank incorporates science and technology into our DRM engagements:

- The World Bank developed Think Hazard, an open-source tool, launched in mid-2016 that provides an overview of the hazard profile for countries and has been used by more than 125,000 users during its pilot phase.
- The web-based online tool has been used for activities ranging from the initial project screening for disasters and climate risks, to more complex disaster and climate risk analytics for Afghanistan, Kyrgyz Republic, and Southwest Indian Ocean Island states.
- The bank has also developed and delivered disaster risk assessments and national-level risk profiles in more than 50 countries that are informing action on the ground. For instance, in Madagascar, risk modelling was harnessed to provide rapid damage estimates to the Ministry of Finance within two weeks of the Tropical Cyclone Enawo in March 2017.
- In Tanzania, a water sector project piloted the use of emergent, low-cost ICT solutions to address the disaster risk reduction challenges posed by changing hydrological hazards in Tanzania's water sector in both urban and rural contexts.

These examples showed not only the importance of science and technology in DRM, but also the importance of various stakeholders including research institutions coming together to tackle complex challenges to reduce human and economic impacts of disasters.

In this sense, GADRI and its institutional members forming international collaborative networks from investigating global disasters and implementation to new research methodologies are monumental in achieving global targets acting on the priorities of the Sendai Framework.

He highlighted two examples of how the collaborative efforts between the World Bank Tokyo DRM Hub and GADRI member institutions have integrated science and technology in DRM engagements:

1. The partnership among Kyoto University, the Research Center for Urban Safety and Security at Kobe University, and the Tokyo DRM Hub supported the implementation of a ten-day science cooperation and learning event in Japan. This event invited technical experts from Ghana, Togo, and the Volta Basin Authority working on strengthening flood forecasting and monitoring in the Volta Basin region. The event provided the visiting technical experts with an opportunity to observe, learn, and experience first-hand cutting-edge Japanese technologies, research, good practices, and lessons learned related to weather and hydrological observations, flood forecasting, and flood risk management through lectures and field-based exercises conducted by Kyoto University and Kobe University. The experts were particularly impressed by how research institutions and universities collaborate to mitigate flood risk and how research conducted by the universities has a direct impact on solving problems communities and populations vulnerable to flooding are facing. The training enabled the experts to improve forecast accuracy as well as the seamless integration of forecast data into the Web Emergency Operational System of Ghana's National Disaster Management Organization.

   This is a great example of how knowledge based on extensive research and best practices shared by universities contributed to the establishment of an operational platform focused in Ghana and Togo while also strengthening the capacity of these governments to provide timely early warning systems.

2. Another example he shared was the contributions from the specialists from Kyoto University in the collaborative work with the World Bank in the development of the case study on "Incorporating Resilience in Infrastructure Prioritization". It was part of the Quality Infrastructure Investment Partnership established by the Government of Japan and the World Bank with the objective of raising awareness and promoting enhanced attention to quality dimensions of infrastructure projects in developing countries. On this project, the Tokyo DRM Hub partnered with GADRI, the Disaster Prevention Research Institute (DPRI), and Kyoto University to develop a framework designed to inform upstream infrastructure project selection that considers resilience to hydrological, meteorological, geophysical hazards such as floods, cyclones, landslides, earthquakes, in developing countries and other factors.

In conclusion, he affirmed that the World Bank collaborations have shown that major advances in technology, information, and research can provide opportunities to policy-makers, development practitioners, and communities to understand disaster risks and equip themselves with information on how best to prepare for hazards in order to save lives and assets. In order to support developing counties and cities in becoming more resilient, the World Bank and the Tokyo DRM Hub will continue to support initiatives that advance disaster resilience and keep fostering collaborations with international research communities to tackle disaster risks by incorporating elements of science and technology in World Bank investments. Through dialogue and great collaborations, our cities and countries will be able to adapt to changes in economic, social, and environmental conditions and mitigate the risks caused by these challenges.

1.1.4 *Prof. Toshio Koike, Director, International Centre for Water Hazard and Risk Management (ICHARM) Under the Auspices of UNESCO, Japan*

Prof. Toshio Koike, Director, International Centre for Water Hazard and Risk Management (ICHARM) under the auspices of UNESCO, Japan—taking notes at a group discussion session

Prof. Toshio Koike, Director, International Centre for Water Hazard and Risk Management (ICHARM) under the auspices of UNESCO, delivered a speech on the "Nation's Synthesis—Concept Design and Implementation". He explained in detail the concept design of the nation's synthesis. Through follow-up discussion for realizing the activity, careful consideration was placed on the relationship between the national platforms and the science and technology synthesis for DRR as each country is expected to lead the implementation of the nation's synthesis.[5]

The national platform should develop a mechanism that allows all stakeholders to share information of science and technology for DRR in its own language. How can the science and technology community contribute to these activities? The science and technology community needs to provide science and technology knowledge to national governments as well as the local governments and civil society and local community and people working as facilitators. For providing effective and usual public support, people should take action to proactively plan and take action for their self-help and mutual support and instead of sole dependency on national government efforts and local government efforts. National platforms could work together with local communities through well-informed facilitators.

The Science Council of Japan in cooperation with UNDRR organized the Global Forum for Disaster Resilience in November 2017 to pursue the study of implementation of the 4 Priority Actions of the Sendai Framework Agenda and concluded with the discussion to develop the following two documents:

- Guideline for strengthening national platform for disaster risk reduction (DRR) and coordination mechanisms through enhanced contributions to science and technology.
- The periodic synthesis reports on the status of science and technology for reducing disaster risk.

During the follow-up discussions to realize this activity, the relationship between the national platforms and the synthesis and science and technology synthesis for DRR were considered. The national platform should undertake to develop a mechanism that allows all stakeholders to share information of science and technology for DRR in its own language.

The tasks of the national platform will entail:

- Review of the status and issues of current DRR efforts that the national platform has implemented based on science and technology.
- National platforms to discuss how DRR should be carried out for the country, and design practical measures to be implemented from a holistic point of view.
- These series of the action should be taken by national platform of each country.

[5] The report is based on the presentation by Prof. Toshio Koike at the 4th Global Summit. It has been reviewed and approved by Prof. Koike.

- Nation—state—consists of three components. Land, sea and atmosphere, and sovereign/governance and people. Nation's series of action will focus on all areas covering the above-mentioned points. This series of action is called as nation's syntheses.

As suggested by the Sendai Framework and Hyugo Framework of Action to identify multi-stakeholder coordination, the importance the nation's synthesis is the coordination and recommendation to establish an infrastructure to support the national platforms for providing effective and usual public support. People should take action to proactively plan for their self-help and mutual support. As Ms. Saya introduced in her talk, not only the national government efforts, and local government efforts, but the people or the residential community effort is strongly expected in this initiative. How can science and technology community deliver and contribute to these activities? The national platforms of the government in collaboration with local governments or civil society, and individual householders and residential communities, and local organizations should take action to support strengthening of the self-help and mutual support before the disaster event and follow up the processes to reduce disaster risk.

How could this kind of activity be promoted? Here, Prof. Koike tapped into what was mentioned by Ms. Setsuko Saya and how can the science and technology community promote and support these types of actions?

As mentioned by previous speakers, the international communities such as the World Bank and the UN and donor countries support national platforms. Local support needs including life safety, asset protection, business continuity, and other various aspects should be taken by local people by themselves. The science and technology community needs to provide science and technology knowledge to national government as well as the local governments and civil society and local community and people working as facilitators and identify the problem structure and possible solutions to make sense and provide governance structure at each level. For effective use and implementation of this activity, it is necessary for each country's science community and international science and technology networks to collect the scientific knowledge from all over the world and convert the data from English or in other language to local languages. This should be facilitated by the local science and technology facilitator, and this is needed for effective integration of information and data to work with each level.

Prof. Koike shared an example of one of ICHARM's projects in Asia to establish a multi-stakeholder coordination mechanism. First, by archiving damage, hazard socioeconomy data, and based on such integrated datasets, it is possible to do integrated risk assessment. Risk is changing from past to the future due to the societal and climate change, and it is needed to identify, monitor, and predict the changing risk. This action provides usable information for policy-making and actionable information for community. To accomplish this, it is necessary to data archive, model development, and societal barrier creation through the interdisciplinary collaboration and transdisciplinary collaboration between the science and technology community and stakeholders. Capacity-building programme is another very important activity to be supported through international cooperation. This formulates the concept of the support of the national multi-stakeholder mechanism.

Another example sited was the case of Philippines—he introduced the implementation mechanisms used after the three plenary meetings for the collaborative work with 20 governmental organizations and three universities also including the department of public work and highways, departments of science and technology, and the secretariats of these departments' co-chairs. Once the requirement for obtaining data was mentioned, the participating organization raised hands to say that they could provide the necessary data. Within one year, the necessary data was obtained from the Philippines. It was not only statistic data but included real-time data from the PAGASA the Philippines hydro-met bureau which used a satellite to produce date on hydrological model and early flood forecasting.

Another example shared was the case of Sri Lanka's Kelaniya river basin. Here, damage data was also collected. There was a higher mortality rate due to heavy rainfall, and a larger number of people affected by the typhoons. It also affected the agriculture production, and the largest damage was caused by the typhoon in 2018. Data from Sri Lanka is still being collected.

This type of information can be shared depending on the collected and collaborative activities among different agencies and department. In addition, work is being carried out with local communities to make contingency plans by formulation of a plan for documentation and sharing. The first three steps—understanding the condition, identifying the risk, and analysing—this should be done in close collaboration between the local community and the science and technology community step by step. Flood diagnostic chart can be developed, and through collaboration, it is possible to identify flood hotspots. Once the local hotspots are clearly identified, it is shared with the community. ICHARM team is working as facilitator in these projects. The local leaders chair the project and lead the discussion, as well as development of the contingency plan. Each residential community plan will be shared among the 29 residential communities and municipalities.

Prof. Koike concluded by stating that in order to accomplish the goal of the nation's synthesis, it is necessary to take action collectively. In Japan, the Science Council of Japan formed an alliance in collaboration with the Japanese ministries, local governments, private sector, UNDRR, GADRI, and international community to promote the nation's synthesis in Japan and in many other countries. Without such strong support, it will be difficult to science and technology community alone to implement the action for a disaster-resilient world.

1.2 Plenary Session II: Fostering Uptake of Science in Governments and Industries

1.2.1 Ms. Shoko Arakaki, Chief of Branch, Partnerships, Inter-governmental Process and Inter-agency Cooperation, the United Nations Office for Disaster Risk Reduction (UNDRR), Switzerland

Ms. Shoko Arakaki, Chief of Branch, Partnerships, Inter-governmental Process and Inter-agency

"The role of science and technology to enhance risk informed development for a resilience and sustainable future" was presented by Ms. Shoko Arakaki, Chief of Branch, Partnerships, Inter-governmental Process and Inter-agency Cooperation, UNDRR, Switzerland. Ms. Arakaki shared her appreciation and offered congratulations to the organizers and GADRI for the timely approach taking to promote the targets of the Science and Technology Roadmap for the implementation of the Sendai Framework and the coherent approach to disaster risk reduction and resilience.

She continued to state that as clearly stated in the Sendai Framework, it is the role of the science community to help understand the risks and assist the decision-making and policy-making processes. Yet adequate financing for DRR strategy as well as for Sustainable Development Agenda is really a problem. Discussions must go beyond traditional short-term "risk financing" to the more long-term integration of risks into investment decisions. Building political will to invest in a more risk-sensitive manner needs to be supported by clear science evidence. Now, it is more important for science and technology to play a key role to better understanding and addressing risks.

The role of the science and technology community is clearly defined by the Sendai Framework in particularly for understanding disaster risk. GADRI's strategic and long-term approach to build relevant scientific evidence and drive innovation was thereby appreciated. This included the organization of the 2016 Science and Technology Conference which was held in Geneva, which adopted the Science and Technology Roadmap and established the science and technology partnership which subsequently led to the Science and Technology Conference in November 2017 in Tokyo led by Prof. Toshio Koike, ICHARM. These achievements fed into 2019 as the year of resilience. Resilience, climate change, and DRR were increasingly recognized in the political realm including the currently ongoing discussion of the G20. Financing for DRR remained an issue, and discussion must go beyond traditional short-term risk financing to long-term risk informed investment decisions. Building political realm to invest in a more risk-sensitive manner needed to be supported by clear science evidence. In this regard, presenting the outcomes of the GADRI 4th Global Summit into the Global Platform for DRR in May 2019 and feeding them into other risk-related policy processes in 2019 were important.

Ms. Arakaki mentioned coherence among DRR, climate change adaptation, mitigation, and development which played a critical role for the effective implementation of risk informed development at the national and local level. Enhanced action on coherence was especially important in light of the 2020 deadline for both Target E of the Sendai Framework and national adaptation plans under the Paris Agreement. Nearly 84 countries had started to report through the Sendai Framework Monitor. However, this number highlights the difficulties many countries are facing with implementation of the Sendai Framework including lack of data and information for all agreed indicators. Countries were also developing their own national custom indicators, national loss databases, and the local strategy and plan as part of Target E. Currently, 109 national databases exist globally, but only 59 are aligned with the Sendai compliance. A total of 107 national DRR strategies is existing, but again not all are aligned with the Sendai Framework requirements. She continued to highlight a few expected milestones to bring the discussions forward from the UNDRR perspective:

1. Contextualization of the 2016 Science and Technology Roadmap which is included in the GADRI 4th Global Summit as a discussion topic.

Despite the Sendai Framework, SDGs, Paris Agreement, and new Urban Agendas, the core part of the agreement was not quite mature enough to be able to inform on the Science and Technology Roadmap. Therefore, the contextualization of the Science and Technology Roadmap was very important. It is hoped that an updated roadmap would be adopted at the Global Platform 2019 in May 2019 and commitment and contribution under the roadmap be reported through the new Sendai voluntary commitment online platform.

2. The effectiveness and current silo-thinking mentality. Silo mentality is a huge challenge among the national ministries, and the G20 will table this as an agenda item too. It is not only within the government, but also between the private sector and public sector, among stakeholders, the science and technology community, civil society, and also in vulnerable groups like women, children, youth, disability, displacements, etc., and it is challenging to implement risk informed development in a comprehensive way. Integrated science across all science fields is very important and welcome leadership and innovative thinking from attending organizations in the GADRI meeting to identify opportunities to fill in the gaps.
3. Data—accessible and applicable data is still big problem. This was addressed during the first meeting on the Global Risk Assessment Framework, an UNDRR initiative to address the questions of data. Prof. Andrew Collins, who is a member of GRAF I, was supporting this through work on "data as a voice and data for action". In this context, disaggregated data was also very important and the science and technology community can play an important role to address specific vulnerabilities to ensure inclusiveness and that no one is left behind.
4. Hazards—Sendai Framework comprised not only natural hazards but also technological, biological, health, pandemic hazards, and related systemic and cascading risks—all these pose significant challenge to tackle, and the science and technology community can play a major role to contribute in this area. The discussions on NATECH were an opportunity to look into this expanded scope of hazards.

In closing, Ms Arakaki reiterated the important role of the science and technology community for better understanding and addressing risks. Ms. Arakaki concluded by stating that GADRI leadership role in this regard was much encouraged and commended the role taken to move forward to implement the revised Science and Technology Roadmap.

1.2.2 Dr. Jörgen Sparf, Mid Sweden University, Sweden, and the European Science and Technology Advisory Group (E-STAG) of the UNDRR

Dr. Jörgen Sparf, Mid Sweden University, Sweden, and the European Science and Technology Advisory Group (E-STAG) of the UNDRR

The presentation by Dr. Jörgen Sparf,[6] Mid Sweden University and the European Science and Technology Advisory Group (E-STAG) of the UNDRR was on "Navigating the European Landscape of DRR—The Case of Data and Socioeconomic Factors".

The E-STAG is an expert group composed of appointed academics from fourteen countries. The group includes five additional thematic experts and two excellent young scientists. The regional office of UNDRR supports E-STAG's operational work, and the European Commission's Joint Research Centre provides essential input.

Before continuing with the presentation, Prof. Sparf shared background information on the studies conducted to date and some of the challenges and dynamics of the two target areas identified by E-STAG: data for resilience and socioeconomic aspects

[6] The report is based on the presentation by Dr. Jörgen Sparf at the 4th Global Summit. It has been reviewed and edited by Prof. Sparf.

of disaster risk reduction. In his presentation covering the European continent, Prof. Sparf shared information particularly on three areas: demographics and economic activity; culture and ethnicity and how people relate to places; and corruption.

Prof. Sparf focused on some of the contemporary challenges in Europe, guiding the audience through three kinds of multiplicities.

1. Population and population density. Despite the small size of the European continent, there is a big difference in population density and economic activity throughout the continent. The population density ranges from more than 500 people per square kilometre to less than one person per square kilometre. The population density is the highest in a broad belt from the Central UK, throughout Belgium, the Netherlands, Germany, and France, down to Northern Italy. Other areas, such as Iceland, Northern Sweden, Norway, and Finland, are very sparsely populated. These differences indubitably pose significant challenges for DRR.

   Sparf also reminded the auditorium about the blue banana and red octopus concepts. They illustrate clusters of industrial economic activity with dense populations living under very tight infrastructure realities.

   The prognosticated demographic changes in Europe are quite different throughout the continent. The prognosis from 2017 to 2050 (source from the CIA, USA) is that, while the western half of Europe will gain population by ca 5–30%, the eastern parts will decrease its population roughly equally. This change in balance is mostly related to internal migration patterns, an increased influx of migrants from outside Europe, and an increased share of elderlies.

   Interestingly enough, the economic prognosis is the reverse to the population. In one of the presentation slides, Dr. Sparf showed the debt-to-GDP ratio in the European countries. While the western countries of Europe have a very high debt of GDP, the eastern countries do not have such a high debt ratio. The combination of an increased population density and a high debt rate puts a hard strain on the economy. This strain is crucial to recognize when discussing investments in disaster risk mitigation and funds allocation for infrastructural investments.
2. The next dynamic was about culture.

   Although Europe is a small continent, the population is significantly heterogeneous in terms of cultures, norms, and values. In many parts of the continent, people are firmly rooted in the local culture, based on customs, traditions, and historical contingencies. Equally, religion varies markedly throughout the continent in terms of religious affiliations and levels of conservatism. Whether people focus on survival issues or self-expression values significantly influence household preparedness for disaster and the need for information on such topics.

3. The third dynamic was corruption. Not only in Europe but on a global scale, this is a growing concern. In recent years, we have witnessed increased corruption rates in some European countries, partly related to cultural norms but mostly to economic aspects. A struggling economy always incentivizes the black market and organized crime. This effect means that countries might face not just recession in itself but also corruption, bribery, et cetera in large-scale investments.
4. The last multiplicity—risks. Just like any region in the world, Europe is facing a wide range of hazards and risks. The natural hazards differ markedly over the continent—earthquakes, volcanoes, droughts in southern parts of Europe. In the northern areas, there are more storms, forest fires, and floods. The climate change effects are shared between the countries, with severe impacts, referring back to blue banana and red octopus concepts of urbanization. Even though city planning is quite heavily regulated, urbanization comes with many challenges. And pandemic is a global issue.

Prof. Sparf stressed that given this background as context, better data is necessary to build resilience. Better data might save more people's lives. One of the problems is that data from different sources is incompatible. Academic disciplines and data collecting entities apply inconsistent standards and different nomenclatures. Work is continuing to identify barriers that affect disaster risk-related data inter-operability and built-in probability so that many stakeholders can use the data for resilience.

In conclusion, Prof. Sparf described how another group in E-STAG is working on the socioeconomics for disaster risk reduction, specifically looking at the vulnerability component. The level of study is the community or city level rather than a household or individual. Four specific resources are of particular interest for this study: economic, political, social, and human. The point of departure is the well-established understanding that disasters do not create inequalities but reveal and amplify existing subsurface ones. The study discusses the fact that resources cannot do anything by themselves. It is always people who make use of resources or decide how to put them into effect. The human and social aspects are thus crucial for the status of social systems and DRR. These actions may also amplify existing vulnerabilities or possibly create new ones related to corruption and inappropriate investments.

The analysis of this empirical data and the report was presented at the UNDRR Global Platform 2019 in Geneva in May 2019.

1.2.3 *Prof. Virginia Murray, Global Health Disaster Risk, Public Health England, UK*

Prof. Virginia Murray, Global Health Disaster Risk, Public Health England, UK

Prof. Virginia Murray, Global Health Disaster Risk,[7] Public Health England gave keynote speech on "The Sendai Framework and Science Advice to Governments to Support Implementation—Suggestion on What Works".[8]

At the start of the presentation, Prof. Virginia Murray congratulated the GADRI secretariat and expressed her appreciation for providing an opportunity to share and reflect on DRR issues. She further stated that science has been absolutely critical in the delivery of the Sendai Framework and it should be useful, usable, and used. The Sendai Framework reflects science throughout the framework, and the challenge for the science and technology community is that there is a job to do but much more to the point, to deliver it.

For further details of the presentation by Prof. Murray, refer to papers under Part 4—Plenary Sessions.

---

[7] The report is based on the presentation Prof. Virginia Murray at the 4th Global Summit. It is completely revised and edited by Prof. Murray.

[8] Public Health England—https://www.gov.uk/government/organisations/public-health-england.

1.2.4 Prof. Andrew Collins, Leader, Disaster and Development Network (DDN), Northumbria University, UK; and Chair of the GADRI Board of Directors; and Member of the Global Risk Assessment Framework (GRAF), UNDRR

Prof. Andrew Collins, Leader, Disaster and Development Network (DDN), Northumbria University, UK; and Chair of the GADRI Board of Directors; and Member of the Global Risk Assessment Framework (GRAF), UNDRR

Prof. Andrew Collins, Leader, Disaster and Development Network (DDN), Northumbria University, UK[9]; and Chair of the GADRI Board of Directors presented on the "Global Risk Assessment Framework (GRAF), UNISDR—"Emphasising the Relevance of Disaster Research Institutes". He highlighted the work done by expert group of GRAF which is a framework for assessing risks.[10]

For further details of the presentation by Prof. Collins, refer to papers under Part 4—Plenary Sessions.

[9] The report is based on the presentation by Prof. Andrew Collins at the 4th Global Summit. It has been reviewed and edited by Prof. Collins.

[10] Global Risk Assessment Framework (GRAF), UNDRR—https://www.preventionweb.net/disaster-risk/graf.

1.3 Plenary Session III: Bridging Science and Decision-Making Through Sharing of Knowledge

In this session, the four speakers enlighten the information further on various tools available for disaster research activities and draw our attention to knowledge gaps in science and technology capacity while showing us how to bridge science and evidence-based decision-making through sharing of knowledge and lessons learned. The session was chaired by Prof. Tomoharu Hori, Vice-Director, DPRI, Kyoto University, Japan, and Prof. Paul Kovacs, Executive Director, ICLR, Western University, Canada.

1.3.1 Prof. Eiichi Nakakita, Disaster Prevention Research Institute (DPRI), Kyoto University, Japan

Prof. Eiichi Nakakita, Disaster Prevention Research Institute (DPRI), Kyoto University, Japan at a group discussion session

Prof. Eiichi Nakakita, DPRI, Kyoto University, Japan, presented an interesting and informative presentation on the "Perspectives on Climate Change Impact Assessment and Adaptation".[11]

As global warming progresses, heavy rainfalls increased across the globe, and Japan was especially impacted with many flood-related disasters making the hearsay of global warming a reality. Enhancements to improve the scientific knowledge on global warming are much more needed now more than ever. Importance of a framework for impact assessment and adaptation methodology for global warming are much required. This emphasizes the importance of collaboration among climatologist, meteorologist, and the academia for disaster-related climate change impact assessment and adaptation as well as the implementation authorities which have responsibilities to the public directly.

Prof. Nakakita continued to explain that the climatologist group provides the scientific basis projection of climate future for disasters related to climate change impact assessment, while the adaptation group provides the scientific basis projection of the future changes in the social impact as well as hazard itself. It is important to create the basic idea of no-regret adaptation policy as well as the development of an evaluation methodology for regret adaptation policy. Prof. Nakakita's presentation introduced a project funded by the Japanese Ministry of Education to strengthen the relationships between climatologist and meteorologist and the relationship between the practitioners.

The collaborative research project which has been carried out for over 15 years included climatologist and meteorologist. DPRI is responsible for the impact assessment and adaptation for natural hazard and water resources aspect under this framework. The basis of this programme is output for future—future projections, for example, output until the end of the century. For example, 4 degree increase in the world temperature is compared to a pre-industrial area. Work involved to obtain these kinds of physical-based projection for future in the world. Not only in the world but also for Japanese areas. Based on this information, evaluations are done for future standards which should adapt in the framework of the disasters and water-related disaster issues.

[11] The report is based on the presentation by Prof. Eiichi Nakakita at the 4th Global Summit. It has been reviewed and approved by Prof. Nakakita.

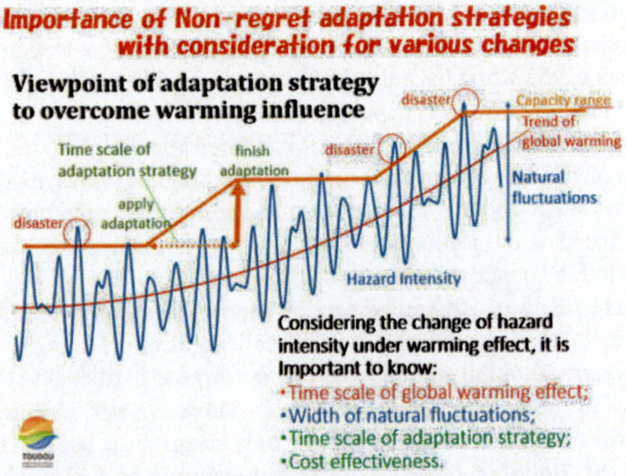

Copy of slide from Prof. Nakakita's presentation.

As part of the project, a design value is implemented. Design value means, the design value for the infrastructure is defined based on some return period to such as 200–100 years. In the case of the global warming, of course, it may increase for some certain value. But still, there are some uncertainties for the projection of this design value itself. This is an important thing which we should take into account.

Another issue to take into account is the worse hazards which may obtain a lot of candidates as a worst hazard scenario. This can be defined that one as a kind of survivability, or critical survivability. Adaptation to new design value and worst case is very important, and projection of the design value and the worst hazard is a question of the projection. How much change there will be? Where will be the relocation be? And when does it become severe? Scientific answers should be provided to these questions.

Another important thing for the future is adaptation strategy, a no-regret adaptation. Because the effect of the global warming comes in gradual stages and on looking back, it proves a lot of changes have occurred in the past. Adaptation countermeasures should be proceeded step by step. Arrangement of the process of taking some of the adaptation measure should be taken into consideration. Or, there could be a new kind of countermeasure. Based on seamless, gradual, increasing of the outer force, it is necessary to think of the no-regret way of adaptation. To realize those no-regret adaptation methodologies, the communication between scientists and the practitioners or authorities is very important. By practitioners, here it means for example, Ministry of Land, infrastructure and Technology, Ministry of Agriculture, and Ministry of Environment. Those are authorities that are responsible for the safety of the public in Japan. In taking into consideration of the worst hazards and no-regret adaptation, cooperation among these three ministries with science and technology and policy-makers is very important.

And the programme corresponding to these three is trying to estimate the higher classes of probability of future corresponding design value as well as the greatest external forces as the worst hazards. Also, it is important to develop a kind of way of rational to make a no-regret adaptation strategy.

To support the above theory, Prof. Nakakita shared an example—currently there is collaboration on the assessment and adaptation methodology with many researchers not only from DPRI, Kyoto University but involving many other universities. The emphasis is placed on the seamless hazard prediction until the end of the century and the non-regrettable adaptation strategies definition.

As there are many expert researchers in natural hazard and social and human sciences, it is possible to conduct this research collaboratively. For example, to prediction future hazards, through simulations, a storm surge is predicted with many other candidates for future occurrences depending on the return period. Based on this, it is further storm surges are simulated and not only storm surge but also the depths of inundation areas. Based on those kinds of computation, a sort of graphic with table is created to show to the public for their understanding.

On typhoons, Prof. Nakakita shared a simulation of the global warming based on actual past severe typhoon cases. With the simulation, it is possible to shift the course itself to make worser and worst condition for specific typhoons. The simulation was based on Nagoya area. With the simulation, civil engineers can computate the height of these storm surges and the inundated areas. Depending on the historical situation and future climate condition by the assumption of global warming, it is possible to shift the inter-condition of the typhoon and thereby get the worst situation.

Another use of these simulations is the public surveys. Based on these situations, a questionnaire is developed to interview the public, and people made their disaster mitigation decision by giving 10% weight on the worst prediction of storm surge. And 90% weight on average prediction. They were presented with both worst and average predictions. As for the target areas, they increase their willingness to pay by 20% for bolstering the worse prediction in addition to the average extreme cases anyway.

In addition to this, we are now undergoing to incorporate the concept of the gradual increase of the outer force by the global warming. So that, we could made no-regret adaptation strategy. In this case, this framework, we must incorporate or introduce the concept of time in our new defined cost function.

Prof. Nakakita concluded by stating that risk management will be a phenomenon of the risk hazards. As global warming continue to increase, there will be many natural hazards and especially an increase in rainfall and to take into account the result from the worse-case classes as one of the forces. It is also important for climate change adaptation to discriminate between the planning with the uncertain development and risk managements in the worst-case hazard itself. To note that the increase in uncertain number is important to estimate the risk of the worst-class hazard itself. As stated in[12] ***Sustainable Development Goals (SDGs)*** adopted at the United Nations Conference held in 2015, there is a need to address the global warming issue together

[12] UN Sustainable Development Goals—https://sdgs.un.org/goals.

with other goals in a comprehensive manner rather than independently. In a similar fashion in Japan, the authorities concerned will need to closely cooperate with each other in promoting the necessary measures to take with regard to adapting to climate changes. In this connection, the Minister of Land and Infrastructure—MLIT—of Japan has decided to introduce the concept of risk management with worst hazards into the official adaptation strategy. The ministries waiting for some establishment of methodology could be uniformly applied nationwide for estimating the worst hazard. Addressing the impact of uncertain climate changes in the appropriate manner therefore makes *improving scientific knowledge of global warming effects.*

1.3.2 Dr. Jean-Paul Pinelli, Department of Mechanical and Civil Engineering at the Florida Institute of Technology (FIT), USA

Dr. Jean-Paul Pinelli, Department of Mechanical and Civil Engineering at the Florida Institute of Technology (FIT), USA

Dr. Jean-Paul Pinelli presented the "DesignSafe: a Cyberinfrastructure Available to Disaster Research Institutes". The DesignSafe cyber-infrastructure is available to disaster research institutes throughout the world (https://www.designsafe-ci.org/)[13] (Rathje et al. 2020; Pinelli et al. 2020).[14] When dealing with natural hazards, whether it is for hazard assessment, or the assessment of the community response, or for the assessment of the infrastructure response, any analysis involves very large datasets which come in very different formats, shapes, and contents. The gathering, storage, and handling of these very large datasets are a critical issue. To provide a solution to that issue, the National Science Foundation (NSF) in the USA funded Design-Safe. DesignSafe is a cyber-platform in the cloud that accompanies the efforts of researchers from inception to the final stage of the research, including publication. Whether the data originates from experiments or simulations or field reconnaissance, DesignSafe can host the data in its data depot. But, DesignSafe is much more than a data repository. It provides the researchers with multiple tools and access to high performance computing to analyse, visualize, manipulate, share, and publish the data in the cloud.

NFS funded the platform to facilitate the interaction between different natural hazard experimental facilities throughout the USA and to provide a way to disseminate the research funded by NFS, but it is not restricted to people funded by NSF or in the USA. It is available for free to anyone in the world, to facilitate and improve disaster risk research throughout the globe. Registering to the DesignSafe platform is quite straight forward and easy.

---

[13] The report is based on the presentation by Dr. Jean-Paul Pinelli at the 4th Global Summit of GADRI in 2019 and it has been reviewed and edited by Dr. Pinelli.

[14] Ellen Rathje, Clint Dawson, Jamie E. Padgett, Jean-Paul Pinelli, Dan Stanzione, Pedro Arduino, Scott J. Brandenberg, Tim Cockerill, Maria Esteva, Fred L. Haan, Jr., Ahsan Kareem, Laura Lowes, Gilberto Mosqueda. “"Enhancing Research in Natural Hazards Engineering through the Design-Safe Cyberinfrastructure," Frontiers in Built Environment, section Wind Engineering and Science, December 2020, https://doi.org/10.3389/fbuil.2020.547706.

Jean-Paul Pinelli, Maria Esteva, Ellen M. Rathje, David Roueche, Scott J. Brandenberg, Gilberto Mosqueda, Jamie Padgett, Frederick Haan, “Disaster Risk Management through the DesignSafe Cyberinfrastructure,” International Journal of Disaster Risk Science, November 2020, https://doi.org/10.1007/s13753-020-00320-8.

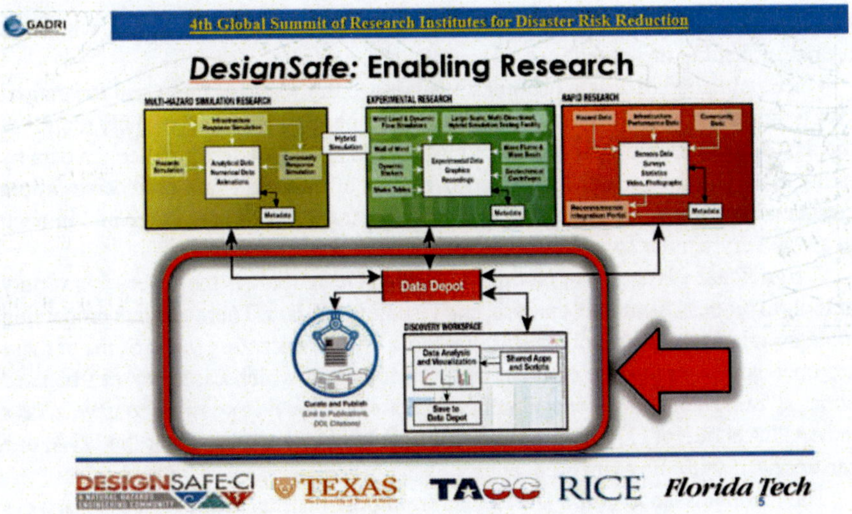

A copy from the presentation by Dr. Pinelli

Flexibility governs the design philosophy of DesignSafe. It is not a one-size-fits-all for the data. The data can come from different sources: experiments, hybrid simulations, reconnaissance, or it can be engineering data, social data, and so on. DesignSafe provides data models to facilitate the curation of the data, and these data models adapt to the specific needs of different types of data. If none of the data models fits a given dataset, there is an “other” data model alternative, which is a bare-bone data model applicable to any kind of data.

The core of DesignSafe is the research workspace, which has several components, including the Data Depot where the data is stored. There are different options to choose from for data sharing, with different levels of privacy, such as to control the level of data access to the public. The data can only be accessible by the user. It is also possible to create projects associating any number of persons who need to be involved in a specific project, and the project data will be available to all of the members of the project. Finally, after curation and publication, the data is accessible to the world at large.

The objective of the data curation process is to share data and ultimately to publish the data to make it available to the worldwide community. The curation process ensures that the data is accessible and searchable. As the research progresses, the researcher can curate the data and thereby assign different attributes to the data in order to organize the data logically according to different objectives. The data curation takes into account that most researchers organize their data using hierarchical file systems and file naming conventions. Therefore, three basic steps govern the curation process. Users select the files or folders they want to categorize and link that data to the main categories identified in the data model. They can then tag descriptions to the files or folders using drop-down menus with terms contributed by the different domain specialists, and they can relate the categories to different processes. As the users create relations, the metadata and the tags coalesce into logic trees, which the users can display on screen.

Once published, the data becomes available to the world at large and the dataset gets a digital object identifier (DOI). The digital object identifier uniquely identifies the data so that people who access and use the data are able to reference the data by citing the DOI number. The owner or originator of the data can then receive proper credit and recognition for the data and can keep track of the data usage and impact, in a way very similar to a journal paper.

A significant advantage of having the data on DesignSafe is the access to a variety of tools to process, simulate, analyse, and visualize the data. There are data processing tools, simulation tools, machine learning tools, and visualization tools. Some of these are open source and some of them are proprietary, in which case they can be used under a "bring your own license" scheme. For example, it is possible to have access to MATLAB free of charge if the user is in the USA. From outside of the USA, one can access it with the owner's licence.

The reconnaissance portal stores data from post-disaster reconnaissance surveys for any natural disaster that occurs all around the world. These include reports, maps, pictures, videos, etc.

DesignSafe already affects disaster research. For example, it enhances the damage assessment workflow when a disaster occurs. DesignSafe facilitates the activities for pre-deployment, during the deployment, and after the deployment. Starting at the pre-deployment stage, DesignSafe provides access to dedicated channels on Slack, which is a communication app, where people involved in a particular reconnaissance effort can organize themselves and communicate effectively. During the reconnaissance, researchers can post and store data on DesignSafe in quasi-real time. Once the reconnaissance field data is uploaded, the researchers have access to a variety of tools to work within the platform to do curation and quality control and the analysis and visualization of the data. Finally, they are able to publish the data on DesignSafe in record time.

These accelerated workflows contribute to speeding up the resilience of a given community. Traditionally, there is going to be a significant time lag between the start of an event and the final recovery of the community. The field assessments will take some time to take place followed by post-processing and traditional analysis. That could take anywhere between one and five years. Only after that lengthy period, the data can be disseminated and become available. Often times, it is too late as the community may have already rebuilt and they might not benefit from any lessons learned from the event. The DesignSafe accelerated workflow shortens the time to post-process and analyse the data so that the data can become available much quicker in less than half a year or between half a year and two years. This will accelerate the rate of learning from natural disaster, and the data will be available to the people that need the data much sooner than before.

The DesignSafe platform is now (2021) in its sixth years of development. It is in constant evolution and constantly updated with new capabilities. For example, a new social science data model was recently added to address the need of social scientist.

The learning centre on the platform provides all kinds of tutorials and resources to help users navigate the platform.

1.3.3 Dr. Elisabeth Krausmann, European Commission-Joint Research Centre (EC-JRC), Italy

Dr. Elisabeth Krausmann, European Commission-Joint Research Centre (EC-JRC), Italy

Dr. Elisabeth Krausmann's presentation was on "Disaster Risk Management Knowledge Centre: A Collaborative Approach to Foster Resilience." The[15]Joint Research Centre (JRC) is part of the European Commission (EC) and is its science and knowledge service.[16] The JRC works on knowledge production and management and technology transfer. JRC's role is to provide scientific support to the EU policy cycle, including policy design and development up to its implementation. The Disaster Risk Management Knowledge Centre (DRMKC) at the JRC was initiated four years ago, and Dr. Krausmann introduced its concepts and examples of concrete outcomes and achievements to date.

[15] The report is based on the presentation by Dr. Elisabeth Krausmann at the 4th Global Summit. It has been reviewed and edited by Dr. Elisabeth Krausmann.

[16] EC-JRC website—https://drmkc.jrc.ec.europa.eu/.

Dr. Krausmann stated that climate change has not spared Europe. Many of the disaster-related issues faced by Europe are due to climate change. For example, in August 2018, temperatures in Spain and Portugal were predicted to exceed 48° which is an unprecedented scenario for Europe. Consequently, there were severe wildfires. In January 2019, there was flooding in various places in the EU. With specific climate change scenarios and simulations about how the temperatures and precipitation levels are going to change in the European Union, vast problems are foreseen in future due to climate change.

Another concrete example shared was of different critical infrastructure sectors looking forward to the 2020, 2050 and 2080s in terms of economic damage due to climate change. For example, in a 2018 study, coastal floods were predicted to cause 26 million EUR in damage per year by 2020 due to impacts on transport infrastructure.

There are other threats and vulnerabilities of importance in the EU, such as natural hazard risks, technological hazards, and Natech accidents heavily influenced by climate change. Then, there are other risks that have crystallized in our societies that might impact the economy, trust in institutions, in governments and might eventually even affect democracy (e.g. security-related threats).

It is necessary to address all these issues and to reduce the risks. The JRC produces knowledge on disaster risk management and has launched major knowledge management initiatives. More specifically, JRC looks at the whole policy cycle with a multidisciplinary approach. It covers disaster risk management under the EU civil protection programme, and it includes natural and man-made hazards risks, pandemics epidemics, and also other security threats like terrorism, cybercrime, or hybrid threats. To deal with the information overload, the JRC aims to make sense of information from different disciplines and filters and distils knowledge out of that information. The messages created in this way can then be transferred to and understood by the policy-makers to act upon. This is not a one-way road because the policy-makers need to tell the JRC what kind of information they require. This consequently led to the establishment of the Disaster Risk Management Knowledge Centre to be able to cope with the multidisciplinary and multi-stakeholder challenges related to disaster risk management.

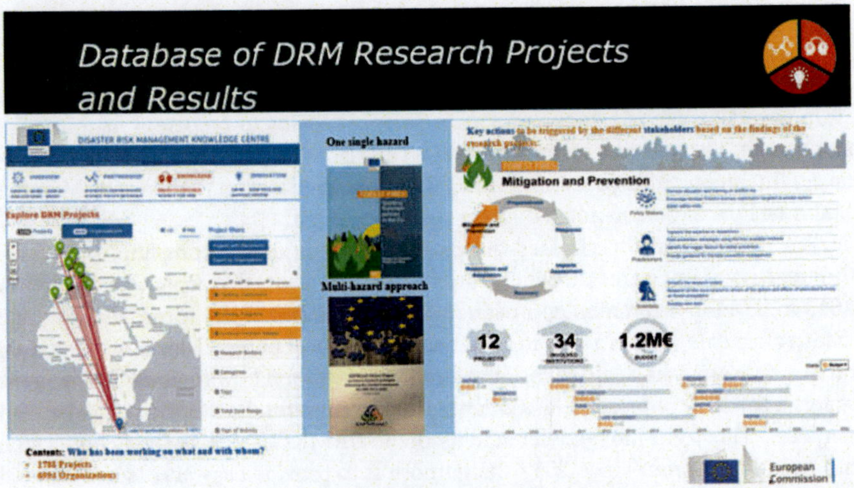

Copy from presentation by Dr. Krausmann

She continued to say that the DRMKC operates within the three pillars partnership, knowledge, and innovation and addresses three main stakeholder groups which are policy-makers, scientists, and practitioners. The DRMKC is supported by several Directorates General within the EC which have a stake in DRM, and it collaborates strongly with the EU Member States. This enhances the effectiveness of the work done by the DRMKC. It manages and transfers knowledge across sectors. The approach is by integration and consolidation of existing multidisciplinary knowledge to co-design and co-develop innovative solutions so that it can address the different needs of disaster risk management.

In practical terms, this is achieved by engaging policy-makers, scientists, and practitioners in the DRMKC's activities, e.g. in multi-stakeholder workshops or in training events. In the frame of the DRMKC's support service, it is possible to invite experts, EU Member State authorities, etc. to help them in case of problems related to achieving DRM objectives, e.g. if they need technical support with risk assessment. Through the JRC, the DRMKC is at the EU science-policy interface which gives it a privileged position to interact directly with the policy-makers and EU Member State authorities.

One work aspect of the DRMKC is to identify gaps in DRM. To avoid duplication of efforts, the DRMKC maps which data and tools are already available and what is still missing. Dr. Krausmann shared two examples of existing DRMKC outputs. The projects Explorer collects information on DRM projects to provide information to researchers on who has been working on what and with whom. It is also open to non-European projects. Another example is a report on the State of Science in Disaster Risk Management which is a 500-page book, which contains the state-of-the-art knowledge on disaster science in Europe. It has also become the basis of 1-week training courses on disaster risk management at universities.

Another book with over 200 authors is in progress which looks at the impacts of disasters. This report is going to be released at the end of 2020. Most of the authors are from Europe with a few from outside of Europe who are contributing to the report. It is felt that it is important to also capture experiences from outside of the EU in case they have relevance for European affairs or concerns.

In the EU, there is a requirement for Member States to prepare a national risk assessment. They need to identify and report all risks that could have national importance. To support this process, the DRMKC has prepared guidelines for Member States on how to do a national disaster risk assessment.

The DRMKC also addresses the problem of loss data incompatibility due to different standards across different disciplines. Work on this topic was started in 2013 to develop a common approach on how to systematically record and manage disaster loss data. This is a very important topic as a lot of political decisions are based on loss data analysis. Guidelines have been developed on how to do that. The latest report from 2017 proposes a loss database architecture for disaster risk management purposes. This feeds into another activity of the DRMKC, the Risk Data Hub, which has been in existence since 2017. Its purpose is to provide easy access and curated risk data in the EU. It provides risk data as datasets, methodologies, or tools. The data is accessible to everyone who has an interest in all hazards and risks, including technological hazards.

Another initiative of the DRMKC is the integration of different EU early warning systems. For example, there is the European Flood Awareness System (EFAS), the European Forest Fire Information System (EFFIS), the European Drought Observatory (EDO), or the Global Disaster Alert and Coordination System (GDACS). The idea is to integrate all these systems and to make it a more powerful tool with different linked functionalities. The tools should also include RAPID-N, used for Rapid Natech risk assessment and mapping. Plans are underway to link RAPID-N to GDACS, and when the alerts go out to the emergency responders, it can include Natech warnings as well. There is also the Copernicus programme that supports emergency responders by increasing the timeliness of satellite mapping production.

In closing, Dr. Krausmann reiterated that for increasing the resilience of society, knowledge transfer, the co-design and co-development of solutions, and the uptake of innovative solutions are key. This needs to be supported by risk awareness, communication, training, and education.

Note: Further information can be found at EC-JRC website https://drmkc.jrc.ec.europa.eu/. Also refer to the DRMKC paper of M. Marin Ferrer et al. in the proceedings of the 4th Global Summit".

1.3.4 Dr. Craig Davis, Founding Executive Committee Chairperson, Infrastructure Resilience, American Society of Civil Engineering (ASCE), USA

Dr. Craig Davis, founding Executive Committee Chairperson, Infrastructure Resilience, American Society of Civil Engineering (ASCE), USA

Dr. Craig Davis, founding Executive Committee Chairperson, Infrastructure Resilience, American Society of Civil Engineering (ASCE), USA, on “Infrastructure Resilience: A Framework for Assessment, Governance and Management”.[17] Dr. Davis presented the development framework for assessment, governance, and management by ASCE and JSCE. The partnership between ASCE and JSCE started in 2017 with the purpose of creating a global infrastructure resilience framework to improve global understanding of resilient infrastructure systems and aid institutions in focusing towards common goals for research, development, and implementation and to fill existing gaps.

Dr. Davis explained in detail how the framework is structured and its operation. The framework is divided to assessment, governance, and management, and it is

---

[17] The report is based on the presentation by Dr. Craig Davis at the 4th Global Summit. It is reviewed. It has been reviewed and approved by Dr. Davis.

broken down to eight elements of processes supported by data needs, sources, and expert opinions.

The framework fulfils an existing gap. It supports the Sendai Framework targets in many different ways. Most significantly, by substantially reducing the disaster damage to critical infrastructure and disruptions to basic services which is very much this framework is all about. It also helps to reduce the direct disaster economic loss; substantially reduce global disaster mortality; and substantially reduce number of affected people globally.

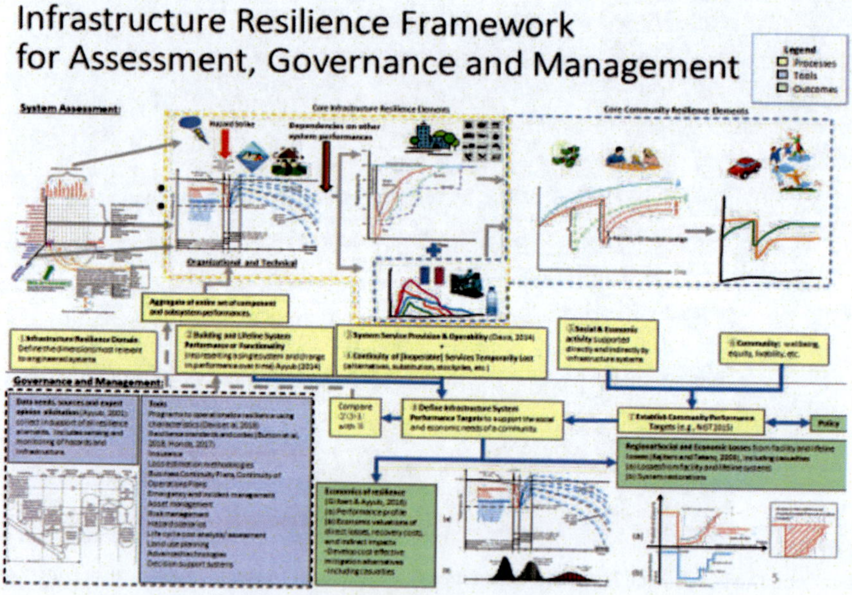

A copy from the presentation by Dr. Davis

He stressed that although many other similar resilience frameworks exist, they are not sufficiently comprehensive to fully address infrastructure resilience in support of the community resilience and the Sendai Framework.

The framework diagram consists with two parts—the upper half on assessment to management and governance and management in the bottom half. He continued to explain the eight elements that represent the process of the framework.

First element 1—**Infrastructure Resilience Domain**—Define the dimensions most relevant to engineered systems—it is important for people to know what is meant by infrastructure resilience. Resilience which is about the infrastructure system, building cluster in all of the life cycle systems covering transportation, water, sewer, floods, flood control management, communication, power, oil and gas as well as all hazards and urban hazards.

Element 2—**Building and Lifeline System Performance or Functionality**—function of the organizational and technical aspects of each of these systems.

You can look at that when hazards strike, there is deterioration over time, and there are different types of losses and ability function.

Element 3—**System Service Provision and Operability**—this is about the dependencies on other system performances—how multiple services and services to customers are affected or damaged by hazards.

Element 4 is—**Continuity of [inoperable] Services Temporarily Lost** when you lose the services, there needs to be quantification massed out in what is lost in what needs to be provided to the customers. Using water is an example here. When you lost the water services, then the emergency organizations and agencies themselves need to have an understanding how to provide the water to the customers. Because, it cannot be handled by the system itself.

Element 5—**Social and Economic activity** supported directly and indirectly by infrastructure systems.

Element 6—**Community:** wellbeing, equity, liveability, etc.—livelihood of people supported by their economic and social activity.

Note: Elements 2–4 are elements within the core infrastructure resilience elements, and 5–6 are core community resilience elements.

There is an overlap of the elements—Elements 4 and 5—so the core issue of continuity of services—when lost, still has the ability to support the community and livelihood through all these sources until the systems are recovered.

Element 7—**Establish Community Performance Targets**—which is related to Elements 5 and 6.

Element 8—**Define Infrastructure System Performance Targets**—to support the social and economic needs of a community—this is supported by Elements 2, 3 and 4. When this is compared with the earlier Elements 2, 3 and 4 with Element 8. Element 8 also supports the outcome for economic to resilience and regional and social economic losses.

He concluded by stating that all is supported by data needs, sources, and expert opinion elicitation (Ayyub 2001)—it also collects in support of all resilience elements including sensing and monitoring of hazards and infrastructure.

Note: For detailed information, refer to the paper by Dr. Craig Davis under Part 4—Plenary Sessions.

1.4 Plenary IV: New Scientific Challenges to DRR and DRM

1.4.1 Prof. Ortwin Renn, Scientific Director, Institute for Advanced Sustainability Studies (IASS), Germany

Prof. Ortwin Renn, Scientific Director, Institute for Advanced Sustainability Studies (IASS), Germany and Prof. Hirokazu Tatano

The presentation under the final plenary session on "Scientific Challenges for Disaster and Risk Governance" was presented by Prof. Ortwin Renn, Scientific Director, Institute for Advanced Sustainability Studies (IASS), Germany.[18]

Prof. Renn shared his thoughts about the challenges for science. Either coming from a practical background or from an academic background, it is very important to understand what the major contributions of science and technology development are, and assessment for disaster research, even more, for disaster management and governance. Three major types of transformation experienced are global in nature, i.e. globalization, digitalization, and sustainabilization. Transformation is the transition that feels partially pushed by people or by nations or by major actors to make sure

[18] The report is based on the presentation by Prof. Ortwin Renn at the 4th Global Summit of GADRI in 2019. It has been reviewed and approved by Prof. Renn.

that things are changing. The first one, globalization—it has all its imperfections and also its conflicts that go along with it and is part of what have been experienced for the last three decades that have tried to globalize. For example, globalize commerce, globalize communication, and a part of being here at the 4th Global Summit in a global community is certainly also a very good sign for this kind of globalized activities.

The second major transformation is digitalization which is a part of the technical modernization development that is facing. Specifically, automation and all the algorithms that come in and that partially have designed our daily lives all the way up to mobile phones and everything else. The last stage of digitization becomes autonomous. Autonomous driving, Internet, and all kind of new developments in which the technology itself takes proposition of properties.

The last one though does not exist, but it fits well with the other two is sustainabilization—that is our acknowledgment that there are limited resources. Not only natural resources, social and economic resources as well. One must make sure to handle these resources prudently overtime in order to have a durable development that creates humane conditions from humankind.

He stated that these three global transformations, perhaps other, are specifically important and have a lot of unintended side effects. Apart from globalization, industrialization, and definitely global environmental changes, there are lot of environmental changes, climate biodiversity, environmental health, and others that are related to modern lifestyles to a very strong industrial base in consumption insulting in increased vulnerability. That is specifically between the interactions among the technological, social, and natural risks.

All types of different hazards experienced today are because they have been interacted by technology development, industry development, settlements, land use patterns, interaction with all kinds of natural processes, and these create a specific composition of habits. All of that is partially due to organization and demographic changes. There is an ageing population in China and Europe, and a very young population coming out in Africa and other countries. There is the strong tendency towards worldwide migration and land use planning. All of those are kind of issues with new risks. Specifically, the intensity of risk has increased because there is more density, more urbanization, more people, and they are all affected. At the same time, there are wildfires and other disasters which weigh huge damages. Because there are very close settlements, and if one is being affected, it has major impacts on everybody else.

There are governance deficits too. The governance of these systems is not working well and perhaps caused by due to corruption; there is the tendency to renationalization; there is anti-movement to globalization and a lot of new authoritarian leaderships. That is an issue of political judgement. Nevertheless, it is something that creates new risks, political risks. Populism on one-hand side and isolation are on the other. Then of course, it creates a lot of tensions which can also lead to violence and lead to other kind of problems.

Another very important element is that although absolute poverty levels have decreased and become improved; there is even more dramatic equity problems with respect to vulnerability between and among people of different nations. This in return creates tension as the least income environments have the highest vulnerability to be affected by either natural or technological risks.

This is a challenge then to risk sciences. It is being mentioned several times that there is a major increase in scientific studies and analysis. Yet it seems that it does not have a lot of impact on the success of risk reduction and building resilience structural changes.

Science is not the only knowledge provider that counts when we talk about disaster and the talk about risk management and governance. There is a lot of other types of knowledge, for example, tacit knowledge, indigenous knowledge, and if these are to be integrated, it will play a much for sophisticated and enriched role in management. There are missing bridges between knowledge and action. There is a lack of convincing concept in science that could make that bridge become more effective. It is called formative research, transdisciplinary, and many other different words to mean the same thing.

To understand the causes of systemic risks, it is the manifestations of the three major transformations of globalization, digitalization, and sustainabilization which tend to be transboundary, highly interconnected, and complex. They are stochastic and no deterministic relationship and very often nonlinear. They have tipping points, nothing happens before the tipping points, and everything collapses afterwards. It is difficult to deal with and to quantify. Then, there is the plurality of knowledge claims and assessments about them. A lot of counterintuitive implications that science is really necessary because of what people think about them may not be true. One important thing is that all humankind learns by trial-and-error. If one has risked with a tipping point, trial-and-error is not a good learning method as nothing happens before the tipping point is reached. It is necessary to anticipate the error beforehand, and this is part of science.

To bridge the gap between knowledge and action, what is important is to say is that there is knowledge and there was action. If steps are taken between passivity from the perspective of what science can do, while it is starting with the knowledge, it is important to investigate what is true and what works and then form the judgement. Very often in science, it is only looked for the knowledge and perhaps monitoring. Everything else is outside of reach and that is part of the problem facing by all.

The transdisciplinary concept of science which is basically are composed of three major elements. One is very classic research—it is more necessary than ever if to think about complex and stochastic processes. For example, curiosity-driven science with lot of methodological rigour, with open questions, almost value free, just telling what is true. This kind of classic research can enlighten to make sure that the situation is understood.

Secondly, goal-oriented investigations—to have coherence strategies to reach a predefined objective. For example, to erase poverty, or to reduce the global temperature increase by 1.5 degrees. All of these objectives need investigation on how to reach those objectives. This is very different from classic research. It is really saying which strategies would lead to the goal one wanted to accomplish and what are the unintended consequences of each of these strategies. That is very important for policy-makers. For example, there are four options to get to point A. Each option has its own benefits, but it also has its own risks and own advantages and disadvantages. Now, it is up to the person to decide what to do and that is something that decision-makers actually do need. And it really believes in and it refers to judgement and to good judgement.

And last, a novel suggestion is catalytic expertise. Catalytic expertise is a chemical expression to say, is the sum of a substance that helps other substances to have combined and to come up with this new substance, and this is used as an analogy. It is analysing, designing, and facilitating processes or institutional arrangements to initiate constructive and productive learning among and between different knowledge gaps, interest groups, and value orientations.

For example, specific social scientists are asked to provide better architecture in models to reach consensus, and on how to improve the political will to do something better, or how to get the stakeholders together. Here, the important element is to research, and intuition is not the best guide for giving the kind of expertise on how to make processes work, which are rational, that are in value, and evidence informed to reach the kind of goal one anticipates to achieve.

This leads directly to four functions of scientific evidence. It is necessary to enlighten policy-makers and inform them about complex relationships. Very often, they have very simple solutions to complex problems. As scientists, it is necessary to provide them with orientation assistance for foresight, vision, and planning scenarios. This is something that they can understand as how to model into the future. What is needed here is instrumental and strategic planning to predefine goals, strategy elicitation, strategy assessment, rate of analysis, and prioritization that relate to this kind of a goal-oriented science. It is important to find good methods, and this is a scientific challenge for co-creating with others. Also with policy-makers, the kind of knowledge and judgement is necessary to make the changes that we want to make.

Therefore, this calls for an integrated approach. Policy-makers need scientific support in all substantive fields, enlightenment, orientation, strategy, and co-creation. These different needs require very special discourse formats that have their own normative rules and implementation criteria. That is where the catalytic science comes into play. There are the normative rules, as one's own implementation criteria. These four different functions need to be addressed in a way that is most effective. And the complex risks and disaster problems demand all four combinations of discourses. They may start out as a more epistemic discourses, lead to orientation where do we want to go or what are the best strategies, and how can we use these strategies in a way that all cooperate together.

In conclusion, there is a new role for science. Just providing casual or functional knowledge is not sufficient. There should be a focus on the combination of curiosity driven, classic concepts, or functional knowledge. It needs to be goal-oriented, strategic, instrumental concepts and requires a process-oriented catalytic concept. All of that need to be synthesized in an integrated governance approach, specifically when talked about systemic risk that can provide understanding, orientation, good strategies, and co-production of knowledge and actions.

1.4.2 *Prof. Mohsen Ghafory-Ashtiany, Earthquake Engineering and Risk Management, International Institute of Earthquake Engineering and Seismology (IIEES), I. R. Iran*

Prof. Mohsen Ghafory-Ashtiany, Earthquake Engineering and Risk Management, International Institute of Earthquake Engineering and Seismology (IIEES), I. R. Iran

Prof. Mohsen Ghafory-Ashtiany, Earthquake Engineering and Risk Management, International Institute of Earthquake Engineering and Seismology (IIEES), I. R. Iran presented on "Closing Gaps Between Scientists and Decision Makers by Creating Demand for Safety as a Basic Human Right".[19]

In his presentation, Prof. Ashtiany discussed about the different types of disaster reduction frameworks and platform at national and international levels but with one common message in different languages which says saving of human lives and resources. If it conveys the same message, why is it necessary to have different platforms? Perhaps, it is because when one system appears to fail, another framework or a different language is brought forward with the same one objective to save human lives and resources. If the language has to be changed throughout the past 30

[19] The report is based on the presentation by Prof. Mohsen Ashtiany-Ghafory at the 4th Global Summit. It has been reviewed and edited by Prof. Ashtiany.

some years, the fact remains that the existing knowledge that has been created especially after mid 60 and 70 s, technologies, standards, guidelines, strategies, policies, according to international levels, one expects to see change with these policies and initiatives, and it would make our built environment safer. The built environment would be expected to or would have at least met the minimum safety and quality standards. However, it has not achieved this goal. All global, national reports these days show that the world is still living in a vulnerable environment with high risk. Disaster reduction strategies and plan have not yet effectively implemented around the world. And the proof of this statement is the casualties and losses that are seen or heard of after each disaster event.

Given the above background, Prof. Ashtiany endeavoured to talk on the topic of demand for safety as a basic human right. Going back to the history of disaster risk reduction and disaster response, it was science development introduced by Frank Press in 1984 at the World Conference. This was followed by other proposals in and current by the Sendai Framework. As initially mentioned, it all has the same message in different language, i.e. saving human lives and resources.

The world experiences an increased number of natural disasters either in the form of hurricanes in the USA or man-made or natural disasters in the developing country. The world is more prone to care and invest extra on response rather than prevention. At a recent meeting, gaps between the science and policy were discussed. The reasons given were the challenges and difficulty to find solutions. Possible reason for this is due to not efficient use of know-how and/or safety measure-related issues that are not integrated into the development processes in each country. There is knowledge, but development is not in conformity with it and as it is the situation in most of the cases but not necessarily in everywhere. The main challenge remains in implementation of safety and disaster prevention and management of know-how in the development. Steps are needed to bring this knowledge into development process. It is necessary to make a bridge between science and development policy for it to be fully operational and apply mainstream science into policy.

In other words, it is essential to see and feel the footprint of science and technology in the decision-making process in the cities. This task should become the future challenge or challenging role of disaster risk institutions, and it is to be one of the challenging roles of GADRI.

He further emphasized that existing structures of research institutions, which are mainly knowledge development bases, should be changed towards people-safety-demand-oriented and provide more doable and affordable technical solutions instead of focusing on numerical modelling, etc. This is not negating supercomputers. It is a positive advancement. But how much can the people feel the result for themselves. For example, how can the cities be made safe unless the people ask for it. Does one wait until the people ask for safety? It is important to create demand for safety.

If one looks at all of these in process in policies and strategies, special emphasis can be given to public awareness and preparedness, public health, safe housing, schools, hospitals, infrastructure, and good governance. But, these operations require implementation oriented, demand-driven science and technology, and good decision-making processes. For example, the mobile phone—it is a demand-driven science application to which requires a payment. When something works, and useful, people invest on it. However, people hesitate to pay for housing safety as it is feared that it may not work.

Another method to make the decision-makers to understand is by changing the language. Should safety be referred as a basic human right, or perhaps, there could be a way to draw the attention of the government and decision-makers to make one's right to be safe and build sustainable cities. Demanding for this basic right requires a risk-aware society. And the reason, risk-aware society requires an effective risk communication. The public should demand for safe housing, safe city, safe infrastructure, reliable energy, good water management, clean air, correct usage of water, and water resources from the government.

People are always been told by the governments, to follow the rules and guidelines, without them providing the infrastructure, and systems that building codes and standards that could be easily implemented. It should be a two-way street between the people and the government.

How does society understand earthquake threat and what can be done about it? What are the factors and ultimate results in behaviour change that reduces the risk and bring safety and sustainability? How existing scientific and technological knowledge can convince the public and decision-makers for effective implementation? Prof. Ashtiany stated that these are things that need to be done in future. How can we close the gap between the science, policy, action, science, and governance? The researchers need to be in dialogue with the policy-makers. Implementation science is different from the evidence science. This should be doable, usable, affordable, adoptable, and be simple solutions. The policy-makers need a special or a different framework of language for them to be able to understand the disaster risk situations. Each of us have a role, for example, scientist have a role, decision-makers have a role, and engineers have a role. However, each person is playing each other's role instead of playing one's own role.

In concluding, Prof. Ashtiany stressed that a paradigm shift is needed in international funds from post-event to real prevention and mitigation. GADRI could take a step forward to create a decision-making toolbox which incorporates all sectors such as governments, private, insurance, public, and science.

1.4.3 *Prof. Desmond Manatsa, Full Professor, Geography Department, Bindura University of Science Education, Zimbabwe*

Prof. Desmond Manatsa, Full Professor, Geography Department, Bindura University of Science Education, Zimbabwe

Prof. Desmond Manatsa, Full Professor, Geography Department, Bindura University of Science Education, Zimbabwe presented on "Taming New Scientific Challenges to Disaster Risk Reduction (DRR) and Disaster Risk Management (DRM) in Africa".[20]

Presenting on new scientific challenges to DRR and DRM: Taming new scientific challenges to DRR and DRM in Africa, Prof. Manatsa stated that when faced with the new scientific challenges all over the world, and the way these affect everyone is felt quite differently. In a developed world, the effects are more positive, but in Africa, it is more of a challenge. His message articulated, how as a global village, to term these challenges in Africa to become positive.

He started with disasters in Africa. Africa is the only continent whose share of reported disasters in the world total has increased over the past decades. This

[20] The report is based on the presentation by Prof. Desmond Manatsa at the 4th Global Summit. It has been reviewed and approved by Prof. Manatsa.

message resonated throughout his message. While disasters in other parts of the world are decreasing, disasters in Africa, they are increasing. The disasters in Africa are hydro-meteorological, with draughts on top followed by floods, and now, it is encroaching to even flash floods, because of settlements which are near rivers and also in urban areas. These disasters are bound to increase in frequency due to climate change. Climate change is compound or already compounding the manifestation of these major disasters. The economy of Africa, as a developing nation, the resources or finances are quite limited, and climate change is going to further reduce the economic activities of Africa due to available resources that mobilize to reduce the damages caused by these disasters. Hence, these disasters are becoming an impediment to sustainable development in Africa. The disaster profile of Africa is closely connected to the vulnerability of its population and economies that are exacerbated by minimum coping capacities. Most countries have limited resources invested in disaster risk reduction or actually do not prioritize investing in DRR, or they have minimal physical space to live in until recovery efforts are in place after a major disaster. However, disaster risk reduction policies and institutional mechanisms do exist in various degrees of completeness. Even though they exist, their effectiveness is limited. This requires a strategic approach that is informed by new scientific knowledge.

Through examples of the extent of the disasters, Prof. Manatsa continued to share information on various disasters impacting Africa. For example, if it is drought which is more concentrated by the eastern parts of Africa, then closely connected to drought is famine. Another is floods, and closely connected are diseases. Drought, famine, and floods, all these three disasters are intrinsically linked to climate change or there being exacerbated by climate change. Disaster risk management challenges in Africa are mostly related to the country's level of development which is characterized by limited fiscal resources, low resilience of economies, limited infrastructure, weak governance, institutional capacity, and limited knowledge base. Emphasis is placed on limited knowledge base. What are the challenges to new scientific knowledge for Africa?

The new global scientific knowledge has become increasingly complex. Scientific knowledge is being produced now more than any other time in history. It is available at just a mouse click, and you are exposed to unlimited amount of scientific knowledge. For example, earth observation is now producing more than 10 terra bites of open data per day. This dimension and this demand approaches for data handling and processing of which Africa has limited capacity. If all of these data that is being generated is needed, in Africa, there is no capacity to harness it to its advantage. More so, the availability of data on the Internet is also created more actors in disaster management who do not need or consult experts from other fields. Even if they do not have the expertise to analyse that data. This situation has put disaster managers at risk in their own decision-making process.

For example, when the tropical cyclone Idai hit Mozambique, instead of disaster managers going to the Met office, or to the experts to solicit for information on what to do or to lean the characteristics of this tropical cyclone, they just go on the Internet to look at the trajectory of the tropical cyclone. They were not interested in the model or its efficiency. Instead, greater trust is placed on the information in the Internet which leads to creating a disaster. People are more reliant on the Internet and believe it has all the solutions. It needs an expert to tell them where the tropical cyclone is going to affect, how many days it is going to be there, and the amount of rainfall that it is going to deposit on that region. This type of challenges is not only among institutions but in society where the Internet is creating a problem in making the people believe that everything on Internet is true and can be acted on it even if it is risking one's life to do that.

Now, the message that is resonated through this is that new scientific knowledge, which abound, which is everywhere in the Internet, and it is being misused. It has created institutions which are now facing their decisions on monitoring disciplinary, rather than performing a transdisciplinary matrix in Africa. A transdisciplinary and holistic approach in science is the prerequisite for effective DRR and DRM, yet many are actually diverging from the core of effectively tackling DRR and DRM because of this huge amount of data suddenly being made available.

In conclusion, Prof. Manatsa requested the GADRI members who are from the developed world and who are very much experienced in DRR and DRM institutions, to chip-in and assist Africa. How could this be done? Through thriving to capacitate the institutions from Africa, to enable them to more systematic in knowledge management or by organizing funded webinars on different topics to AADRI members. It is time to rethink that Africa do not have sufficient funds but need the knowledge to be hand-in-glove with the rest of the world, reduce the risks, and manage disasters. It is possible to train or do joint research publications on DRR and DRM. Especially through the platforms created by GADRI and its regional alliances.

Note: Also refer to the paper by Prof. Desmond Manatsa et al. under Part 4—Plenary Sessions.

1.4.4 Prof. Rajib Shaw, Keio University, Japan; and Chair, Scientific and Technical Advisory Group (STAG), UNDRR

Prof. Rajib Shaw, Chair, STAG, UNDRR on "Advances and potentials of science and technology in implementation of Sendai Framework".[21] Prof. Shaw shared some examples of the work that is underway at the global levels with more focus on the regional levels, and gaps or issues, especially on the science technology in DRR which are more regional specific, to Asia. For example, the 2016 UNDRR's, Asia Science and Technology Advisory Group did an analysis of eleven different countries on how science and technology are trying to influence or contribute to different levels of decision-making in disaster risk reduction. Few countries in Asia were taken, and

[21] The report is based on the presentation by Prof. Rajib Shaw at the 4th Global Summit of GADRI in 2019. It has been reviewed and approved by Prof. Shaw.

three different sets of parameters were checked. Where is science into decision-making; where is investment in science and technology; and the link of science and technology two people. The survey was done by the A-STAG group members in collaboration with a disaster focal point as well as the science technology focal points in the specific location. It showed that some countries take a very strong focus of science in decision-making; and science and technology investment. But when it comes to the link of science and technology to the people, it is below the expected targets. In spite of a good mechanism for science and technology for decision-making, in spite of a good investment, it not possibly reaching the last mile which is the link to the people.

But on the contrary, in some countries which has possibly not that much high scores in the science technology into decision-making, nor high in the investment, but the link of science technology to people appeared to be quite high. One of the main reasons for that could be that there is a very strong legal systems on disaster management laws. The point stressed here is how much a legal system is very important in some cases to influence science technology to this particular last mile which is the linking to the people.

UNDRR's 2016 Asia Science and Technology Advisory Group conducted an analysis of eleven different countries to study how science and technology influence or contribute to different levels of decision-making in DRR. This particular research focused on the issue of gaps to learn which type of research studies was undertaken on disaster risk reduction, irrespective of the language. Yet, science in national decision-making or science investment does not necessarily link to the people's needs. That is the major challenge possibly recognized and thought earlier, as a strong existing gap in what is done and what is required.

There are different regional frameworks and global frameworks, but the most important thing is the national and local linkages. The governance system, or the legal framework, is also very important for science-policy direction. However, it is necessary to possibly go beyond, as scientists' approach and bring other stakeholders into the process.

On the subject of issues in research gaps, Prof. Shaw shared light on the analysis done using the Scopus database of the World Bank database. The analysis conducted was to see which type of research studies has been done on the research of disaster risk reduction, irrespective of the language. It showed that 0.22% of the global scientific research is on the disaster risk reduction. Whether this a low or a high value depends on the way people perceive it. However, in Japan, interestingly it is 0.6%. If this is compared with the global average, Japan's disaster forecast or disaster research forecast is almost three times than the global average. But, these two top points are debated on data versus publication. While the maximum number of data is there, the publication is very, very limited. The maximum number of publications comes, where the economic loss is very high, simply a north–south division. Many of the northern countries are focusing more on the economic losses and possibly not reaching the actual people or saving lives.

When looking at the cluster in Europe, it actually showed the importance of policy research in Europe compared to some parts in Asia. Perhaps, one of the major reasons behind this is the education system. There is hardly any interdisciplinary and multidisciplinary master's programmes. The higher education programmes are very compartmentalized. There is the need for a higher education system which contributes to types of research in the global level.

With those two backgrounds, on the point of the Science and Technology Milestone on global and regional levels, after 2015 March Sendai Framework, in January 2016 at the first global Science and Technology Conference, the Science and Technology Roadmap was adopted followed by several global and regional events organized to follow global science technology. There are ongoing events at regional levels. For example, Asia is the regional platform for organizing the inter-ministerial conference. It is very important to bring the science voice into inter-ministerial discussion. There were also similar types of regional platforms in Arab states, in Africa, in Latin America, in Europe. Prof. Shaw encouraged the audience to have more interaction with regional bodies, or regional platforms, or regional science and technology platforms and make an effort to influence them.

Prof. Shaw continue to describe priority actions for Asia Science and Technology Advisory Group. He described the 12 actions initially identified for Asia, which was agreed in Bangkok and the addition of two more actions in Beijing in 2018. A survey conducted in Asia with the help of the Integrated Research on Disaster Risk (IRDR) among a group of mainly consisting of scientists below 30 s or below 40 as well as some senior researchers, to see their needs, what do they think are their priority research areas, resulted in a very interesting finding. Almost 70% of the participants being in the academia or stakeholders, it was recognized that science is not only scientist's business.

It was interesting also to that the answer to as to which type of resources is more challenge, whether it is technical, governance or legal, human resources or financial resources. But, the result was government or the legal resources seem to be key issues seems to be more challenging to enable science-policy interaction.

With examples, Prof. Shaw portraited how science and technology can bridge the gap between scientists and other different stakeholders. Science in national decision-making or science investment does not necessarily link to the people's needs. The major challenge is to recognize that there is a strong gap exists between what we do and what is required, and there are different regional frameworks and global frameworks. Out of all this, the most important thing is the national and local linkages. In addition, he reiterated the importance of the governance system or the legal framework for science-policy direction. The system should go beyond the only-scientists approach and draw other stakeholders into the process.

1.5 Panel Session 1: Group Discussion Session I: Contributions to Increase the Effectiveness and Relevance of Our Institutes in DRR

Panel discussion session 1

Chaired by Dr. Srikantha Herath, Professor and Director, Center for Urban Water (CUrW), Min. Megapolis and Western Development, Sri Lanka, and Dr. Yuichi Ono, Professor, International Research Institute of Disaster Science (IRIDeS), Tohoku University, Japan.

1.5.1 Group 1-A: Water and Weather Related—Rapporteur Prof. Tetsuya Takemi, DPRI, Kyoto University, Japan

Group A focused on water and weather-related disasters from the scientific, engineering, and implementation perspectives with emphasis on the state-of-the-art concepts, methodologies, and approaches.[22] Evidenced-based research results, best practices, and approaches were shared and discussed during the session. Highlights

[22] The report is based on the rapporteur presentation by Dr. Tetsuya Takemi at the 4th Global Summit of GADRI in 2019. It has been reviewed and edited by Prof. Takemi.

showed the interaction between natural events and human-related events. Discussions also focused on understanding risks and dealing with emerging risks in each region.

As an important element, Prof. Takemi emphasized the importance of communication with technical development organizations for weather forecasting and other water-related warning as well as with local agencies and local governmental officials or other stakeholders.

Prof. Takemi emphasized the importance of technical development and communication for weather forecasting or other water-related warnings and the importance of communicating with local agencies and local governmental official or other stakeholders. He highlighted particular points that were of concern as a whole for the group in particular based on past research implementation and experiences.

- Understanding of risk and transfer of science and engineering knowledge on hazards and risks from one region to another.
- Dealing with emerging risks in each region and inter-relationships between academic, decision-makers, and stakeholders.
- Filling the gaps between academic sector and decision-makers.
- The above points were discussed with examples specific to water and weather-related disasters.

He reiterated that the scientific, engineering knowledge, and human society need to forge cooperation between designers, protected infrastructure, and analysis on behaviour responses in crisis situations by understanding cultural diversity around the world. It is important to produce information intelligence for end-users, and well-structured information requires diverse demands.

In addition, education is vital not only for researchers but also for facilitators in developing countries. Non-structural measures should be incorporated into disaster mitigating measure.

It was also noted that integration of technological and social sciences from the beginning of the analysis on design in structural and non-structural measures is needed. Once again emphasis was placed on the importance of education, for example education for mass media and local society.

The importance of creating some mechanisms for communication between science and technology community and policy-makers to understand the stakeholders or decision-makers perspective for increasing the effectiveness of academic sectors.

The warn-on-forecast, a US government initiative to improve weather forecast from observational-based to model-based (i.e. data assimilation of multiple, heterogeneous observations), is an ongoing challenge and has an important issue in conveying warning to the public for them to take proper actions.

Discussions focused on how to understand risk in general sense. Increase in capacity for disaster management is important to reduce existing risk and to prevent risk creation and to build resilience in recovery. Risk creation may emerge from the trade-off between the human interferences and upcoming challenges to the socioeconomic factors at longer timescales. Not to create new risks under urbanization and climate change is an important issue.

Climate change at a local level is not well known by decision-makers and general public especially in developing countries. For example, the impact of extreme weather on agricultural and its climate impact should be conveyed to the general public and decision-makers. Specific to the risk from water and weather-related phenomena, sharing weather climate data among countries is important.

Business continuity plan must be well recognized based on the experiences of damaging tsunami as an example.

It is important to recognize that water resource management should take into account risk especially in urban areas. Green infrastructure or eco-disaster risk reduction should be taken into account and explored to achieve mitigation risks in urban areas in various regions.

GADRI as a platform should prioritize how non-structural measures incorporated in disaster risk reduction, and GADRI can be a platform to transfer scientific and engineering knowledge from one region to the other by understanding cultural and societal diversity in each region.

Commentator Prof. Srikantha Herath commended Prof. Takemi for his very comprehensive and excellent presentation. He stressed that there were a number of very important points made from that group discussion. Probably, some of the points could be common to other groups as well. Especially the group emphasized the interface between natural and human and the importance of communication. It is important to recognize how one can learn how best to communicate of intelligence and also transfer of experiences across regions.

The take away points from the comments received from the audience are as follows:

- Importance of combining structural and non-structural measures—how non-structural measures could be incorporated with structural measures in DRR.
- Understanding cultural and social diversity. Add to that experiences on not only knowledge but also implementation practices and then incorporating the communication to make the last mile.
- Developing countries need to prioritize non-structural counter measures and the necessity to incorporate it. Another point of opinion shared was that it might be equally important to combine structural (resources) aspect as well. Structural should be complemented by non-structural measures. Such level of coordination might be very important in science.

1.5.2 Group 1-B—Earthquakes and Volcanoes Group—Rapporteur Dr. Craig Davis, Founding Executive Committee Chairperson, ASCE Infrastructure Resilience Division, USA

The Earthquake and Volcano Disasters group discussion session was chaired by Prof. James Mori, Prof. Yoshiki Ikeda and Prof. Hiroshi Kawase and co-chaired by Prof.

Khalid Mosalam and Mr Soichi Nakajima.[23] The discussion session summary was presented by Dr. Craig Davis.

He stated that although the session title included earthquake and volcano disasters, the group specifically focused on earthquake disaster-related issues with specific focus on:

- understanding disaster risk;
- strengthening governance;
- investment in risk reduction;
- enhancing disaster risk preparedness.

Using a matrix for each of the above-mentioned topics, discussions focused on short term which meant 1–3 years and long term up to 2030 earthquake disasters. Some difficulties were encountered defining whether the event is short term, long term, or continuous.

As an important element, Dr. Craig mentioned, he would strongly encourage to have all of the universities within GADRI and beyond to create courses and educational materials for students, for politicians and others. If it is to be taken to a second level, it would be to have that warranted culturally specific, and culturally meaning, and to a broader perspective. If that could be accomplished, the world would change in a significant way.

Understanding disaster risk. On reviewing short-term aspects, the need to open sharing of data and identify how and what tactics and what types of projects might be undertaken was discussed. The best method of accomplishing this recommendation is by sharing data. For example: use of modern artificial intelligence and machine language to analyse big data, and use of satellite data for reconnaissance.

Other suggestions included of developing methods for regional site characterization, which are useful when looking at a large region like city-wide or portions of city-wide to capture ideas on what the subsurface looks like to micro-zone for ground shaking or for liquefaction or for fault rupture. These are not well characterized and were highlighted as a problem in both, modern and developing countries.

The need is to promote success stories such as volcanic predictions. He stated that general physics for volcanic predictions is understood; deaths related to volcanic eruptions are small, and it is very difficult to control lahar or lava flow. In addition, the types of mitigation activities are different and smaller in land use planning and evacuation.

GADRI could promote interdisciplinary studies on especially on monitoring the landslides, environmental, ocean studies—getting this large group of multiple disciplines of experts working together to better understand how to help with the disaster risks.

In the short term, they suggested carrying out detailed post-earthquake surveys, investing in monitoring for buildings, micro-zonations or regional seismic networks, undersea tsunami monitoring which are very costly. If, at least, some plans were being

[23] The report is based on the rapporteur presentation by Dr. Craig Davis at the 4th Global Summit of GADRI in 2019. It has been reviewed and approved by Dr. Davis.

made, then maybe some of the investments could be undertaken by the government or another funding.

In the long term, one of the biggest problems in developing countries is non-engineered structures. To understand the situations and to assist with future disaster reduction, it is proposed to create an inventory for buildings, developing retrofit methods, carrying out historical studies on past earthquakes specific to regions.

Under strengthening of governance, he stated that the main focus was on communications. Promoting effective evidence-based science and engineering reports that would be more of a consensus base. Maybe, they have third-party reviewers and written to communicate to wide writing just like technical people in a very concise way.

He further stated that using the GADRI platform, it should:

- encourage teaching of students to communicate to government officials;
- encourage teaching of policy-makers to communicate with technical people;
- encourage teaching of technical people to communicate with the public;
- encourage more disciplinary communication among different technical disciplines;
- facilitate social sciences to teach technical people how to better communicate.

With regard to investing in risk reduction, in the short term, it was proposed to slow down bad practices as simple building and engineering practices can provide effective low-cost solutions. Though this is not a problem, it is not necessarily easy either to stop it. However, GADRI could take up the initiative to slow down some of these bad building practices.

Education and training are a very important aspect for distribution of information to many people about simple hazard and risk information. There is a lot of information now but if we looked at culturally specific issues in different regions, could we mass distribute this information and give at least a limited amount of initial teaching that would carry on progressively through the long term and hopefully by 2030 catch on and be self-perpetuating? We can start now and have this group through GADRI that discontinue this type of work.

There are long-term effects on recurring hazards and the risk associated. Having to deal with the difficult problem of low probability but high-risk hazards. How does that fit compared to high probability lower risk that you can expect every day? Events that happen once every thousand years or more, but when happen damage will be very severe. It needs to be better understood on the whole.

Setting priorities for implementing scientific results. We have heard this through our discussions today. That there is a lot in the protocol or the framework, the Sendai Framework, and you have to prioritize to get it done. Not everybody can do everything and basically nothing done.

He stated that it was suggested that GADRI could develop a series of Mass Online Open Courses (MOOCs).

Under enhancing disaster preparedness, we had the least amount of discussion on this yes but in the short term and long term encourage setting priorities for choosing efficient solutions and promote early warning systems for earthquakes and volcanoes which are very dependent on resources of the country. Here, emphasis was on the need to focus on actions. When the early warning systems are in place according to country-specific and regional specific, what actions do you actually take? Another long-term aspect would be to encourage and help undertake comprehensive studies on a response and recovery to help learn from these and to send that back to emergency responders, to make better decisions. This will, eventually, create great real-time decision support systems.

Prof. Herath commended Dr. Davis for his excellent and comprehensive summary. He stated that there were lot of references to communication and also some common elements are available for open data sharing and early warning systems.

- A comment included a case study in Indonesian village stressing the importance of local knowledge and whether discussions explored around local knowledge and how it impacts preparedness and how it alerts communities, specifically related to volcanic disaster. It is not only the social scientists helping the natural scientists to communicate with people but also learning together from them.
- Earthquake- and volcanic-related disasters have the highest impact on built-in homes and critical infrastructure compared to any other hazard. Given that, in many places, land use changes take place. On monitoring issues, it is not only monitoring, but also related to early warning issues which are very limited in the case of earthquakes. Given this situation, what scientists can convey to the policy-makers, or what are the short-term measures it should take account to evaluate how affordable and accessible they are.

Group discussion in progress

1.5.3 Group I-C—Geohazards—Rapporteur Prof. Michel Jaboyedoff, University of Lausanne, Switzerland

Although landslide-related loss of lives is high, it is yet the most underestimated hazard among natural disasters.[24] Statistically, landslide-related hazards are underestimated, and the perception of geohazards especially, landslides, is localized by people. Unlike earthquakes, landslide occurrences are fairly evident.

Discussion focused on improvement and assessment of vulnerabilities of geohazard and the need to push for global coordination of projects.

Following highlights are the important elements of the discussion:

- Establishing early warning systems and flexible solutions by finding ways to improve community participatory tools. This could be the way to manage to go through early warning systems and monitoring and inform people that they have to develop these tools and systems together with the help of the scientific community.
- Improving the perception of geohazard, and especially, landslides as it is localized. For example, it is possible to know where the landslides will occur unlike an earthquake occurrence where the location cannot be exactly pinpointed although it is possible to know about an earthquake going to take place but not exactly where the epicentre will be in advance. Such is not the case with landslides.

How to solve the problem with the available means.

- It requires multi-level communication and exchange by insisting on exchanges of dialogue with the politician and all stakeholders. Often, there is a problem between the local people in the decision-makers, and they have different perceptions of the consequences of hazard depending on the interest.
- It is necessary to improve the exchange between scientists and other stakeholders often as there are several types of knowledge.
- It is important to get everybody to understand each other and how geohazard disasters work.
- Geohazard can be dangerous because of different perceptions. People need to be informed about the real risks and see how they accept the situation. This means introducing participatory approaches to teach about existing risks and how to live with risks.
- One of the major issues is lack of participatory approaches. When a disaster occurs, it is not cost-effective and logistically not possible to remove people after the disaster occurs. There should be an acceptable risk threshold.

Further, the discussions had recommendations for GADRI.

It could be to bring in the good and bad experiences about geohazard mapping. Geohazard mapping is really important for the disaster risk reduction strategy implementation. When comparing issues and extracting the main processes, it is possible

---

[24] The report is based on the rapporteur presentation by Prof. Michel Jaboyedoff at the 4th Global Summit of GADRI in 2019. It has been reviewed and approved by Prof. Jaboyedoff.

to see how best to implement disaster risk reduction to obtain accuracy of different situations.

It could set up recommendations on how to implement disaster risk reduction by means of sharing knowledge and improvement of assessment of the vulnerability of geohazard which is up to now has proven to be very precise.

It is important to push for global coordination of projects, push for new technology, and provide recommendations for agencies with special needs in terms of resolution and characteristics of the data that requires an update.

What is really important is to develop those high-tech–low-tech flexible monitoring and early warning systems as it depends on the situation where people are able to locally implement. It is a simple means in some cases, and it is also possible to develop very high-tech monitoring systems.

Another point is to adapt communication to transmit the good knowledge and explore good practices for human-induced landslides due to the problems faced with human-induced landslides. For example, in Nepal which is a hilly country, there are many landslide disasters due to roads being cut through mountains.

Other issues to anticipate are new risks and especially cascading effects of different natural hazards. Cascade effects are becoming a global phenomenon in the global changes.

A proposed and viable solution is to disseminate exchanges around GADRI by creating an exchange platform for GADRI remembers. Through the platform, they share experiences and especially mapping and DRR experiences and their implementation. These could include good and bad experiences as well as sharing of data, recommendations, etc.

For example, phase one will be implementation or putting together guidelines. And proceed to add implementation tools in practice. The goal is to have simple documentation but comprehensive. Very often, the documentation is too complicated with different levels of reading for different levels of stakeholders although all the topics link to DRR, vulnerabilities, and so on.

The goal is to improve resilience and to create a panel of experts in multidisciplinary that can be dedicated to answer questions coming from various stakeholders. For example, using mobile tools to take a picture of a hazard and then sending it to the platform and asking their opinion as to what can be done for the current situation. It is important to have pilot studies where people can share experiences. The platform will be multilingual progressively.

Through examples, the need to communicate and exchange of activity were stressed. It is necessary to consider the problems and issues and link them to communication tools and documents.

Listening to local people and learning from local knowledge were highlighted as a tool for the scientists. For example, what are the quality requests in the format and at the end of the exchange between the stakeholders? There needs to be an assessment on the quality of the request and exchange between different partners.

Prof. Herath, one of the chairs for the session, thanked Prof. Jaboyedoff for the wonderful presentation and very structured set of recommendations. He further stated that the exchange platform captured many of the issues that needed to be further strengthened and discussed. Use of local knowledge and local intelligence is one of the topics widely expressed by the previous presenters with examples on earthquakes and early warning systems at the village level.

Comments from the audience included:

- During the discussion session, many concerns on communications and early warning systems which focus on how high-tech measures were expressed. The role that can be played by the local communities is often considered as these people are able to provide early warning intelligence.
- The application of high-tech measures, in many developing countries, has a lot of limitations due to communication constraints in areas that do not even have access to different types of networks.
- Local communities, most of them in developing countries have their livelihoods on land farming. Some landslides are slow onset hazards. They develop overtime. These communities see these cracks developing. If they are well educated and they are informed, they can act as early warning systems to warn people that this is what is happening.
- While discussing a lot about high-tech measures, it is also necessary to consider the role which the local communities can play in providing early warning intelligence. That can even help mitigate the impacts of geo hazards.
- On the point of low-tech–high-tech flexible monitoring system—It is usually the local people who are aware of disaster when they see cracks and know what may occur during the rainy season. For example, in Nepal where the people living by mountain sides, they leave their houses when it rains. While taking advantage of the local knowledge, it is better to provide them with some very simple comprehensive manual, but with simple instructions in the right language.

1.5.4 *Group I-D–I and II—Social Dimensions of Risk and Health in Disaster Risk Management—Rapporteurs Dr. Ram Sateesh Pasupuleti, Indian Institute of Technology, Roorkee, India; and Ms. Robin Eve Miller, Northumbria University, Newcastle Upon Tyne, UK*

This group was divided to two parts: one on social dimensions of risk and the other on health in disaster risk management.[25]

The purpose and objectives of the two sessions on Social Dimensions of Risk and Health in Disaster Risk Management were:

- to identify how research on the construction of risk can be used for effective DRR;

[25] The report is based on the rapporteur presentations by Dr. Ram Sateesh Pasupuleti ad Dr. Robin Eve Miller at the 4th Global Summit of GADRI in 2019.

- to consider how the understanding of risk can be better circulate it among communities and discuss among stakeholders also;
- to work out how the understanding of risks can help to design the countermeasures of DRR;
- to reflect on future research directions relating to the social dimension of disaster risks.
- Selected key questions addressed were:
- to what extent these risks and disasters within society currently well known, and we went through a lot of case studies which will really interesting;
- how can the social construction of risk be effective for DRR;
- how can an improved understanding of risk be communicated around varying DRR stakeholders;
- in what way does understanding the social dimensions of risk help us drive DRR forward.

The two groups discussed the issues separately and combined to discuss both topics on health as well as a social dimension to compile the outcomes of the discussions.

Social Dimension of Risk presented by Dr. Ram Sateesh Pasupuleti—The discussions focused on the context of risk and living with the risks. To define risk, risk itself is a social construct due to the nature of unequal distribution of resources, skills, and abilities, and also due to the fact that victims are from different genders, differently abled individuals and social setups, and so forth. When it is about the social dimension, it is thought about even before it addresses the society. It starts with "I". The concept of I is accountable for we, you, and ours, and theirs, and it continues to grow. Further, the accountability of I is very important. How does one look at the risk? How is it not only considered in one's own context rather it can be looked in a holistic manner.

The concept of "I" although it is discussed within the social dimension, it is taking the inner spiritual dimension of the social dimension. It would be good to see if stakeholders consider the spiritual dimension of such social dimension as part of DRR.

In terms of knowledge, much emphasis was on local knowledge systems. How and what experts understand about risk is very different to how the local knowledge person understand and perceives the risk. There are not sufficient and adequate inquiries pursued to ascertain on how the local knowledge can help in reducing vulnerabilities to disaster risks. It is worthwhile to take a step back and think about how the local knowledge understands risk. It is difficult to assess how the local knowledge helps in reducing the vulnerability or cope with the acceptable and unacceptable risks. Further, very less is known about the acceptable and unacceptable risks from a local perspective and in this regardless evidence is documented. Much of scientific knowledge dealt with risk studies must be transferred to local levels as in most cases such immense knowledge is mostly reposited in libraries. We need to explore and develop certain mechanisms to interface the scientific knowledge and local knowledge.

It is also necessary to point out legal frameworks, implementation challenges, and existing gaps as well. At the policy level, there are gaps especially for instance in coastal regulation zones. Although the regulation (for instance CRZ in India) was formulated in 1991, but was not adequately implemented until the Indian tsunami in 2004. There are huge gaps existing within the policy and practice not only in local levels but as well as at national levels. How can one direct from the policy-level decisions to the practice orientation? One of the possible solutions is by creating a culture to introduce DRR at school curriculum and exploring ways to integrate that in an early learning stage. This would not only help to disseminate the knowledge but also in producing evidence-based knowledge.

From a contextual perception, it is necessary to realize the dependency on the built-in environment which poses more risk. The more dependent on the built-in environment, we become more vulnerable encountering more risks. The perception is too difficult to narrow down as it is very subjective and it depends onto whom the risk is addressed to, and to what is needed, and what sort of risk is at hand. Another aspect is the development of communication and how risk is communicated to the community. Different agenda and information can be perceived differently by the educated and uneducated, and it is necessary to bridge the gap through proper channels of communication between society and the scientific community.

There is also an interface where education and research and policy and practice come together. Coordination at different levels is needed despite of different actors in national or regional levels, different actors in the local-level communities, and even with various action plans. This is where it must develop strategic platforms to strengthen to achieve the global, national, regional, and local collaboration, cooperation for increasing self-reliability rather than the dependency. It is necessary to build and enhance the trust among the communities through various partnerships and community participatory activities.

Health and Risk presented by Dr. Robin Eve Miller—It is very difficult to isolate health-related risks from the disaster as there will be consequences to health after a disaster. Health and safety are an integral part of DRR. One may recognize that there is no upper limit of the health, and a lower limit is to be safe and alive meaning to be safe in DRR with associated direct and indirect risks and known and unknown risks.

With other new risks emerging, health risks will always increase and become severe. The disciplinary orientation gaps in understanding risks related to health and DRR come from unequal resource allocations in terms of medical infrastructure, or targeted communities, or the professionals available in the rural areas due to unequal resource availability which requires advocacy to go deep into the bottom level of building partnerships across various stakeholders.

Other factors that contribute to health risks are:

- Direct and indirect risks.
  - Associated with certain risks as well as uncertainty related to some known and some unknown risks.
  - There will be new risks as well as health risks.

- The disciplinary orientation gaps in understanding risks related to health and DRR and the unequal resource allocation.
  - It could be in terms of medical infrastructure, resources allocated to target communities, and professional human resources availability in the rural areas.
  - Facilities such as good doctors and medical services available in developed countries may not be available in developing countries.
- Market-driven risk
  - How different drugs are encouraged in the market; and how the farmers manage such risk management due to lack of frameworks.
  - Quality and the legal enforcement of drugs—in the immediate onset of a disaster, who will be responsible to look at such controllers.
- Lack of basic hygiene conditions
  - The above facts are all contributing to the health and safety in the DRR, and specifically:

    Expressing from a disaster context—there is the time dimension of health disasters—is it immediately after the disaster?
    It is a difficult situation to assert.
    Sometimes, there are different situations and sometimes the epidemic and endemic sprung after a longer period of time after the disaster. Even a different and unexpected new health risk may emerge.
- Accessibility is another important point—how communities have access to medical needs during the emergency situations.
- Accountability is where and how the supply chain processes and to ensure necessary legal frameworks and protocols are followed.
- Affordability—this is one of the hardest of the factors in developing countries—sometimes, it is not possible to have affordable medical treatment due to high cost.
- Advocacy of right practices is how to disseminate information to follow the right approach. Advocacy has to go into the bottom level through building partnerships across various stakeholders.

Comments:

Prof. Yuichi Ono thanked the two presenters for preparing charts overnight and for a comprehensive and succinct presentation. There were three previous groups and some common points on hazards approach, and health was added under the health and social and human dimension of it.

- The point on communication issue was common among the previous three groups. In addition, the concept of "I"—which is person centred was a very good illustration. When scientists construct models, frameworks, or synthesis or whatever the tools they prepare, sometimes the "I's" are lost. Now, it is easy to hold on to this "I" concept for social and human dimension.

- From the audience commended the presenters for an impressive presentation on both topics—disaster risk management and health risks. One very important point is the access to health based on gender. In certain developing countries, the access is not equal for men and women due to cultural problems. For example, a woman cannot visit a clinic if there is only male doctors, and this should be taken into consideration too.
- From the audience, a comment was that there is another point which is probably close to what is called cooperative reclusive style. That it is very crucial to have cogeneration of knowledge between the local and the international community as well as science and administrative unit is. What is very often seen is a conflict between local knowledge and universal knowledge being brought in by scientist or between administrative knowledge thinking that they know how to do it and the local traditions. These kinds of struggles can only be overcome; there is a process in which the kind of knowledge is code generated. Very often, the native or indigenous knowledge is also not always superior—that is kind of romanticizing it sometimes. But, it is very important that it is a part of the knowledge generation and that this is really a learning exercise rather than an exercise of teaching people how to behave. This kind of mutual learning things is somethings that should be emphasized as they are part of the collaboration co-relation scheme that was presented, and it could be more explicit.
- Another comment from the audience—Whether the presentation is on risk or risk communication or risk measurement, and health and safety, all these things are integrated to one specific thing called culture. The culture, how to develop this culture? Culture development is basic to all mankind. Whether it is started with education at the school-level developer culture or at home. It could divide this cultural development to two sectors—structural and non-structural. Structural ways and non-structural ways sometimes blend with each other. There are different examples to show the non-structural way of developing culture is far more deeper and integrating and sustaining than this structural side. This is an area for research to find out—how to develop the culture in the whole process. Whether within the area of health, safety, or even risk. Risk acceptance, risk understanding, risk perception are part of cultures. It varies from different societies, and this is a very crucial point that should be underlined.

1.5.5 Group I-E—Natech (Natural and Technological Accidents) and Cross-Cutting Issues—Rapporteur Dr. Maria Suarez from DPRI, Kyoto University, Japan

The Natech and Cross-cutting Issues group was chaired by Professor Ana-Maria Cruz and Dr. Elizabeth Kraussman.[26]

---

[26] The report is based on the rapporteur presentations by Dr. Maria Suarez at the 4th Global Summit of GADRI in 2019. It has been reviewed and edited by Dr. Suarez.

Natech, natural and technological accidents, entails huge consequences and impacts on losses of human lives and economic losses. Although, learning from lessons of past accidents is possible, further research is much needed to support and develop Natech accident modelling and simulations for risk assessment, addressing cascading events, and considering risk reduction issues. Social impact of Natech is a topic to be further developed.

For example, what is the difference of impact between people and research community; how are they prepared or how are they impacted in the long term? Research on Natech and economic impacts are lacking, and it is always very difficult to retrieve data from the previous accidents in terms of economic damage. There is limited awareness and understanding of Natech scenarios among all stakeholders. It is a topic that is not so well known among different stakeholders and the industry over technological risk itself.

What happens in a situation where there is release of a chemical substance? Or what happens if a natural hazard impacts the industrial facility? There is also a lack of systematic collection of Natech data, and much constraints are observed in collecting details of previous accidents as these only exist in the databases directly related to Natech. There is a need to establish and create reconnaissance team for damage and impact assessment following disasters.

To strengthen the idea of the extent of the consequences that entail, Dr. Suarez used the cited examples of the incidences that took places in Turkey earthquake that affected the refineries, hurricane Katrina in 2015 which affected and caused a huge release of petroleum and oil; and the Great East Japan earthquake and tsunami which caused the nuclear disaster in 2011. They even had huge losses not only human but economic as well.

What are the main challenges for Natech research collection?

More research is needed to support and development of Natech accident modelling and simulations, risk assessment, addressing cascading events and considering risk reduction issues. Although lessons learned from previous accidents can be visited, there is much more to do regarding these aspects. It was also pointed out that social impact of Natech is a topic to be further developed. A recent interview conducted showed that there are a very few studies that have touched upon this topic. Much more need to be studied to know the effects of long-term impacts of Natech in the society.

For example, what is the difference of impact between people and research community? How are they prepared or how are they impacted in the long term?

More research is need regarding risk communication and evacuation planning. Understanding how best to release the information to the community; consider how the interaction between the industry and the community and the local government be regarded and how society can be physiological impact of the society upon receiving an evacuation warning or order to evacuate. It requires multi-stakeholder interactions for Natech risk governance issues and the need to consider interactions between all the stakeholders not limited to the industrial facility itself but the surrounding community and the local government with the development support needed to form strategies from academia.

There is limited awareness and understanding of Natech scenarios among all stakeholders—the way of technological risk is perceived. For examples, what happens if a chemical substance is accidentally released or what is the impact of natural hazard industrial facility. The communities surrounding the industrial facilities are not aware of the imminent risk they are exposed to, or the local government does not know how they can limit or reduce the impact of potential Natech. Industries are hesitant to share information for various reasons including potential security reasons, and this is a fact in every country.

There is a need for a paradigm shift to expand traditional technological risk management to a holistic systems perspective.

There is a need to establish and create reconnaissance team for damage and impact assessment following disasters. The information and date collected during such missions should reach the databases that are willing to develop further.

In terms of the contributions to the Science and Technology Roadmap, it was proposed to build awareness of Natech among all stakeholders. It was noted that there are many natural hazards that have not studied and recorded—like research and natural hazards triggering Natech. For example, hydro-meteorological hazards. A few counties like Japan have started to study tsunami impact on Natech, and the lightning hazards have been studied in Italy. Further developments are needed on natural hazards impact on Natech.

An interesting point noted was the Natech narrative knowledge. This is based on a collection of residents' voices and based on their experiences, and thereby creating stories and build on storytelling. This can help raise awareness among society and community and assist them to learn from previous experiences. This has been practiced and used in Japan related or focused on natural hazards such as earthquakes or flooding.

Other points of recommendations included:

- Development of performance indicators for Natech in industries to evaluate how prepared industries are for risk management and which measures are in place to combat them and how they are implemented to reduce potential risks.
- Development of Natech framework for resilience including testing and validating is needed to pool different stakeholders, their interaction, and to integrate their knowledge and their expertise to make society more resilient to Natech risks.
- Training and allocation of industrial staff and workers, government officials are needed to as there is a lack of understanding and awareness on the specificities of Natech scenarios.
- It is also very important to structure the data collection and narrow the gap of data availability on Natech.
- Participatory disaster risk reduction strategies are needed.

On long-term goals—It was proposed to include Natech as part of educational programmes and curriculum on disaster risk management. This of course will broaden the scope and knowledge of people. It is important to establish minimum risk management standards, risk information disclosure, and risk communication across all countries so people know what are the risks they are facing and act accordingly.

In concluding, Dr. Suarez stressed on the importance of raising awareness through education, training, and engaging all stakeholders in discussions on Natech risk factors. It is important now more than before to improve research on Natech and its economic impacts.

1.5.6 Recommendations and Wrap-Up from the Panel Chairs

The session brought forward good, constructive comments.

- It showed one-way forward is by creating exchange platforms. It could be a dynamic system, bringing in new developments on communications and collecting data from different sources to provide a platform to have a dialogue instead of one-time reporting.
- The platform will provide a portal to exchange of practices, good and bad practices, and education. This is recommended to GADRI to take it up as a first step.
- Another way to bring forward sharing accumulated research knowledge would be to publish through Disaster and Risk Research: GADRI Book Series.
- One common topic in all discussion group sessions was the use of local knowledge and its application in combating disaster risks and risk management. This is highly recommended to researchers to take advantage of local knowledge and practices to enhance their research towards DRR and DRM.

1.6 Panel II—Group Discussion II—Effectiveness and Relevance of Disaster Research Institutes in DRR—Contributions to World Stakeholders

Group II—A—discussion session in progress

Group II—D—discussion session in progress

1.6.1 Group II-A GADRI Contributions to the 2016 Science and Technology Roadmap—How Best Could GADRI Promote SFDRR 2030 Agenda?—Rapporteur, Dr. Annisa Triyanti, Postdoctoral Researcher, Utrecht University, the Netherlands

The session was chaired by Prof. Hirokazu Tatano, Ms. Shoko Arakaki, and Prof. Rajib Shaw and co-chaired by Prof. Ortwin Renn, Prof. Paul Kovacs, Prof. Irwan Meilano, and Dr. Annisa Triyanti.[27]

The group discussions focused on the contextualization of the Science and Technology Roadmap; discussed GADRI's involvement and contributions to promote the Sendai Framework Agenda; and the monitoring and evaluation plan.

With an introduction to the Science and Technology Roadmap which was launched in 2016 under the panel led by Prof. Virginia Murray, and the Sendai Framework 2015–2030, the presentation highlighted the role of the roadmap which in itself is to promote the role of science and technology activities and to realize the Sendai Framework for disaster risk reduction goals and targets. The roadmap is a guideline and a checklist stressing the importance of decision-making and evidence-based policies. Each of the priority for actions has a component of assessment or updates of data and knowledge, a component of dissemination, a component for monitoring and review, and capacity building. It is aiming towards the realization to consolidate scientific effort for collective impact and ensure that the impact is not scattered. With the same direct or same objective of reducing disaster risk, it is also important, at the same time to realize the need to achieve sustainable development to mitigate and adapt to climate change and global environmental challenges. Another outcome of the 2016 Science and Technology Conference held in Geneva was the Science and Technology Partnerships which continue to grow with a membership of 211 members from 58 states. With the launch of the Science and Technology Roadmap, it was also realized that there is a need to contextualization. The motion started in 2017, and in 2018 during the conference held in Chengdu, it is specific details and the process started.

The group discussion session focused to build a matrix covering the four different Priority Actions 1—understanding disaster risk; 2—risk governance; 3—investment; and 4—enhancing of disaster preparedness. The matrix will indicate the expected outcomes.

On interlinkages and interconnections among stakeholders including the science and technology community, the importance to open up to communicate, learn from each other, and other sectors, to be transdisciplinary and ensure transformation were stressed.

By dividing the group discussion into three parts, various questions were approached. For example, what are the concerns on the current version of the Science

---

[27] The report is based on the rapporteur presentation by Dr. Annisa Triyanti at the 4th Global Summit of GADRI in 2019. It has been reviewed and edited by Dr. Triyanti.

and Technology Roadmap, the role of GADRI, and the monitoring and evaluation plan?

The Tokyo Statement which was an important outcome of the Science Council of Japan Tokyo conference in 2017 was integrated to the 2016 version of the roadmap. A gap analysis was conducted, and missing points were added and restructured the roadmap actions and implementation strategy.

Using a matrix, priority for actions were discussed to have a component of assessment or updates of data and knowledge as well as a component of dissemination, monitoring and review, and also capacity building.

Key concerns of the current version of Science and Technology Roadmap discussed were:

- Focus on assessment—to fill the aspect of assessment of data on activities of the science and technology community, it requires partnerships and working together with other actors not only among the science technology community but also with the civil society and private sectors.
- To ensure that this process is inclusive and include all stakeholders, it is important to have both, system and action orientation if it is to be strengthened and evidence based.
- Concern of the feasibility of the Sendai Framework compared to other frameworks such as SDGs and climate change processes. That is where the synergy is needed as the science and technology community has a role in ensuring there is coherence among the processes.
- Emphasis on the importance of the local knowledge and the mechanisms of utilization. For example, using local knowledge is a validation of what is currently being done in the science and technology community and what other people, for example, like governments and private sectors are undertaking. It goes to capture real life, the action, and understanding of the implementation of the action.
- How and by whom will the implementation process of this roadmap be monitored and evaluated?
- There is a need of monitoring mechanism. That can be a product like report, document, etc. One of the things that is most important is also make sure that this monitoring mechanism realize the multidisciplinary approaches of the products and also the process is well and to link with the current platform like Sendai Monitoring and voluntary commitments.
- The other ideal comes to creating Science and Technology Roadmap supplementary sheet.
- It is a template for institutions to report like physically, to be actually disseminate in the next GADRI Summit. Because we feel like this summit happens every two years, and this is a very good momentum to showcase and also do monitoring and evaluation and tracking of what we have been doing. Is it actually giving an impact or not?
- To contribute to stocktaking in other processes like the global platform and regional platforms to monitoring updates at the implementation of roadmap, I heard from presentation yesterday GADRI has been developing so much with the

new alliances, regional alliances, and it is a very good opportunity to use this for regionalization as well of our activities.

What is the role GADRI can play in the implementation of the Science and Technology Roadmap?

- Focusing on GADRI, dissemination of evidence and building capacity utilize the existing mapping and knowledge management database or inventory of research. This is important to avoid duplication because as Professor Shaw mentioned today as well. Because there are lot of things already going on outside, we have to make sure that our work builds upon each other rather than competing or duplicating each other work.
- To be more inclusive, we think that there is an opportunity for GADRI to actually open up. May be the mechanism, of course, it is based on internal discussion and upon internal discussions, I think it is important to within this, I think it is important to open up the access of Science and Technology outside of Science and Technology Community including this NGOs or civil society organizations and also the private sectors.

In concluding, the following points were highlighted as references:

- Evidence-based policy and decision-making are extremely important, and it should be expanded beyond assessing and explore into a directive for dissemination, communication, capacity building, monitoring, and evaluation.
- It is important to realize that is coordinated effort and consolidate scientific effort for collective impact
- Interlinkages and interconnections among stakeholders including the science and technology community. The emphasize is that science is not the only input. The need to open up to communicate, learn from each other, and from other sectors as well as transdisciplinary. This is important to ensure transformation.

1.6.2 Group II-B—SDGs, Climate Change Adaptation—What Engagement Mechanisms and Research Linkages are Needed to Influence Research Directions Among Policy-Makers, Governments, Localities, Media, and Other Groups?—Rapporteur Prof. Yuichi Ono, IRIDeS, Tohoku University, Japan

The group discussion was co-chaired by Professors Tetsuya Takemi and Yuichi Ono assisted by Professors Kazuyoshi Nishijima, Eiichi Nakakita and Srikantha Herath.[28] The group discussion recognized the importance of evidence-based decision-making

[28] The report is based on the rapporteur presentation by Prof. Yuichi Ono at the 4th Global Summit of GADRI in 2019. It has been reviewed and edited by both Prof. Ono and Prof. Tetsuya Takemi.

concerning SDGs, climate change adaptation to come up with better mechanisms to integrate research outcomes into policy-making so that research outcomes could adequately influence policy-makers, governments, localities, media, and other stakeholders. The group discussion led to identify the importance of better communication mechanisms to narrow gaps between research outcomes of climate science, including climate change adaptation and policy-making. While IPCC reports provide strong basic findings at the global level, actual policy-making happens at the national and regional levels and GADRI has roles there. The group recognized the need for a guideline linking the Sendai Framework for disaster risk reduction, Sustainable Development Goals, and Paris Agreement at the global level as there is a great deal of confusion at the national levels, including government and academia. One proposal was to identify and prioritize some of the most significant indicators from them to monitor as there are too many indicators to follow. Finally, the group discussed how GADRI and its researchers could improve the situations and identified some key issues to be reported at the next GADRI Summits.

Results of discussion:

1. **How researchers in GADRI could better communicate with policymakers on climate change adaptation and SDGs**

More than often, climate change issues become scapegoats for some of the hydro-meteorological and large-scale disasters, instead of recognizing the malpractices and inactions by governments and policy-makers. The group discussed on how this trend could be reversed with a phased approach as below.

Short-term goals

- Understand mechanisms through which policy on climate change is enacted.
- Share experiences to develop mechanisms at national and regional levels.
- Facilitate to adjust the mechanisms at national and regional contexts.

A long-term goal

- Policies on climate change adaptation are enacted at national and regional levels.

A long-term plan

- GADRI members to provide policy advise to high-level government officials.

How to better communicate with climate impacts and SDGs

As the main mode of change, communication was identified as one of the keywords, the following points are considered:

- Disseminate messages to specific audience and consider the communication means through trusted persons or education public awareness in a wider way.
- Integrate climate change issues into formal education system (e.g. school curriculums/syllabus in universities).
- Disseminate messages more effectively through scientific institutes being engaged with communication experts (e.g. media).

- Focus on SDGs and climate change-related questions at the community level so that science and technology community messages are delivered to communities.
- Foster the scientific community to implement long-term projects on climate change in collaboration with policy-makers to build common understandings and trust.

2. **How can researchers/GADRI develop an inventory/atlas/portal of climate change adaptation measures around the world and make them (the adaptation measures) usable/implementable in other locations.**

The following points were proposed by the group:

- To develop a database and interactive web-based platform, including:
  - a list of all those climate change-related hazards,
  - geographic distribution of houses at risk of climate change related hazards around the world,
  - case studies and local climate change adaptation practices and strategies, categorized as cultural, geologic, geographic, and capacity levels, including infrastructure quality, planning, etc.
  - good practices and lessons learnt (may need a scale to measure with a list).
- To disseminate adequate information to certain sectors, including:
  - researchers
  - policy/decision-makers
  - media
  - communities
  - children and so on

3. **How to differentiate/relate Climate Change Adaptation (CCA) from/to Disaster Risks Reduction (DRR)**

The group agreed that climate change adaptation and disaster risk reduction are related and inseparable as disaster risk is a function of hazard, vulnerability, exposure, and capacity. Hazard used to be purely natural, but now is changing due to anthropogenic effects over climate change. With regard to linkages with SDGs, climate change adaptation, and disaster risk reduction, a few recommendations from the group were made as follows:

- There are many confusions among governments and people due to lack of indicators or too many indicators. With regards to the Sendai Framework for disaster risk reduction, there are a handful of indicators. Yet, others have too many indicators. Although they are all interconnected and interrelated, it is not clear as to who is doing what and what is being done are not clear at the governmental level.
- Thus, there should be a guideline linking SFDRR, SDGs, and CCA at the country level. To reinforce the guidelines, the information should be provided by the UN as a blue print in compliance with experts. The GADRI members could help at the regional and national levels.

- Identify/prioritize the most significant indicators—due to numerous indicators, it is necessary to focus or prioritize some of the significant indicators accounting (present and future in terms of SDGs) for risks to enable policy-makers to focus on the implementation more practically and effectively. There are some universities helping the government monitoring of Sendai and other frameworks.

Lastly, group discussions identified following points for GADRI members to consider and report the progress back to the next GADRI Summits.

1. How can researchers in GADRI communicate with policy-makers on climate change adaptation and SDGs? How can researchers in GADRI communicate with communities on climate change adaptation and SDGs?
2. How can GADRI develop inventory/atlas of climate change adaptation measures around the world to share our knowledge and to adapt them to individual situations?
3. How to differentiate between climate change adaptation and risk reduction (DRR)?

1.6.3 Group II-C—Knowledge Management and Science Synthesis—Nation's Synthesis: Online Synthesis System and Pre-synthesis—Rapporteurs Dr. Rodrigo Cienfuegos, Centro de Investigación para la Gestión Integrada del Riesgo de Desastres (CIGIDEN), Chile; and Dr. Indrajit Pal, Asian Institute of Technology (AIT), Thailand

The group discussion kicked-off by sharing different experiences from different countries on gaps for data issues, gaps in science and policy coordination, interdisciplinary research, and how the nations in a concept of synthesis could help to reach those gaps.[29] It is desirous of various stakeholders in the government, business, industry, civil society, and the social science and technology community to work together through national platforms for disaster risk reduction. While building the system to be effective, it needs to be facilitated and coordinated for the proper functioning between different stakeholders. A framework is needed to bridge the gap and foster science and policy coordination at local and national levels.

The group was chaired by Prof. Toshio Koike and Dr. Elisabeth Krausmann, Dr. Indrajit Pal, and Dr. Rodrigo Cienfuegos. Group was divided to two groups.

Group discussion focused on interdisciplinary programmes on disaster management to develop an integrated scientific knowledge base through by promoting interdisciplinary research methods.

[29] The report is based on the rapporteur presentations by Dr. Rodrigo Cienfuegos and Dr. Indrajit Pal at the 4th Global Summit of GADRI in 2019. It has been reviewed and edited by both Dr. Cienfuegos and Dr. Pal.

The first group focused by sharing different experiences from different countries on various data gap issues, science and policy coordination, interdisciplinary research and to reach those gaps through the nations synthesis concept.

Gaps or problems identified for data issues:

- There are gaps in unified online systems for data sharing. It is common around all different experiences.
- There is a need to establish guidelines and institutional framework for data sharing and data usage.
- Another problem identified was that local governments do not efficiently translate/utilize the available data. Often there is data available, which is part, which is distributed, but there is a gap on its efficient use.
- Last mile connectivity with the simplified information flow which could be useful for communities in local levels.

On science and policy coordination, the following recommendations were issued:

- Necessity to identify and address the policy needs.
- Specific gaps need to be bridged for implementation between science and policy.
- Promote dialogues between the scientific community and policy-makers.
- Risk communication needs to be strengthened and incorporated in a better way for policy formulation

Gaps in a research perspective of interdisciplinary brought the following suggestions:

- There is a lack of incentives especially for researchers in the academia. Often academia silos system dominates, and the intensives are not well aligned with the interdisciplinary work. To avoid the frustrations faced by young researchers, it is necessary to promote to foster interdisciplinary work.
- At national levels, funding mechanisms are need to be encouraged for interdisciplinary work. In some countries, this is already initiated but not well established globally.
- Need to join natural disaster research with technological accidents➔Natech—which is also going into interdisciplinary work.

Specifically, on the nation synthesis:

- Proposed on a framework to build an online system that could help bridging the gaps between the science and technology community and the national levels with different stakeholders.
- Stakeholders need to work together for disaster risk reduction.
- The second group discussions focused on short-term plan (3-year plan) and long-term plan up to 2030.

Under short term, the following were the recommendations:

- Conducting a questionnaire survey to ascertain the situation among different institutes/countries, within a short-term basis. Propose to share the questionnaire among GADRI members (after GADRI BoD endorsement).
- Develop a strategy for promoting national-level synthesis.
- Try to build standards for data sharing processes and mechanisms.
- Learning from the lessons set out by the European Commission and Japan, adapt good practices on data sharing mechanisms.

On long-term plan up to 2030, the following were proposed:

- Contribution to the Science and Technology Roadmap—promote contribution of data sharing.
- Foster the establishment of science and policy coordination framework at the local and national level.
- Bridge gaps between science and policy to communities.
- Develop the integrated scientific knowledge base through the promotion of interdisciplinary research method.
- On interdisciplinary programmes, there is no sufficient data as to the existing programmes on the disaster management in the region and as to whether these programmes are following the interdisciplinary process.

A short comment related to data accumulation also was received from the audience. It was emphasized that data promotion should be encouraged as it helps reducing loss of lives, assists humanitarian relief work, helps other stakeholders. Also stated was that accumulating data systematically, and archiving data contributes to prosperity.

1.6.4 Group II-D—Research Funding—Where is the Funding for Disaster Risk Reduction Activities Coming From and is this Invested to the Right Causes and the Areas that are Most Needing Support?—Rapporteur Dr. Subhajyoti Samaddar, DPRI, Kyoto University, Japan

The group was chaired by Prof. Kaoru Takara, Dr. Takahiro Sayama, Dr. Subhajyoti Samaddar and co-chaired by Prof. Charles Scawthorn, Prof. Mahua Mukherjee, and Dr. Sameh Kantoush.

The group discussion focus was on research funding sources, nature of funding, research areas in need of funding, current sources of funding, and the priority areas that receive funding.[30] The discussion also focused on the provision and distribution of funding by sector and region and appropriate usages of funding. Funders or donors

[30] The report is based on the rapporteur presentation by Dr. Subhajyoti Samaddar at the 4th Global Summit of GADRI in 2019.

are apt to fund soon after a catastrophic event to attract media attention. However, funding for the DRR component is ignored or not considered by most policy-makers or funders. As a result, most of the funding targets short-term response and relief operations. Fewer fundings are available for prevention and preparedness. There is a need to advocate for DRR funding, encourage governments to allocate a certain percentage of GDP for DRR research activities, and perhaps show costs and benefits for funders in the DRR sectors. Another possibility to encourage donors would be to promote their awareness of priority areas of DRR.

The main topics of the discussion covered the following:

- Gaps in current research funding
- Key priority areas of research
- The value of fundamental science versus need-based research
- Information provision
- Distribution of funding
- Appropriate usage of funding
- Research collaboration and advocacy

On sources of funding, it was observed that the primary sources of funding for DRR are from the national government agencies. At least, this is the case for certain developed and industrialized countries. International funding is least available to developed countries as well as for some developing countries. The private sector is least interested in DRR as it could prove to be a risk on the return on investment. An area to be explored is the Corporate Social Responsibility (CSR) funding which could prove to be interested in funding for DRR efforts.

The nature of funding focus was on hazards and aids for response and relief, particularly on engineering-based solutions or engineering-based research and funding on fundamental science. For DRR social science, it was noted that the difficulties faced in obtaining funds for research. Funding, in general, is very few for DRR activities as most of the funding is allotted for disaster relief and response; afterwards, only a few budgets get sanctioned for preparedness, rehabilitation, and resiliency.

In developing countries, there is a lack of funding for DRR and certain independence to conduct own research. Depending on the funding agency, it was observed that most funds are spent on post-disasters in developing countries rather than pre-disaster initiatives. For example, funding is used for catastrophic events by donors as propaganda to promote the business and attract the media. DRR as a component of the development is ignored or perhaps not considered by the policy-makers or funders. Most of the fund goes to response and relief operations or this kind of short-term benefit research.

As the role of GADRI, the following highlight a few suggestions:

- GADRI to advocate for DRR funding. For example, request for a certain percentage from GDP to be given to DRR research activities. For that, maybe, we can show some examples of what are the costs and benefits if the funders are funding the DRR sectors.

- Expand focus on institutional capacity building such as knowledge-based development or implementation research.
- GADRI may encourage crowd-funding and initiate building partnerships with other crowd-funding agencies to raise the budget for DRR research.
- GADRI would promote and showcase the importance of DRR research through adopting selected cost–benefit analytical research on disaster and development.
- GADRI may carry out a comprehensive survey on funding availability for various DRR activities in different geographical regions. The survey may include the process and criteria of the funding.
- GADRI Database: The participants propose to develop an international online cyber-base. GADRI would promote its networks for training and academic exchange through regularly held workshops and seminars involving donors and policy-makers to discuss and exchange ideas and expose the priority areas in DRR that they should fund.
- GADRI can promote and endorse selected training courses or online DRR courses as GADRI approved projects that may in future encourage and influence the donors to render funds.

1.6.5 Group II-E—Data Working Group—What are the Real Strengths and Weaknesses of Data and Risk Management? Rapporteur—Dr. Sameh Kantoush, DPRI, Kyoto University, Japan

The group was chaired by Prof. Andrew Collins, Ms Irina Zodrow, and Dr. Jean-Paul Pinelli and co-chaired by Dr. Jörgen Sparf.

The data group discussed the real strengths and weaknesses of data and risk management.[31] There are different types of data, and the concept relating to the data is—data types—raw data, treated data, synthesized knowledge or structured/semi-structured/unstructured, or data beyond scope. There are large volumes of data and storage varieties. At the storing stage, data is saved as tables, counting numbers, or statistics of mortality or other numbers, names, graphs, documents locally distributed or concentrated, or hard or soft data which are in hard discs. Also discussed were variability, speed and accessibility of the data, flow rate of data and the changes in time, and real time or refresh rate of the data formats and protocols. The Science and Technology Advisory Group (STAG) and UNDRR are conducting qualitative survey to get more data. The discussion summarized the issues and gaps in data, risks associated with "big data" as well as challenges.

There were many questions about the data itself how to organize, data to measure indicators, and how it should confirm as far as possible to the following principles. These as you can see as data, useful, relevant, complete, timely, accessible, and many

[31] The report is based on the rapporteur presentation by Dr. Sameh Kantoush at the 4th Global Summit of GADRI in 2019.

different criterions and indicators should be confirmed and discussed the building standard for data sharing processes and the mechanisms and other information.

The life cycle of processed data itself will start with specific goals or purposes. To start to monitor or collect these data and do some treatment or statistics for the data, the collected information required processing. It also needs analysis for building knowledge, taking actions, and monitoring again for the benefits of specific needs and finally evaluate and process life cycle.

The group focused on the following four main questions, mainly used for the STAG survey:

- What data do you consider already contributing to the progress of the Sendai Framework Priorities for Action?
- What other existing data is there that could contribute to the progress of the Sendai Framework?
- What cooperative actions enable data to progress the Sendai Framework Priorities?
- How is data a voice for those exposed to disaster risks and for facilitators of the Sendai Framework Priorities?

Further discussion also covered the current situation about the characteristics of big data that included several present situations such as:

- Scientific papers may not have complete data but need to provide frameworks for data
- Value is produced when the data output is put into action
- Quality, decentralized, and diversity issues need revisiting:
- Well-formed data
- Missing values
- Cleanliness of the data itself

In each country, there are networks collecting data, and it varies from country to country with type of data:

- Office of national emergencies and of statistics can be political sometimes and very much linked with science information.
- Minimum standards for data sharing are important if donors and research funders have an impact

Data characteristics issues discussed included:

- Vulnerability indices are often most effective at local level not national level—This is the area of work for action data theme
- Building data comes from building tax data and maintained by every government, by city
- Boarders between privately owned data
- Building data comes from building tax data
- Maintained by local government from every county or city
- Combine data, interpreting, share with water companies and have feedback

Also discussed were:

- Who is responsible for local warning if nothing happens or someone dies in the evacuation? Who is responsible?

- Given so many ways to gather data from satellites and other sources, what applications are there for data mining and processing the good from the bad data?
- Summarizing the issues of gaps in data on the one hand and risks associated with "big data" on the other hand.
- What does the policy-maker to determine how it will affect people?
- There are tools for people to express themselves, so we need to think about how we communicate better—e.g. e-health has transformed health—citizen science is a huge movement in itself.
- Networks that people belong to—if someone is ordered to evacuate—if the network is local such as Hurricane Katrina, then people may not evacuate—may have no impact on those at risk?
- Is soft data being pushed out by hard data?—how should soft data be included and gender relevant—a million women is very different to men—this is from our colleague from Chile.
- An apparent difference between "developed and developing" world—implications for growing data inequality.

What are the challenges?

- Deep learning and machine learning rapidly developing—in five years, things have evolved very fast
- How to get the influence of indigenous knowledge and how to get engagement of local resilience in all different disciplines?
- How to coordinate local data—be it territorial or indigenous knowledge++?
- Migration may make data collection more complex
- Strategy and intervention do not construct wadi or dam in this place, and indigenous knowledge is key
- Type of soft data that links to memories for migration or staying and trying to use past data, and this is difficult to find and is not collected routinely

There are several information from the local communities which are very important to be included from the beginning. However, local populations might not have knowledge of the local hazard risk: they might include people moved to the area and a majority of recent immigrants; return period of events might be long; there might be a large number of tourists.

The presentation was concluded with following suggestions to GADRI:

- Promote the flow of data between all parties including academics, local communities and governments;
- GADRI to provide a platform for sharing information for the alliancing context

- The platform needs to link/partner/coordinate with other database initiatives (DesignSafe, ISC CODATA and other networks)
  - Agreement on classifications and taxonomies
  - Agreement on principles and standards for data loss collection, recording and reporting
  - Need for comparability of disaster loss data
  - Agreement on data exchange and standards databases

1.6.6 Group II Panel Discussion Session

The chairs Prof. Charles Scawthorn and Ms. Shoko Arakaki identified the following two topics as commonly expressed by all groups:

- Data—the need, and the usefulness, the utility, and how to process the data;
- Research funding—how to pay for it?

Comments from the rapporteurs and the audience included:

- Establishment of a platform within GADRI for data sharing and information flow may provide a chance to share case studies, reports, pre- and post-event report information, etc., which can be interactive and useful to all alliance members and counterparts.
- It is also necessary to talk about the data policy of the individual institutions.
- Prior to embarking on creating a platform for data sharing, it is important to work on the terminology of data. There are hazards terminology, adaptations in the disaster world, and the risk world. The mitigation world is quite different too. It is equally important to have some translation of those terminologies.
- It is important to see that data is not just a numeric but also qualitative data which includes important case studies. It is important to understand the mapping of the inventory and how to use that to maximize the utilization.
- Each person has their own data, and that data is important to them. Most likely, it is important to some other people, but some people's data is not important others.
- Regarding creating a GADRI platform, one option is to make use of an existing platform such as DesignSafe instead of duplicating efforts.
- Attention was also drawn the streaming of fake data, and this could opt people to hold on to their data in the hard discs. To avoid such situations, perhaps a control system for data management could be created. Something similar to what is done with chemicals, where it is possible to say the origin of the chemical is known and how it is changed. Otherwise, it is extremely difficult to make a difference between fake data and original data.
- There is an overloading of data many times, and this has created an apartheid, not to use so much of data. Whoever is responsible for using the data, and storing the data, should also be responsible for bringing down data to utilizable levels for actual users and actual implementers. The challenge remains in how to narrow

down data and make it reliable, adaptable, and acceptable data, to prepare a toolbox for the doers.

- With projects funded by local governments, it is not easy to access data and sometimes results in researchers purchasing the data from governments.
- While agreeing with the challenges for data gathering and collection, it is really important to define it. What is the true importance or mission to gather data among the alliance? It should eventually contribute to a safer world. By collecting data and gathering, the purpose is to develop a GADRI global knowledge on a common system. Data needs to be differentiated from common data, background data, raw data or basic data, and to what extent the data should be shared.
- The role for GADRI might be creation of a global databases, and the contribution of the data, with provenance of the data by a member institution, and cross-validation is done by another institution. The data that is in GADRI archives can be trusted and has complete provenance. There will be quality assurance, to some extent, and this could be a way to fight back against fake data.
- There are very good existing data collecting initiatives with very little incentive for researchers for contributing the data or using the data collecting systems. There need to be a strong motivation or benefit for researcher for sharing the data or for using a particular data storage system.
- The role for GADRI is to create a standard of ethics for disaster research that GADRI disaster research institutions both create that standard and adhere to it. One of the standards for example is that GADRI members commit to open source the data. From an ethical position for research funders, in order to fund research, it requires to be open sourced. Not GADRI fund it but the people fund it. For example, that the nation governments commit to a certain fraction of the GDP to fund disaster research.
- A survey or research should be conducted to this effect to find out as to what kind of data is favourable. For example, should one pay for the data; if the data is free, what is the quality and level of data or other conditions.
- In terms of role for GADRI and also from the point of view of Data Working Group, UNDRR, the work on data will be an iterative process. Nonetheless, it is a very strong theme which is coming through which have not being emphasized. What is the impact of one's data? How is the data contributing? Picking up that theme might actually may be the way to ask some of the other questions. How to incentive for people sharing the data? How to improve the quality of data? Is it related to the end product? The path is probably moving towards some sort of idea or framework where to work from the end result, and work backwards to identify, which data and what type of data, from where and whose data is actually affecting this change the Sendai Framework is looking for.
- This view is of the policy practice end of the equation. From a science basis, it is a bit different as the data may be used for many other purposes than just impact. This could be the bridge to be built with UNDRR interest, and this can give a potential role for GADRI to fill in that space.

The chair drew precaution that there has to be balance between the demand for data and the results. The opportunity is for pure research that will lead in directions that cannot be foreseen for which, once gone down that road, may discover great value. It is important to have balance in between. Of course, while the desire is to spend money wisely, at the same time ensure the human genius is not suppressed.

1.7 Wrap-Up Session

The wrap-up session panel members were Prof. James Mori, Prof. Andrew Collins, and Prof. Hirokazu Tatano. The session was chaired by Prof. Hiroshi Kawase.

Prof. James Mori at the 4th Global Summit draft resolution session

Prof. Hiroshi Kawase chaired the wrap-up and the draft resolution session

1.7.1 *Resolution of the 4th GADRI Summit—15 March 2019*

The member institutes of Global Alliance of Disaster Research Institutes (GADRI) fully support the United Nations Office for Disaster Risk Reduction (UNDRR) Sendai Framework for disaster risk reduction 2015–2030.[32] Recognizing the urgency of meeting short-term goals, to support implementation of this plan, as described in the Science and Technology Roadmap, GADRI advocates:

- Evidence-informed technical reporting for topics in interdisciplinary disaster risk reduction sciences, which currently include natural, engineering, social, and health sciences.
- Development of effective interfaces between technical information and policy-making.
- Proactive implementation of policies that will reduce human, economic, environmental, and cultural losses from disasters.

[32] The final Resolution is published separately in the Proceedings of the 4th Global Summit of 2019.

GADRI promotes changing roles of science and reflective practices of implementation that will contribute to the effectiveness of disaster risk reduction. This may include development of tools to support a new generation platform.

GADRI will contribute to the actions of the roadmap by providing a common voice and identifying priorities on important issues, such as:

- Contributing to the revision of hazard terminology
- Contributing to understanding risks and vulnerabilities
- Syntheses for all types of risks on a country or regional level, including examples of successes and contributions of local knowledge, to expand sponsor and user groups (e.g. DRRKC, Nations Synthesis Framework)
- Effective education programmes about disaster risk reduction for scientists, government and community officials, media, and the public
- Improved research and practice on communication and interactions among and between technical experts, policy-makers, implementation groups, and stakeholders
- Development of new methodologies and technologies for data analyses, promotion of action data and encouragement of open data sharing, for research and risk reduction applications

GADRI institutes will report their achievements at future Global Summits of Research Institutes for Disaster Risk Reduction.

- The 4th GADRI Summit—Institutes of GADRI fully support the Sendai Framework for disaster risk reduction 2015 2030 and support the implementation of the plan as described in the Science and Technology Roadmap,
- GADRI advocates evidence-based technical reporting for topics of natural sciences, engineering, and social sciences,
- development of effective interfaces between the technical information and policy-making,
- proactive implementation of policies that will reduce human economic and cultural losses from natural and man-made disasters.

Next section identified a few key issues which have been a common objective of the 4th Global Summit and which GADRI work towards the progress of the roadmap.

- GADRI will contribute to the actions of the roadmap by providing a common voice and identifying priorities on important issues such as
- clarification of hazards and risk terminology
- synthesis for all types of risks on a country and regional level including examples of the successes to expand sponsor and user groups
- Effective education programmes about disaster risk reduction for government and community officials, media, and the public
- improve communication and interactions among and between technical experts, policy-makers, implementation groups, and stakeholders
- development of new technologies for data analysis, promotion of action data, and encouragement of open data sharing for research and risk reduction application

- GADRI institutes will report their achievements at future Global Summits of Research Institutes for Disaster Risk Reduction.

In concluding the session, Ms Shoko Arakaki introduced the Global Platform for Disaster Risk Reduction to be in May 2019 in Geneva. Dr. Elisabeth Krausmann announced that the 5th GADRI Global Summit will be held at the European Commission's Joint Research Centre in Milan, Italy, with support of the Northumbria University in the UK. Mid Sweden University in Swede and GADRI in March 2021.

1.8 In Conclusion

The contextualization of the Science and Technology Roadmap played a major part in the 4th Global Summit of GADRI. GADRI deliberations on the contextualization were submitted at the UNDRR Global Platform for DRR held in March 2019.

Most of the group discussion sessions pointed to the necessity for change in science orientations and language, including hazard and risk terminology, to one that policy-makers, government officials and local groups, be these communities or other assemblages of people, could understand in relating to disaster risks. Streamlining different hazard languages in different frameworks such the SDGS, Paris Agreement, Sendai Framework, and other aspects of the 2015–2030 agenda and beyond is important to convey the science message to different stakeholders.

Data management, research funding, and local knowledge were other prominent topics variously debated with both contrasting and convergent opinions throughout the preparation and delivery of this event.

The importance of listening to local people and making use of the local knowledge, which can be vastly ignored by scientists, are a few of the points emphasized by most discussants. If a local knowledge has been considered in current research, it often will add impetus and provide much more practicable solutions while involving communities in a process of science findings that in return contributes to greater effectiveness during pre- or post-disaster. Local communities experience fatalities of disaster risks increasing the imperative of their rights to engage in analytical processes. Their awareness and understanding of risks with which they are living will further assist researchers to understand and enhance the quality of research work in a way that also opens opportunities to better reduce casualties.

Data management as an iterative process needs to be further expanded and discussed across contributing groups to enable evaluation of its importance, value, and usefulness in disaster risk reduction and response. While data sharing, including inter-operability, has currently become a more-debated topic, the means to more impactful data remains underexplored.

Research funding remains difficult to accommodate in DRR policy and practice activities. While most donors in the private sector are willing and interested in post-disaster-related projects, to be able to boost their own agendas in a more immediately visible way, donors for pre-disaster prevention and management activities, which can be less apparent in the short term, are scarce and need greater incentives to motivate donors to secure funds into DRR research. A poignant suggestion emanating from the discussion sessions was the contrasting funds received for climate change initiatives, but how researchers could astutely combine disaster risk reduction and climate change crisis into projects to tap into these, particularly where they remain unspent. Another suggestion is to request national governments to contribute a minimum sum from their GDP towards DRR-related activities. GADRI members need to collate these ideas and act on them, such that research funding problems could be reduced to some extent and at the same time to draw in the government, policy-makers, and other stakeholders to this agenda. The need to bridge the existing gaps between scientists and policy-makers in the critical current dependence of disaster risk reduction and response on science knowledge has been heavily confirmed and elaborated in the summit. It remains and ongoing driver of the GADRI purpose.

References

4th Global Summit of Research Institutes for Disaster Risk Reduction (4GSRIDRR)—https://www.dpri.kyoto-u.ac.jp/news_en/11487/

EC-JRC website—https://drmkc.jrc.ec.europa.eu/

Rathje E, Dawson C, Padgett JE, Pinelli J-P, Stanzione D, Arduino P, Brandenberg SJ, Cockerill T, Esteva M, Haan FL, Kareem A, Lowes L, Mosqueda G (2020) Enhancing research in natural hazards engineering through the designSafe cyberinfrastructure. Frontiers in Built Environment, Section Wind Engineering and Science, December 2020. https://doi.org/10.3389/fbuil.2020.547706

GADRI newsletter—GADRI ACTIONS, April 2019—http://gadri.net/resources/pdf/GA_Newsletter_April2019_final_rev2.pdf

GADRI website—http://gadri.net/

Global Risk Assessment Framework (GRAF), UNDRR—https://www.preventionweb.net/disaster-risk/graf

Kyoto University website—WINDOWS concept—http://www.kyoto-u.ac.jp/window/en/

Pinelli J-P, Esteva M, Rathje EM, Roueche D, Brandenberg SJ, Mosqueda G, Padgett J, Haan F (2020) Disaster risk management through the designsafe cyberinfrastructure. Int J Disaster Risk Sci . https://doi.org/10.1007/s13753-020-00320-8

Public Health England—https://www.gov.uk/government/organisations/public-health-england

The report is based on the presentation by Ms. Setsuko Saya at the 4th Global Summit

The report is based on the presentation by Mr. Soichiro Yasukawa, UNESCO, France at the 4th Global Summit of GADRI in 2019 and it is reviewed and edited by Mr. Yasukawa

The report is based on the presentation by Mr. Masato Miyazaki at the 4th Global Summit. It is reviewed and edited by Mr. Miyazaki's Office

The World Bank Tokyo Office—https://www.worldbank.org/

The report is based on the presentation by Prof. Toshio Koike at the 4th Global Summit

The report is based on the presentation by Ms. Shoko Arakaki, Chief of Branch, Partnerships, Intergovernmental Process, and Inter-agency Cooperation, UNDRR, Switzerland at the 4th Global Summit. It is reviewed and edited by Ms. Irina Zodrow

The report is based on the presentation by Dr. Jörgen Sparf at the 4th Global Summit

The report is based on the presentation Prof. Virginia Murray at the 4th Global Summit

The report is based on the presentation by Prof. Andrew Collins at the 4th Global Summit

The report is based on the presentation by Prof. Eiichi Nakakita at the 4th Global Summit

The report is based on the presentation by Dr. Jean-Paul Pinelli at the 4th Global Summit of GADRI in 2019a and it is reviewed and edited by Dr. Pinelli

The report is based on the presentation by Dr. Elisabeth Krausmann at the 4th Global Summit. It is reviewed and edited by Dr. Elisabeth Krausmann

The report is based on the presentation by Dr. Craig Davis at the 4th Global Summit. It is reviewed and approved by Dr. Davis

The report is based on the presentation by Dr. Ortwin Renn at the 4th Global Summit of GADRI in 2019b and it is reviewed and approved by Dr. Renn

The report is based on the presentation by Prof. Mohsen Ashtiany-Ghafory at the 4th Global Summit. It is reviewed and edited by Prof. Ashtiany

The report is based on the presentation by Prof. Desmond Manatsa at the 4th Global Summit. It is reviewed and approved by Prof. Manatsa

The report is based on the presentation by Prof. Rajib Shaw at the 4th Global Summit of GADRI in 2019

The report is based on the rapporteur presentation by Dr. Tetsuya Takemi at the 4th Global Summit of GADRI in 2019a

The report is based on the rapporteur presentation by Dr. Craig Davis at the 4th Global Summit of GADRI in 2019b

The report is based on the rapporteur presentation by Prof. Michel Jaboyedoff at the 4th Global Summit of GADRI in 2019a

The report is based on the rapporteur presentations by Dr. Ram Satish ad Dr. Robin Eve Miller at the 4th Global Summit of GADRI in 2019

The report is based on the presentations by Dr. Maria Suarez at the 4th Global Summit of GADRI in 2019

The report is based on the rapporteur presentation by Dr. Annisa Triyanti at the 4th Global Summit of GADRI in 2019c

The report is based on the rapporteur presentation by Prof. Yuichi Ono at the 4th Global Summit of GADRI in 2019b

The report is based on the rapporteur presentations by Dr. Rodrigo Cienfugos and Dr. Indrajit Pal at the 4th Global Summit of GADRI in 2019

The report is based on the rapporteur presentation by Dr. Subhajyoti Samaddar at the 4th Global Summit of GADRI in 2019d

The report is based on the rapporteur presentation by Dr. Sameh Kantoush at the 4th Global Summit of GADRI in 2019e

The final Resolution is published separately in the Proceedings of the 4th Global Summit of 2019

The following appeared in GADRI newsletter—GADRI ACTIONS, April 2019—http://gadri.net/resources/pdf/GA_Newsletter_April2019_final_rev2.pdf

UNESCO website—https://en.unesco.org/

UNDRR, Geneva, Switzerland—https://www.undrr.org/

UN Sustainable Development Goals—https://sdgs.un.org/goals

Part II
Plenary Sessions

Chapter 2
Enhancing Disaster Preparedness and Building Resilience: UNESCO's Multi-hazard, Multi-disciplinary, and Multi-stakeholder Approach on Disaster Risk Reduction

Soichiro Yasukawa and Lesly M. Barriga Delgado

2.1 Introduction

Climate-related hazards such as floods and hurricanes and geo-hazards, such as earthquakes and volcanoes, are triggering catastrophes, disturbing, and altering the livelihoods of populations worldwide (UNESCO and UNICEF 2012). From 2000 to 2019, natural hazards caused US$1.23 trillion in damages, claimed 1.23 million lives, and affected over 4 billion people (UNDRR 2020). Fatalities are deepened by the vulnerability of poor societies (UNISDR and CRED 2018). Since the 2000s, people from low-income countries are seven times more likely to die in disaster-affected areas than equivalent of people from high-income countries (UNISDR and CRED 2018). With the frequency and magnitude of occurrence increasing due to climate change, losses associated with disasters are on the rise (UNDRR 2019). It is estimated that by 2050, the number of people in urban areas exposed to cyclones will increase twofold to 680 million, and those who are at risk of major earthquakes will be approximately 870 million people (World Bank 2010).

Over the past two decades, disaster risk reduction (DRR) has caught significant attention as the main approach to reducing vulnerabilities and risks and improve preparedness for protecting regional, national, and local development (Van Niekerk and Terblanché-Greeff 2017). Related to the latter, one of the key international agreements adopted for better and resilient societies has been the Sendai Framework for disaster risk reduction (2015–2030) (UNDRR 2019). The Sendai Framework constitutes a long-term international agreement that draws a voluntary route for the next 15 years of DRR (Kelman 2015). The framework was adopted at the UN Third World

S. Yasukawa (✉) · L. M. B. Delgado
Division of Ecological and Earth Sciences, Section on Earth Sciences and Geo-Hazards Risk Reduction, UNESCO, London, UK
e-mail: s.yasukawa@unesco.org

H. Tatano and A. Collins (eds.), *Proceedings of the 4th Global Summit of Research Institutes for Disaster Risk Reduction*, Disaster and Risk Research: GADRI Book Series,
https://doi.org/10.1007/978-981-19-5566-2_2

Conference on Disaster Reduction (WCDR) held in Sendai, Japan, and endorsed by the UN General Assembly in June 2015 (UNDRR 2019). While the Sendai Framework is the roadmap for DRR, other global agendas—including the Agenda 2030, the Paris Climate Agreement, and the New Urban Agenda—have targets that cannot be attained without DRR. In this sense, there are clear links between those international instruments.

UNESCO operates at the interface between natural and social sciences, education, culture, and communication, playing a vital role in building a global culture of resilience. UNESCO assists countries in capacity building for management of disasters and climate risks, particularly supporting the Member States on (1) Early warning systems; (2) Safe critical infrastructures; (3) UNESCO designated sites risk prevention; (4) Science, technology, and innovation such as artificial intelligence and big data; (5) Built environment; (6) Risk governance; (7) Nature-based solutions, and (8) Post-disaster response. The present review aims to expose some of UNESCO's main contributions to these global major challenges, developed through multi-hazard, multi-discipline, and multi-stakeholder DRR mechanisms to enhance preparedness and build resilience at local, national, and regional levels.

2.2 The United Nations Educational Scientific and Cultural Organization and Its Role on Disaster Risk Reduction

2.2.1 The Multi-hazard, Multi-stakeholder, and Multi-disciplinary Approach on Disaster Risk Management and Reduction

Countries face different challenges and have different perceptions or knowledge of risks. This is reflected in a large diversity of preventive, hard and soft measures, risk communications and policy as well as in their implementation. Through a multidisciplinary approach, UNESCO builds capacities and fosters partnerships so that science and technology can support a holistic understanding of the climate crises and natural hazards, and thus, efficient disaster prevention, preparation, response, and recovery. UNESCO maximizes its comparative advantage by combining its expertise in earth, ocean, water, and ecological sciences with its mandate in education, social sciences, communication, information, and heritage preservation to reach out to communities, the private sector, governments, and other UN agencies for combining effort toward resilient societies. Under its DRR activities, UNESCO assists countries in capacity building for the management of disaster and climate risks. Efforts are made in close collaboration with the overall UN system to support Member States with their commitments and aspirations contained in the Nationally Determined Contribution (NDC) and the UNFCCC Paris Agreement (Fig. 2.1).

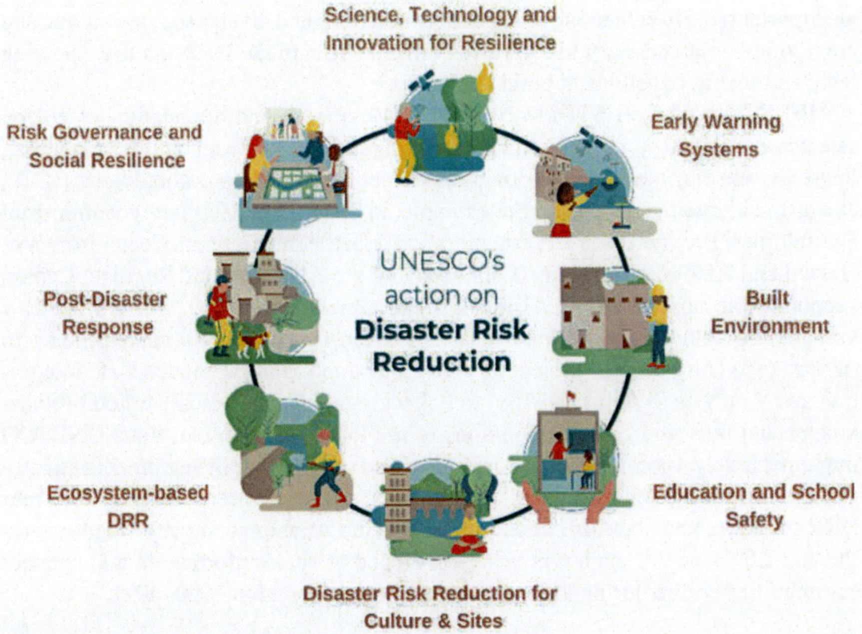

Fig. 2.1 UNESCO's action on DRR. *Source* UNESCO

In addition to the multi-disciplinary approach through inter-sectoral and inter-divisional collaboration, UNESCO takes multi-hazard and multi-stakeholder engagement approach to the challenges of mitigating and adapting to hazards, disasters, and climate change. Multi-hazard approach is necessary as one natural hazard will have cascading effect such as earthquake to tsunami, storm to flood, and climate change to glacier melting. UNESCO is in a good position as it covers both weather and geo-related hazards. Multi-stakeholder approach allows academics, civil society, the private sector, and the public sector to collaborate in establishing risk prevention plans and strategies. The engagement of different stakeholders and organizations working at different scales of governance grants more coordinated and integrated DRR actions and projects.

2.2.2 *UNESCO's Main Contributions on DRR to Major Global Challenges*

2.2.2.1 Science, Technology, and Innovation for Resilience

Science, technology, and innovation (STI) systems are critical contributors to the agility and responsiveness of states to disaster crises (OECD n.d.). These systems play an important role in enhancing linkages between DRR and diverse socioeconomic and environmental processes while securing citizen's role in the DRR inquiry, allowing local and inclusive actions to build resilience.

UNESCO's take on STI involves the enhancement and application of citizen science, participatory research, local and indigenous knowledge and development, low-cost sensors, advanced information and communication technologies (ICTs) and artificial intelligence or AI. For example, in 2019, UNESCO Intergovernmental Hydrological Program (IHP), in collaboration with the International Center for Water Hazard and Risk Management (ICHARM) and the AGRHYMET Regional Center, supported the application of AI in a project aimed at reducing Western Africa's vulnerability to floods. The "Water disaster Platform to enhance climate resilience in Africa" (WADiRe-Africa) project involved the adoption of AI-related Data Integration and Analysis Systems (DIASs) to collect, analyze, and visualize flood-related images and thus find concrete solutions to the local issues in real time. UNESCO implemented a project in Eastern Africa that also supports the development and integration of citizen science, modern technologies, and gender-responsive actions into DRR strategies and plans. In collaboration with the Japanese company Weathernews Inc and LINE, an AI chatbot is being developed to share information and connect communities before/during/after disaster events in the region 2020–2021.

2.2.2.2 Early Warning Systems

UNESCO works to address various hazards—notably tsunamis, earthquakes, floods, drought, and landslides—creating spaces and projects that enhance collaboration and scientific knowledge exchange to establish, maintain, and improve early warning systems (EWSs). EWS constitutes complex people-centered systems that aim to reduce the impacts of natural hazards and increase the resilience of societies to related risks through the provision and dissemination of information and awareness (UNDP 2018). In the last years, UNESCO has been continually active in promoting this global target of the Sendai Framework for DRR, particularly in countries with significant challenges toward EWS development.

For example, in 2019, UNESCO supported the establishment of a Tsunami-EWS in Fort-Liberté, the capital of the northeast department of Haiti. Warning signs and preparedness materials were installed and distributed across the city. Fort-Liberté now has operational tsunami procedures, and 50 local and national focal points were trained on the reception and dissemination of alerts.

Significant efforts have been also undertaken to strengthen seismic observations. In Ghana (2019) and Morocco (2021), for instance, UNESCO supported the donation and installment of equipment for seismic monitoring, provided by the Japanese company Challenge.

2.2.2.3 The Built Environment

Earthquakes are one of the deadliest natural hazards, with more than 80% of the resulting casualties caused by the collapse of buildings. In this framework, UNESCO supports its Member States in strengthening capacities and construction-policy sectors to increase the safety of their built environment and thus reduce the socioeconomic impact of earthquakes.

The project "Capacity Building for Disaster Risk Reduction in the Built Environment in LAC" is the latest UNESCO project concerning this issue. The three-year initiative, also known as the BERLAC5 project, started in 2020 and aims to build capacity for DRR in the built environment against earthquakes. The project has five target countries (Cuba, Dominican Republic, Guatemala, Mexico, and Peru) and four components: (1) secure safer new buildings, (2) school facilities safety, (3) strengthen existing buildings, and (4) develop risk-informed policy making. BERLAC5 is funded by the Japanese government and implemented in collaboration with local and international experts, practitioners, and seismic engineers through knowledge exchange.

2.2.2.4 Education and School Safety

UNESCO is actively engaged in empowering schools and their communities to identify the risks they are exposed to, map their vulnerabilities and capacities, and enhance school safety. To do so, UNESCO promotes a multi-hazard school safety assessment methodology known as Visual Inspection for defining Safety Upgrading Strategies (VISUS). The methodology has been developed by UNESCO in close collaboration with the University of Udine (Italy), and it is based on the use of visual inspections and the application of pre-set algorithms to assess the hazards potentially affecting schools at local or national levels. VISUS provides decision-makers, technical staff, and universities the necessary tools and information to make science-based decisions on where and how to invest their resources for school safety upgrading interventions (UNESCO and Udine University 2019).

The VISUS methodology has been successfully tested in seven countries: Italy (2010), El Salvador (2013), Lao People's Democratic Republic (2015), Indonesia (2015), Peru (2016), Haiti (2017), and Mozambique (2017). Overall, the security of more than 500,000 students and educational staff has been assessed. By 2023, UNESCO is planning to evaluate the safety of 100 schools in the Dominican Republic.

2.2.2.5 Disaster Risk Reduction for Culture and Sites

Cultural heritage plays a significant role in the economic and social development of a country and represents an asset for resilience and recovery after a catastrophic event.

In 2020 and recognizing the value and importance of safeguarding cultural heritage, UNESCO and the Caribbean Disaster Emergency Management Agency (CDEMA) conducted a regional workshop to build awareness and foster synergies on disaster resilience in the Caribbean's Culture Sector. UNESCO is currently supporting the incorporation of the workshop's results in national/local DRR policies and plans for the culture sector and selected World Heritage Sites in the Caribbean.

2.2.2.6 Ecosystem-Based Disaster Risk Reduction

Nature-based solutions (NbSs) constitute actions to effectively and adaptively protect, restore, or modified ecosystems to provide both human well-being and biodiversity benefits (Cohen-Shacham et al. 2016). Being most of disasters often caused or aggravated by environmental degradation, it has been recognized the contribution of nature in building local resilience to disasters and climate change (UNEP 2019).

UNESCO promotes the Ecosystem-based Disaster Risk Reduction (Eco-DRR) approach, where ecosystems, such as wetlands and forests, are sustainably managed and conserved to prevent and mitigate natural hazards and climate change impacts (UNEP 2019). For this, UNESCO works closely with international experts to mainstream Eco-DRR in development planning at global, national, and local levels. For example, UNESCO actively participates in the ongoing activities of the Intergovernmental Science-Policy Platform on Biodiversity and Ecosystem Services (IPBES) and the Partnership for Environment and Disaster Risk Reduction (PEDRR). Likewise, UNESCO is involved in the EU-funded OPERANDUM project that aims to reduce hydro-meteorological risks in European rural territories through co-designed, co-developed, co-deployed, tested, and demonstrated innovative NbS.

2.2.2.7 Post-disaster Response

UNESCO assists Member States in post-disaster response to assess damage and losses, and to identify recovery and reconstruction needs. For this, UNESCO—in collaboration with the International Preparedness and Response to Emergencies and Disasters (IPRED)—has established a system to dispatch engineers and seismologists to earthquake-prone countries for carrying out post-earthquake field investigations and draw lessons toward better preparedness, response, and recovery. With the IPRED expertise support, engineering analyses are applied to collapsed buildings with the aim of understanding the cause of the collapse. The findings are then used to create better building codes and practices. Related missions have been sent to Van in Turkey (2012), Bohol in the Philippines (2014), and Kermanshah in Iran (2017) (UNESCO n.d.).

2.2.2.8 Risk Governance and Social Resilience

UNESCO promotes civil society engagement in the management of disaster risks to ensure the coherence of national and local regulations and policies. Following the premise of "leaving no one behind," UNESCO encourages the participation of vulnerable populations and the youth in DRR planning and implementation at community and policy-making levels.

Concerning the latter, UNESCO encourages the formation of youth networks and platforms to contribute to regional and national DRR through science, engineering, technology, and innovation (SETI). In November 2019 and supported by UNDRR, UNESCO established the Asia Pacific Youth and Young Professionals in Science, Engineering, Technology, and Innovation for DRR (U-INSPIRE). U-INSPIRE advocates and facilitates youth in Asia and the Pacific to strengthen their SETI as well as to use their capacities, capabilities, and initiatives to support DRR and CC.

In January 2020, UNESCO supported the establishment of the African Youth Advisory Board for Disaster Risk Reduction (AYAB-DRR) to institutionalize the African Union working relationship with children and youth as major stakeholder groups in DRR on the continent. A Youth Professional Platform for the Caribbean is currently being established by UNESCO, supported by CDEMA, whose official launch is expected for the summer/autumn of 2021.

2.3 Conclusions

Disaster risks are one of the major threats facing humanity that do not stop at national frontiers. Natural hazards as well as climate change must be dealt collectively, transforming thinking, and behavior toward disaster prevention and the reduction of societies' vulnerability. UNESCO is committed to the United Nations system-wide implementation of the Sendai Framework and thus to a worldwide resilience building. Through a multi-stakeholder, multi-disciplinary, and multi-hazard strategies, UNESCO continues to advocate, through abovementioned eight pillars of intervention, influencing decision-makers in fulfilling their responsibilities to save lives, and making scientific and cultural knowledge integral part of national and regional disaster planning.

It is imperative to highlight that UNESCO's contribution to the implementation of the 2030 Agenda maintains adequate attention to UNESCO's global priorities (i.e., gender equality, Africa, and the Small Island Developing States or SIDS). In this regard, UNESCO continues to encourage men and women to be agents of risk mitigation and climate change adaptation, constitutes a catalyst for DRR cooperation across Africa in line with the Agenda 2063, and calls for increase attention by the international community to SIDS's disaster risk landscapes and the need of regional technology, capacity building, and financial support within a climate change context.

Bibliographys

Cohen-Shacham E, Walters G, Janzen C, Maginnis S (eds) (2016) Nature-based solutions to address global societal challenges. IUCN, Gland

Kelman I (2015) Climate change and the sendai framework for disaster risk reduction. Int J Disaster Risk Sci 6:117–127

OECD (n.d.) What role for science, technology and innovation in building resilience?. OECD, Paris. Recovered from https://www.oecd.org/sti/inno/what-role-for-science-technology-and-innovation-in-building-resilience.htm

UNDP (2018) Early warning systems. Recovered from https://www.google.com.pe/url?sa=t&rct=j&q=&esrc=s&source=web&cd=&ved=2ahUKEwjHl4mO3uDxAhXmsKQKHRfFC0UQFnoECAQQAA&url=https%3A%2F%2Fwww.undp.org%2Fcontent%2Fdam%2Frbec%2Fdocs%2FUNDP%2520Brochure%2520Early%2520Warning%2520Systems.pdf&usg=AOvVaw1EZxADtL2k-NOVsjCq5USA

UNDRR (2019) Global assessment report on disaster risk reduction 2019. Recovered from https://gar.undrr.org/

UNDRR (2020) The human cost of disasters: an overview of the last 20 years (2000–2019). Recovered from https://reliefweb.int/report/world/human-cost-disasters-overview-last-20-years-2000-2019

UNEP (2019) Disasters and ecosystems: resilience in a changing climate Recovered from https://climate-adapt.eea.europa.eu/metadata/publications/disasters-and-ecosystems-resilience-in-a-changing-climate-source-book

UNESCO and Udine University (2019) UNESCO guidelines for assessing learning facilities in the context of disaster risk reduction and climate change adaptation. Recovered from https://unescochair-sprint.uniud.it/en/unesco-guidelines-for-assessing-learning-facilities-in-the-context-of-disaster-risk-reduction-and-climate-change-adaptation/

UNESCO and UNICEF (2012) Disaster risk reduction in school curricula: case studies from thirty countries. Recovered from https://www.google.com.pe/url?sa=t&rct=j&q=&esrc=s&source=web&cd=&cad=rja&uact=8&ved=2ahUKEwimkrSr3ODxAhUI-qQKHWcHA0EQFnoECBwQAA&url=https%3A%2F%2Fsustainabledevelopment.un.org%2Fcontent%2Fdocuments%2F928unesco11.pdf&usg=AOvVaw1sOuqc1-wb6gvEjP56vfJB

UNESCO (n.d) IPRED Post-earthquake field investigation. UNESCO, Paris. Recovered from www.uesco.org/new/en/natural-sciences/special-themes/disaster-risk-reduction/geohazard-risk-reduction/networking/ipred/post-earthquake-field-investigation

UNISDR and CRED (2018) Economic losses, poverty & disaster 1998–2017. Recovered from https://www.unisdr.org/files/61119_credeconomiclosses.pdf

Van Niekerk D, Terblanché-Greeff A (2017) Anticipatory disaster risk reduction. In: Poli R (ed) Handbook of anticipation. Springer, Cham, pp 1–23

World Bank (2010) Natural hazards, unnatural disasters: the economics of effective prevention. Recovered from https://openknowledge.worldbank.org/handle/10986/2512

Chapter 3
Presentation on "THE Sendai Framework and Science Advice to Governments to Support Implementation—Suggestion on What Works" by Prof. Virginia Murray, Head of Global Disaster Risk Reduction, Public Health England, UK

Virginia Murray

This presentation seeks to identify the calls for action identified by the UN member states in the Sendai Framework for Disaster Risk Reduction for science advice to inform governments to support implementation; provide examples of tools currently being used to deliver science to help to inform policy and practice; and to provide an example of how the work has been taken forward in partnership to understand more of the all hazard paradigm as a policy and practice process.

GADRI and its science partners are key to sharing how suggestion on what works for science advice to governments to support implementation of the Sendai Framework. Indeed, the Sendai Framework calls for action.

- *'To promote and improve dialog and cooperation among scientific and technological communities, other relevant stakeholders and policymakers in order to facilitate a science policy interface for effective decision-making in disaster risk management;'* (Paragraph 24 h), and
- *'Academia, scientific and research entities and networks to focus on the disaster risk factors and scenarios, including emerging disaster risks, in the medium and long term; increase research for regional, national and local application; support*

The report is based on information provided in the presentation by Prof. Virginia Murray at the 4th Global Summit.

Public Health England—https://www.gov.uk/government/organisations/public-health-england.

V. Murray (✉)
Global Disaster Risk Reduction, UK Health Security Agency (formerly Public Health England), England, UK
e-mail: Virginia.Murray@UKHSA.gov.uk

H. Tatano and A. Collins (eds.), *Proceedings of the 4th Global Summit of Research Institutes for Disaster Risk Reduction*, Disaster and Risk Research: GADRI Book Series,
https://doi.org/10.1007/978-981-19-5566-2_3

action by local communities and authorities; and support the interface between policy and science for decision-making;' (Paragraph 36b).

Examples of networks that have developed to provide systems for support include the International Network for Government Science Advice (INGSA) which provides a forum for policy makers, practitioners, national academies, scientific societies, and researchers to share experience, build capacities, and develop theoretical and practical approaches to the use of scientific evidence in informing policy at all levels of government.

Other networks that help to inform the science policy interface include:

- The UNDRR PreventionWeb which provides a valuable knowledge sharing platform for disaster risk reduction.
- The Campbell Collaboration, started in 1999, promotes positive social and economic change through the production and use of systematic reviews and other evidence synthesis for evidence-based policy and practice.
- The Cochrane Collaboration, started about 25 years ago, is an international knowledge synthesizer network which gathers and summarizes evidence from research to help make informed choices about health care related issues and management. Cochrane's members and supporters come from more than 130 countries, worldwide. The volunteers and contributors to Cochrane are researchers, health professionals and patients.
- Evidence Aid started in 2004 and has been building on experience in research custodianship, synthesis and communication. It works to ensure that humanitarian actors have access to clear summaries of evidence, presented in clear, simple formats, so that users can design and deliver the interventions that will be most likely to save lives and livelihoods.
- European Commission Disaster Risk Management Knowledge Center (DRMKC) started in 2015. It was developed in part as a result of the Sendai Framework request for enhancing the means of implementation by the call for action for knowledge exchange:
  - *'To promote the use and expansion of thematic platforms of cooperation, such as global technology pools and global systems to share know-how, innovation and research and ensure access to technology and information on disaster risk reduction;'* (paragraph 47c).

    DRMKC identified that better knowledge, stronger evidence and a greater focus on transformative processes and innovation are essential to improve our understanding of disaster risk, to facilitate building resilience and risk-informed approaches to policymaking, and to contribute to smart, sustainable and inclusive growth. DRMKC provides a networked approach to the science policy interface in DRM across the European Commission, Member States and the DRM community globally, engage to

    - support the translation of complex scientific data and analyses into usable information;
    - provide science-based advice for DRM policies;

- provide timely and reliable scientific-based analyses for emergency preparedness and coordinated response activities; and
- brings together existing initiatives contributing to the management of disaster risks.

- UK Government What Works Network started informally in late 1990s, What Works is an initiative that aims to improve the way government and other public sector organizations create, share and use (or 'generate, translate and adopt') high quality evidence in decision-making. It was formally created in 2013 to ensure that the best available evidence on 'what works' is available to the people who make decisions on public services. Five years on, a report looked at the impact of the Network, which has grown to include 10 independent What Works centers.

  It supports more effective and efficient services across the public sector at national and local levels. The network comprises 9 independent What Works Centers, 3 affiliate members and one associate member. Together these centers cover policy areas which account for more than £250 billion of public spending. What Works Centers are different from standard research institutions. Each center is committed to increasing both the supply of, and demand for, evidence in their policy area, and their output is tailored to the needs of decision-makers.

  The centers help to ensure that robust evidence shapes decision-making at every level, by:

  - collating existing evidence on the effectiveness of programs and practices;
  - producing high quality synthesis reports and systematic reviews in areas where they do not currently exist;
  - assessing the effectiveness of policies and practices against an agreed set of outcomes;
  - filling gaps in the evidence base by commissioning new trials and evaluations;
  - sharing findings in an accessible way;
  - supporting practitioners, commissioners and policymakers to use these findings to inform their decisions.

As an example of the science policy issues identified for further work, it was apparent that there was a need to address all hazards paradigm call for action in the Sendai Framework.

- *To strengthen technical and scientific capacity to capitalize on and consolidate existing knowledge and to develop and apply methodologies and models to assess disaster risks, vulnerabilities and exposure to all hazards*; (paragraph 24j)

Following the 4th GADRI Global Summit, there was a compelling case for an all hazards approach to achieve risk reduction as a basis for sustainable development. In the Report of the Open-ended Intergovernmental Expert Working Group on indicators and terminology relating to disaster risk reduction 2016 is the definition of 'hazard' which is shared in Annex 2 of the UNDRR/ISC Hazard Definition and Classification Review:

'A process, phenomenon or human activity that may cause loss of life, injury or other health impacts, property damage, social and economic disruption or environmental degradation'

Annotations: *Hazards may be natural, anthropogenic or socionatural in origin.* ***Natural hazards*** *are predominantly associated with natural processes and phenomena.* ***Anthropogenic hazards****, or human-induced hazards, are induced entirely or predominantly by human activities and choices. This term does not include the occurrence or risk of armed conflicts and other situations of social instability or tension which are subject to international humanitarian law and national legislation. Several hazards are socionatural, in that they are associated with a combination of natural and anthropogenic factors, including environmental degradation and climate change.*

Hazards may be single, sequential or combined in their origin and effects. Each hazard is characterized by its location, intensity or magnitude, frequency and probability. Biological hazards are also defined by their infectiousness or toxicity, or other characteristics of the pathogen such as dose-response, incubation period, case fatality rate and estimation of the pathogen for transmission.

Multi-hazard *means (1) the selection of multiple major hazards that the country faces, and (2) the specific contexts where hazardous events may occur simultaneously, cascadingly or cumulatively over time, and taking into account the potential interrelated effects.*

Hazards include (as mentioned in the Sendai Framework for Disaster Risk Reduction 2015-2030, and listed in alphabetical order) biological, environmental, geological, hydrometeorological and technological processes and phenomena.

Biological hazards *are of organic origin or conveyed by biological vectors, including pathogenic microorganisms, toxins and bioactive substances. Examples are bacteria, viruses or parasites, as well as venomous wildlife and insects, poisonous plants and mosquitoes carrying disease-causing agents.*

Environmental hazards *may include chemical, natural and biological hazards. They can be created by environmental degradation or physical or chemical pollution in the air, water and soil. However, many of the processes and phenomena that fall into this category may be termed drivers of hazard and risk rather than hazards in themselves, such as soil degradation, deforestation, loss of biodiversity, salinization and sea-level rise.*

Geological or geophysical hazards *originate from internal earth processes. Examples are earthquakes, volcanic activity and emissions, and related geophysical processes such as mass movements, landslides, rockslides, surface collapses and debris or mud flows. Hydrometeorological factors are important contributors to some of these processes. Tsunamis are difficult to categorize: although they are triggered by undersea earthquakes and other geological events, they essentially become an oceanic process that is manifested as a coastal water-related hazard.*

Hydrometeorological hazards *are of atmospheric, hydrological or oceanographic origin. Examples are tropical cyclones (also known as typhoons and hurricanes); floods, including flash floods; drought; heatwaves and cold spells; and coastal storm surges. Hydrometeorological conditions may also be a factor* *in other hazards such as landslides, wildland fires, locust plagues, epidemics and in the transport and dispersal of toxic substances and volcanic eruption material. '**Technological hazards** originate from technological or industrial conditions, dangerous procedures, infrastructure failures or specific human activities. Examples include industrial pollution, nuclear radiation, toxic wastes, dam failures, transport accidents, factory explosions, fires and chemical spills. Technological hazards also may arise directly as a result of the impacts of a natural hazard event'*

The all hazards approach was needed to understand the broad range of hazards of relevance to risk reduction and resilience building. It was decided that although understanding of hazards and their related impacts has evolved over recent decades, and lists of hazards are available at many levels from many organizations, a single overview was required to provide a full picture of hazards to help inform the policy.

In May 2019, the UN Office for Disaster Risk Reduction (UNDRR) and the International Science Council (ISC) jointly established a technical working group to identify the full scope of hazards relevant to the Sendai Framework as a basis for countries to review and strengthen their risk reduction policies and operational risk management practices. To implement partnership working the technical working group used an iterative process of developing and reviewing the hazards listed through extensive consultation with over 500 technical experts from relevant science groups, UN organizations, the private sector and other partners.

The technical working group collaborated to identify an initial hazard list which was compiled from existing hazard glossaries and terminologies. To limit the potentially infinite scope of hazards addressed, a hazard was only included if it fulfilled each of three criteria: has the potential to impact a community; has measurable spatial and temporal components; proactive and reactive measures are available.

The outcome of the UNDRR/ISC Hazard Definition and Classification Review published in July 2020, states that:

- *'The Sendai Framework for Disaster Risk Reduction 2015–2030 ('the Sendai Framework') was one of three landmark agreements adopted by the United Nations in 2015. The other two being the Sustainable Development Goals of Agenda 2030 and the Paris Agreement on Climate Change. The UNDRR/ISC Sendai Hazard Definition and Classification Review Technical Report supports all three by providing a common set of hazard definitions for monitoring and reviewing implementation which calls for 'a data revolution, rigorous accountability mechanisms and renewed global partnership'*

In the recommendations of the report, it was documented that these hazards are increasingly interconnected, cascading and complex in nature both in the natural and human-induced hazards, including their potential impact on health, social, economic, financial, political and other systems.

Indeed recommendation 4 of the review states:

- *'Use this hazard list to actively engage policymakers and scientists in evidence-based national risk assessment processes, disaster risk reduction and risk-informed sustainable development, and other actions aimed at managing risks of emergencies and disasters.'*

In conclusion it was stressed that there is a need for partnership to work together to synthesis evidence to inform policy and practice, and to make it accessible to all to achieve the goals and targets of the Sendai Framework.

References

Campbell Collaboration. https://campbellcollaboration.org/about-campbell/vision-mission-and-principle.html

Cochrane Collaboration. https://www.cochrane.org/about-us

Evidence Aid. https://evidenceaid.org/who-we-are/our-strategy/

European Commission Disaster Risk Management Knowledge Centre (DRMKC). https://knowledge4policy.ec.europa.eu/disaster-risk/about_en

International Network for Government Science Advice (INGSA). https://www.ingsa.org/

UNDRR PreventionWeb. https://www.preventionweb.net/english/

Public Health England. https://www.gov.uk/government/organisations/public-health-england

Sendai Framework for Disaster Risk Reduction. https://www.preventionweb.net/files/43291_sendaiframeworkfordrren.pdf

The report is based on information provided in the presentation by Prof. Virginia Murray at the 4th Global Summit of GADRI held at the Disaster Prevention Research Institute, Kyoto University, Kyoto, Japan from 13th to 15th of March 2019

UK Government What Works Network. https://www.gov.uk/guidance/what-works-network

UNDRR/ISC Hazard Definition and Classification Review. https://www.undrr.org/publication/hazard-definition-and-classification-review

Chapter 4
Presenting the "Global Risk Assessment Framework (GRAF), UNISDR—Emphasising the Relevance of Disaster Research Institutes" by Prof. Andrew Collins, Disaster and Development Network (DDN), Northumbria University, UK; Chair of the GADRI Board of Directors; Associate Member of the Global Risk Assessment Framework (GRAF), UNDRR

Andrew Collins

Abstract Prof. Collins shared information on the work done by the expert group of GRAF which is a framework for assessing risks. He presented this also as a significant manifestation of the principles often presented in earlier forums, that understanding risk better amplifies the risk assessment and management cycle. This is a risk governance concern dependent on the application of good data and in-depth analysis of processes of change in reducing major threats, resilience building and moving towards a more well-being orientated emphasis for the future of people and planet. A key point is that there is a common vision of enabling sustainable societies and systems for improved planetary well-being; this requires global risk assessment to help increase resilience in the light of more comprehensively understood risk. The GRAF vision, objectives, principles and processes are based around an appropriately comprehensive theory of change.

A. Collins (✉)
Disaster and Development Network (DDN), Northumbria University, Newcastle upon Tyne, UK
e-mail: andrew.collins@northumbria.ac.uk

H. Tatano and A. Collins (eds.), *Proceedings of the 4th Global Summit of Research Institutes for Disaster Risk Reduction*, Disaster and Risk Research: GADRI Book Series,
https://doi.org/10.1007/978-981-19-5566-2_4

[1]Prof. Andrew Collins, Disaster and Development Network (DDN), Northumbria University, UK, and Chair of the GADRI Board of Directors presented on the "[2] Global Risk Assessment Framework (GRAF), UNISDR—Emphasising the Relevance of Disaster Research Institutes". The message by Prof. Andrew Collins runs parallel and mutually supportive of the message delivered by Prof. Murray where she stressed the importance of evidence data, data sharing and management. Prof. Collins shared information on the work done by the expert group of GRAF which is a framework for assessing risks. He wished to present this also as a significant manifestation of the principles often presented in earlier forums, that understanding risk better amplifies the risk assessment and management cycle. This is a risk governance concern dependent on the application of good data and in-depth analysis of processes of change in reducing major threats, resilience building and moving towards a more well-being orientated emphasis for the future of people and planet. The GRAF vision, objectives, principles and processes are based around a comprehensive theory of change. The draft framework is already in place with a basic governance structure. It will be also presented and then launched at the Global Platform 2019 in Geneva in May 2019. Sciences that are operational in a multidisciplinarity space are fundamental to the GRAF. It is a framework to produce a steer and impact. The influence of GRAF should be to bring about change, not only analyse how it can be understood, guiding pathways to achieving the Sendai Framework priorities for action.

Prof. Collins emphasised that he is speaking on behalf of the GRAF of which the secretariat is UNDRR. He would add some of his own interpretation, but essentially the GRAF was the product of many great inputs from a wide range of experts. GRAF's expert group had requested his input and presentation of the framework. This is a work-in-progress and meant to be an iterative process that is intended to continue for many years. GRAF also supports some of the other discussions presented earlier in the Summit and was seen as complimentary and potentially highly interactive with other work commissioned by UNDRR, including the STAG Data Working Group (DWG) he was currently chairing. This was leading to a specific steer for greater use of "action data" that has subsequently been published at the UNDRR web pages (Collins et al. 2020).[3] The action data concept is complimentary to and further supported by the active risk engagement process GRAF also promotes.

One of the fundamental key points to take away from the systemic representation of risk and risk intervention in GRAF is its orientation from a scientifically verifiable basis. Without investment in science, it does not exist. It is also a form of experiential science because it is science in policy and practice and not just science sitting on its own. He said he emphasises an "s" at the end of science to reflect the fundamental dependence of GRAF on multidisciplinarity.

[1] The report is based on the presentation by Prof. Andrew Collins at the 4th Global Summit.

[2] Global Risk Assessment Framework (GRAF), UNDRR—https://www.preventionweb.net/disaster-risk/graf.

[3] Collins et al. (2020).

This presentation further stressed that if GRAF is a depiction of engagement of actions that bring about change and impact to achieve the priorities of the Sendai Framework, then it must be drawing on science with a voice and with impact.

GRAF vision is to improve the understanding of complex risks behaviours and where relevant and applicable, to transform behaviours and catalyse a proactive decision-making culture by democratising everyone's understanding of the systematic nature of risk through time.

What is recognised through the GRAF process is that the Sendai Framework's Priority for action number 1 concerns the need to understand risks both vast and loaded with action orientations that derive from such a statement. The Sendai Framework priorities a mechanism represented by GRAF references to inclusiveness, governance, democratising of knowledge, access to it and how it is used. This leads to political will and investment in societal, economic and environmental processes that achieve sustainable societies and systems. He directed the audience to the GRAF website to obtain further details on its objectives and principles.

Prof. Collins proceeded to expand further on the process of GRAF, which is about getting the right sort of insights and understanding to generate the tools and the process. Ongoing analyses are the key theme of this taking place within contexts of change that prevent risk creation, mobilises finance investment and communities whilst enabling a transformation of behaviour. Whilst risk assessment is needed to effect good risk management, capacity to implement this is an ongoing learning system through applying risk management in a suitably transformed governance context.

The effectiveness of change in the elements represented by the GRAF can be themselves expanded in terms of more comprehensive and more in-depth considerations of actionable insights and models. The projects informing the framework are those that can be more sensitive to the changing urgency, cultures and the need to progress implementation of the Sendai Framework.

GRAF principles support both transitioning and transformational steps in disaster risk reduction in that the way is stated as open, collaborative, universal, trusted, practical, transparent and capable of living with uncertainty. In addition, there is the use of openly accessible terminology, the key words being widely applied through the GRAF concept also being widely embedded and applied within risk reduction scholarship, policy and practices.

The concept and considered processes of change can be traced back into the emergence of knowledge and how one understands the way the world works, being fundamentally based on systems analysis and adaptation. Interventions in systemic risk are dependent on systemic risk assessment that is cognisant of complex underlying and proximate processes of change in how people alter the way they live. The processes of influence that are needed to implement the Sendai Framework is underpinned by the Science and Technology Roadmap and is part of the Agenda for Sustainable Development. Effective processes require inclusive co-design, that can inform decision-making and transform behaviour, being based on collective and generally multidisciplinary, multi-institutional and multi-sectoral work. There is still much to understand and be better communicated regards the interactions between hazard driven

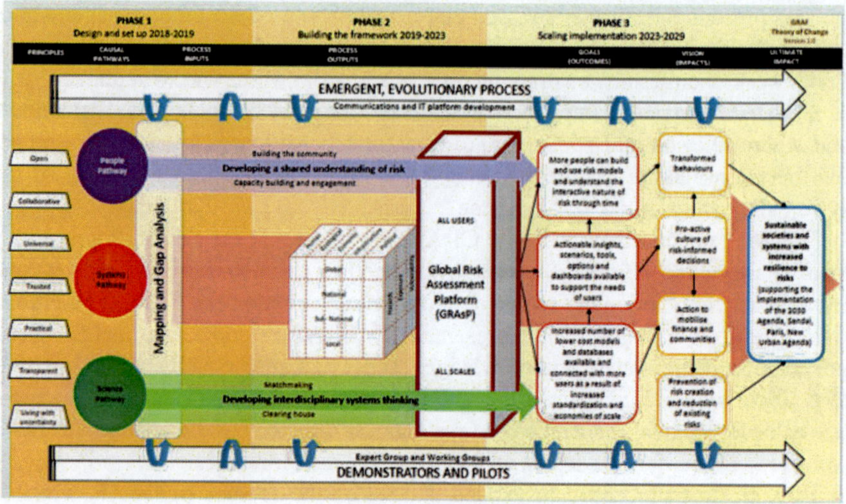

Fig. 4.1 GRAF theory of change as presented by Prof. Collins. *Source* GRAF, UNDRR)

shock exposure and inherent and induced vulnerabilities across human, ecological, economic and financial systems. Critical understanding comes from engaging this process and integrating this knowledge of the risk landscape in more coherent and inclusive decision-making.

Through diagrams used with permission of the GRAF group, Prof. Collins was able to illustrate the process further. Key points from Fig. 4.1 were to flag the centrality of the risk cube in the centre of the analysis, as basic knowledge for GRAF in creating the framework. This mediates the variable situatedness of inputs on one side and outputs produced on the other side. Of relevance to the GADRI membership would be the need to furnace the latest research for the inputs side—where progress in science and academic scholarship is concerned. This however needs to be also informed by indigenous wisdom and multiple levels of learning and understanding in a cultural context. He also pointed out that, although there is a theory of change for GRAF, it is yet unclear how change for many components of risk actually happen.

The iterative process referred to as a feature of the GRAF process involves appropriate consultation. Those present and much wider have an opportunity to comment on the document as it will be presented in many additional forums and via the Global Platform 2019 itself.

It was made clear in this presentation that at the end of the day, action-orientated risk assessment is about people developing and sharing understanding of risks, and this interaction being the critical aspect of implementation, impact and developing future scientifically guided pathways. GRAF is an attempt to bring in all energies possible in contributing to a framework for global risk assessment.

Recognising complexities and barriers to some of this process at the global, national, sub-national and local areas are also the start to being able to address the hitherto less addressed hazards and vulnerabilities. In particular, there is a need to identify actions on vulnerability in a way that it can be offset by capacity; this depends on the different representations of how each of these elements are experienced in varying cultural and economic contexts.

GRAF, he stressed, had been organised extremely well by UNDRR as a natural extension of its promotion of the era of the Sendai Framework. This is also very much a part of the voice of the science community, and of the disaster management community beyond the academy. It comes together as a Global Risk Assessment Platform, though definition of what the platform does is part of the ongoing discussion and debate.

GRAF is currently in its phase 1—it includes a mapping and gap analysis as one of its ongoing activities. There is a working group associated with that and a draft schematic of the risk assessment framework for developing this contribution alongside other working groups is also available.

There is a governance structure organised by UNDRR for the expert group as well as the working groups. The governance and management of GRAF include steering of potentially a lot of activity, such that decisions have to be made as to what projects are particularly relevant and need to be prioritised.

By way of a call to GADRI members, it was flagged that GRAF needs to draw on all forms of risk-related research to foster uptake of science in governments and industries. Action on global risk assessment that progresses successfully is a fundamental role of science. The mapping and gap analysis working group had designed a survey in terms of different components of this challenge.

The link between GRAF and STAG which has been particularly active in the recent years is various, though also through the mapping and gap analysis which is linking to the DWG of the STAG itself. The work of the DWG is wider than GRAF and temporary, supporting GRAF, whilst GRAF is one of the routes of DWG utility. He concluded his remarks on wider points of relevance to disaster research institutes through the approach represented by GRAF and how bringing these various efforts together and further integrating them as much as possible is one of our common objectives. A key point is that there is a common vision of enabling sustainable societies and systems for improved planetary well-being; this requires global risk assessment to help increase resilience in the light of more comprehensively understood risk. GRAF has been organised very well by UNDRR to date, and it is a good reflection of the progress that can be made in concept development with intent to strengthen the effectiveness of the disaster management community.

References

Collins AE, Richardson B, Zodrow I (2020) UNDRR STAG data working group (DWG): report on concept, activity and implication, United Nations Disaster Risk Reduction. https://www.undrr.org/publication/stag-data-working-group-report
Global risk assessment framework (GRAF), UNDRR. https://www.preventionweb.net/disaster-risk/graf
The report is based on the presentation by Prof. Andrew Collins at the 4th Global Summit
UDNRR. GRAF concept note. https://www.preventionweb.net/files/globalplatform/5cd89ca4c27b3GRAF_Concept_Note_2018_FINAL.pdf
UNDRR. https://www.undrr.org/

Chapter 5
Disaster Risk Management Knowledge Centre: A Collaborative Approach to Foster Resilience

Christina Corbane, Montserrat Marin Ferrer, Elisabeth Krausmann, Ainara Casajus Valles, Karmen Poljansek, Charalambos Moussas, Tiberiu-Eugen Antofie, and Marzia Santini

Abstract The risk landscape is changing in the European Union (EU) but also globally. The emergence of new and the intensification of existing risks requires a joint effort to enhance society's resilience to these risks. In support of this goal, the European Commission, in close collaboration with Member States and the disaster risk management community, has launched the Disaster Risk Management Knowledge Centre (DRMKC). The DRMKC is an instrument which provides a networked approach to the science–policy interface in disaster risk management in the EU and for developing and pursuing evidence-based EU policies. True to the provisions of the Sendai Framework for Disaster Risk Reduction, the DRMKC follows an all-hazards, multi-stakeholder approach. Its driving force is collaboration, and it fosters knowledge transfer, the co-development of solutions for existing needs, as well as the continuous uptake of innovative approaches to managing disaster risks. It also reinforces the existing links across Disaster Risk Management- and Disaster Risk Reduction-related policies. At the occasion of the Sendai Midterm Review a number of successful DRMKC initiatives and products are described.

Keywords DRMKC · Disaster risk · Partnership · Knowledge · Innovation

5.1 Introduction

We are coming to the end of the first quarter of the twenty-first century marked by a constantly evolving disaster risk landscape in Europe and the world. The convergence of the COVID-19 pandemic with natural hazards, exacerbated by climate change, brought profound implications to our societies and the way we organize ourselves to protect people, property and economic interests. In the future, more extreme events, sudden and slow onset and more complex cascading disasters with

---

C. Corbane (✉) · M. M. Ferrer · E. Krausmann · A. Casajus Valles · K. Poljansek · C. Moussas · T.-E. Antofie · M. Santini
European Commission Joint Research Centre (JRC), Via E. Fermi 2749, 21027 Ispra, VA, Italy
e-mail: Christina.corbane@ec.europa.eu

H. Tatano and A. Collins (eds.), *Proceedings of the 4th Global Summit of Research Institutes for Disaster Risk Reduction*, Disaster and Risk Research: GADRI Book Series,
https://doi.org/10.1007/978-981-19-5566-2_5

potentially increasing damages are expected. The last two years have shown that managing disasters amid the COVID-19 pandemic is an emerging challenge calling for better coordination and collaboration between countries and communities to address compound risks.

As recently evidenced in the 2021 IPCC report (Coppola et al. 2021), within the next two decades we are guaranteed to face more dangerous and destructive extreme weather events than we are seeing today. Likewise the WMO report (World Meteorological Organization 2021) shows that the expected increasing losses are not only related to the growing intensity and frequency of weather events in many parts of the world as a result of climate change, but also due to the increasing number of people exposed to disaster risk and to population growth in hazard-exposed areas. The report underscores the need for "greater investments in disaster risk reduction following a multi-hazard approach to disaster risk management and early warning systems to build resilience, reduce risks and strengthen preparedness for multiple disaster scenarios". Under the current landscape scenario, the role of science, technology and innovations is instrumental and needs to be further exploited to ensure continuity of attention to causes and consequences across the full disater risk management cycle.

Better gathering, sharing and use of data, information and knowledge in the European Commission (EC) are essential to deliver integrated policy work and overcome silo mentalities. The EC officially endorsed the concept of Knowledge Centres in its Communication on Data, Information and Knowledge Management (Communication to the commission C(2016)6626), as one of the four main actions for improving data, information and knowledge management within the Commission and enhancing synergies with external stakeholders.

The Knowledge Centres are virtual entities bringing together experts and knowledge from different sources inside and outside the EC, hence creating a new flexible instrument to develop and pursue informed and evidence-based EU policies. They provide contextualized evidence, reviews, data analysis and visualization, communicating in a concise way through visual and digital channels. The 2030 strategy of the EC's Joint Research Centre (JRC) has put strong emphasis on the need for the creation and development of Knowledge Centres in areas where there is a clear and specific EU policy demand and where the JRC already has a critical mass of knowledge it can build on.

The Disaster Risk Management Knowledge Centre (DRMKC[1]) is the first Knowledge Centre launched in 2015. It sets out to promote a science–policy interface in DRM across the EC, EU Member States and the Disaster Risk Management (DRM) community within and beyond the EU. The DRMKC aims to facilitate access to and uptake of research developments and results and supports the development of capacities and new technologies, in particular within Member States.

The DRMKC is an effective tool to handle the information overload by simplifying the scientific complexity for policy making (Fig. 5.1).

Within the Union Civil Protection Mechanism (UCPM), the pivotal role of science is acknowledged in Article 3(e) of Decision No 1313/2013/EU which calls for an

[1] https://drmkc.jrc.ec.europa.eu/.

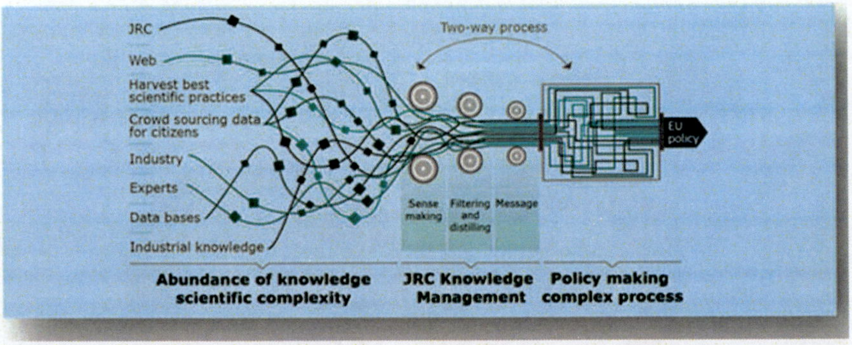

Fig. 5.1 Simplifying scientific complexity for policymaking (extracted from Spirito et al. 2017)

increase in "the availability and use of scientific knowledge on disasters". The role of scientific knowledge is further emphasized through Article 13 establishing the Union Civil Protection Knowledge Network. The Knowledge Network brings together civil protection and disaster management actors to work together, collaborate and learn from each other with the objective of strengthening the EU's overall ability and capacity to deal with disasters.

The EC published on 10 November 2021 the Decision (EU) 2021/1956 which formally established the Union Civil Protection Knowledge Network, with a strong Science Pillar, clearly referencing the DRMKC, as a core building block. This is a major step in the science–policy interface in the disaster risk management domain.

It is also an example of how an EC Knowledge Centre can evolve into a more long-term initiative, gain full trust and recognition by Member States and set the model for a science–policy interface in Europe.

Within the newly established Knowledge Network, the Science Pillar aims at pooling together the scattered high-quality data and analytical tools and putting them at the disposal of the UCPM Member and Participating States and the citizens. It will facilitate access to the tools and systems and the operationalization of knowledge to ensure evidence-based decision-making and operations.

One of the missions of Knowledge Network is to map knowledge gaps, learn from good practices, ensure knowledge transfer and co-design innovative and tailored technologies based on existing needs.

5.2 Developing Collective Knowledge and Innovative Tools to Enhance Resilience

Improving collective knowledge can only be done through collaborative processes. More than ever before, the COVID-19 pandemic has shown that disasters are systemic, and collaboration across policies is essential. Therefore, the DRMKC

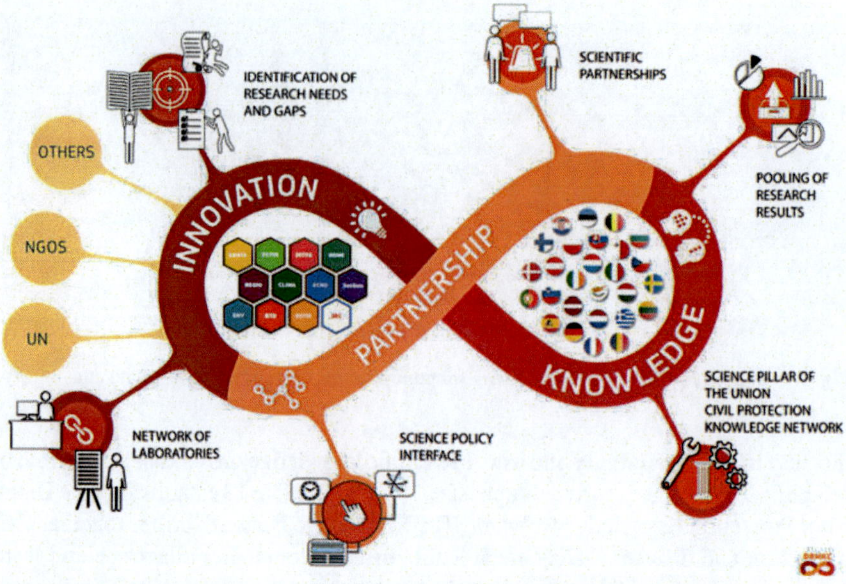

Fig. 5.2 The DRMKC mission: facilitate information and knowledge through integrated scientific solutions while enhancing the connection between science, operational activities and policy

brings together policymakers and experts dealing with disaster risk at the Commission, the EU countries and around the world to.

- build networks and partnerships to transfer science into policy
- integrate, consolidate and share knowledge across sectors
- co-develop innovative solutions for the mitigation of disasters and their impacts.

The DRMKC mission illustrated in Fig. 5.2 is to enhance the knowledge base to support disaster risk management policies and operational needs.

In the following sections, we briefly present the main instruments developed by the DRMKC to facilitate evidence-based decision-making and operations and to enhance resilience to disaster risks:

5.2.1 The Projects Explorer[2]

While the production of knowledge in the field of disaster risk management is abundant, there seems to be a lack of expertise in conducting cutting-edge risk assessments within the new paradigm of vulnerability and resilience, rather than focusing narrowly on hazards (Aitsi-Selmi et al. 2016). The gap between what is known about natural

[2] https://drmkc.jrc.ec.europa.eu/knowledge/PROJECT-EXPLORER.

hazards and disaster mitigation, on the one hand, and how research findings are translated into disaster risk reduction policies and research programs, on the other hand, is still not fully closed. Besides, the current understanding and recent research findings show that dealing with compound disaster impacts, cascading and concurrent hazardous events and factoring them into disaster scenarios remain challenging.

The DRMKC is actively contributing to map these gaps to help shaping the research agendas of the new Horizon Europe and the EU missions, in particular the EU mission for Adaptation to Climate Change.

The Projects Explorer acts as a common repository of relevant research and operational projects along with their results. It allows capturing existing research results to support their operational exploitation, and at the same time, it identifies the remaining gaps, hence promoting a more focused research. The Projects Explorer is developed in close collaboration with the Community of Users on Secure, Safe and Resilient Societies (CoU) of the EC's Directorate General for Migration and Home Affairs.

Currently, it covers more than 2000 research projects involving more than 8000 institutions.

5.2.2 *Science for DRM*

The DRMKC has committed to producing a series of reports to analyse, update the state of the art and identify research and innovation gaps in the field of DRM. Each report is multi-hazard, multidisciplinary and addresses the full disaster risk cycle; it has scientific-oriented contributions presenting the state of science and practitioner-oriented contributions presenting the use of science. This series of reports aims to bridge science and policy as well as operational communities.

The first report, published in May 2017 with the sub-title "Knowing better and losing less" (Poljansek et al. 2017), covered the DRM cycle through a scientific hazard-based approach.

The report "Science for DRM 2020: acting today, protecting tomorrow" (Casajus Valles et al. 2020), is the second flagship report prepared by the Disaster Risk Management Knowledge Centre. In this second volume, the emphasis is on broadening the understanding of current and future risks faced in Europe, focusing on the consequences suffered by different assets, such as population, the environment and economic sectors. By knowing the causes and drivers of risks, it will be possible to move from identifying needs to developing solutions and new approaches.

The Science for DRM series of reports is produced thanks to the collaboration of many partners. More than 300 experts engaged in disaster risk management, coming from academia, governmental agencies, ministries, civil protection groups, research organizations, civil society organizations and the private sector, were involved in the preparation of the Report that was publically launched in March 2021.

Both reports have allowed the launch of initiatives as the one undertaken by the Cooperation Network for Risk, Safety & Security Studies (CONRIS).[3] CONRIS is a network of universities with accredited degree programs in risk, safety and security management. CONRIS aims at increasing safety and security in Europe through collaboration in education and research. This network of universities has adopted the reports as textbooks for DRM training offered in several European universities. The scope is to promote a multidisciplinary, international and harmonized level of understanding of DRM at the university level through online webinars. These open-access sessions provide students studying disaster risk management with international, evidence-based knowledge about state-of-the-art practice, contributing to the quality of DRM provision. Materials from this training are available in the DRMKC learning corner.[4]

The training has been provided twice per year so far in different universities that are part of the CONRIS consortium. Each session has been attended by about 60 students coming from different universities.

5.2.3 Recommendations for National Risk Assessment in Disaster Risk Management in EU[5]

Another line of activity in which DRMKC is engaged to provide scientific advice to Member States is a series of reports on "Recommendations for National Risk Assessment in Disaster Risk Management in EU" produced in collaboration with a large number of scientists. The potential users are the national authorities of the UCPM Participant States that are in charge of the preparation of the National Risk Assessment (NRA) process as well as to link the NRA exercise to the whole DRM planning.

Article 6 of the Union Civil Protection Mechanism Decision No 1313/2013 with amendment Decision (EU) 2019/420/EU requested Participating States to regularly report to the Commission on their disaster risk management activities. With this reporting process, the EU gains a more comprehensive view of disaster risks in the EU and better evidence to support formulating an EU risk management policy framework that would complement and enhance the national efforts.

The purpose of this series of "Recommendations" is to have maximum benefit from the reporting process for all involved. The reporting outcomes are only the tip of the iceberg of the real activities related to disaster risk management in the country. The Reports recommend a common risk assessment approach for existent and new/emerging risks to be addressed and which capacities to develop within risk governance processes at the national level. The goal is to have risks arising from

[3] http://www.conris.eu/.

[4] https://drmkc.jrc.ec.europa.eu/knowledge/learning-corner.

[5] https://drmkc.jrc.ec.europa.eu/knowledge/SCIENCE-FOR-DRM/Recommendations-for-National-Risk-Assessment-for-Disaster-Risk-Management-in-EU.

different hazards in different regions comparable, to understand the risk drivers to act upon and help decision-makers in DRM in their prioritizing of DRM planning and actions. Then, information shared at EU level fosters a shared understanding of risks relevant not only for the national context but also at EU level. These are risks requiring activities at cross-border, inter-regional and international level and include issues such as climate change adaptation, critical infrastructure protection and disaster loss data collection.

The reports explain the purpose and objective of each step of the reporting to give meaning and motivation to demanding risk governance processes. In Version 0 (Poljansek et al. 2019), the focus was on the national risk assessment process from the scientific perspective and how to to provide solid evidence to risk managers. We collected the contributions of 9 expert teams that prepared short step-by-step instructions of disaster risk assessment approaches specific for the risks of different origins usable in the context of a national risk assessment exercise. They addressed also national risk assessment capabilities to bring evidence to the next level. Version 1 (Poljansek et al. 2021) was more about risk governance that facilitates the policy cycle for the implementation of integrated disaster risk management and provided instructions on how to tackle the risk asessment of 6 additionl risks. The risks covered in this document are of natural, anthropogenic and socio-natural origin: floods, droughts, wildfires, biodiversity loss, earthquakes, volcano eruptions, biological disasters, Natech accidents, chemical accidents, nuclear accidents, terrorist attacks, critical infrastructure disruptions, cybersecurity and hybrid threats. Both reports are available on the DRMKC Website.[6]

5.3 Innovation as Key Engine to Face a New Era of Risks

The DRMKC also launched a number of initiatives and services aimed at supporting EU Member States in the collection and analysis of loss data and the provision of risk models and tools for risk assessment.

5.3.1 DRMKC Risk Data Hub[7]

Post-event damage and loss data have an intrinsic key role in all the phases of disaster risk management, from the understanding of the impacts of past events to the calibration and validation of forecasting models (risk models), to the evaluation of progress made in reducing disaster risk.

The DRMKC Risk Data Hub (Antofie et al. 2020) is a multi-hazard Geo-portal aiming at providing a common tool for scientists, practitioners and policymakers to

---

[6] https://drmkc.jrc.ec.europa.eu/knowledge/science-for-drm/nra.

[7] https://drmkc.jrc.ec.europa.eu/risk-data-hub.

support bridging science and policy, past and future, local/national and European scales in the frame of DRM. It is conceived to be co-developed by scientists and end-users looking for crystallizing the science–policy interface.

Working together with the different user communities in the frame of the collaborative approach offered by the DRMKC, we count on facilitating the translation of research results into useful and used inputs for risk reduction.

The concept of the GIS-based platform Risk Data Hub was the result of a needs and gaps assessment performed in 2016, during the preparation of the European Commission Staff Working Document "Overview of Natural and Man-made Disaster Risks the European Union may face" (2021). The existing gap between the data and knowledge available in the scientific community and their accessibility and usability by decision-makers was one evident outcome of the overview.

5.3.2 DRMKC Going Global: INFORM SUITE[8]

INFORM (IASC and the European Commission 2022) is a joint initiative of the European Commission and the Inter-Agency Standing Committee Task Team (IASC) for Preparedness and Resilience, in partnership with many UN Agencies, donors, NGOs and EU Member States. Based on this international collaboration, INFORM provides a quantitative assessment of the risk of humanitarian crises in a multi-hazard context with a major focus on developing countries. JRC is a scientific lead of INFORM developments to provide scientific support to a large number of European and global policy initiatives in the frame of humanitarian aid and development.

Since 2014 INFORM Risk has not only become a global reference for multi-hazard assessment of crisis risk, but INFORM has been developing into a suite of quantitative, analytical products. They are designed to support decision-making at different phases of the crisis risk management cycle: INFORM RISK for prevention and preparedness, INFORM SEVERITY to support response actions, INFORM Climate Change to inform decision-making around the projected risk of climate-amplified hazards and INFORM WARNING, still to be developed, will monitor early warning signals to support preparedness and early actions.

All INFORM products follow the same principles. They are global, open, reliable and flexible. Availability of methodologies and underlying data allows organizations to adapt the INFORM products to suit their decision-making processes.

[8] https://drmkc.jrc.ec.europa.eu/inform-index.

5.3.3 DRMKC Gaps Explorer[9]

In the scientific world, as the flood of knowledge increases, there is a greater need for systematic reviews to guarantee proper exploitation of results along with the definition of the way forward. Directorate General (DG) Research and Innovation (DG RTD) contributes to mapping achievements and needs for further development through their "Projects for Policy" series of reports.

Inspired by this initiative, the DRMKC has developed the Gaps Explorer to provide targeted recommendations, tailored to different stakeholder profiles—policymakers, practitioners and scientists—based on thematic reviews. The first pilot has been co-developed by DG RTD and JRC on forest fires. The second thematic review builds on the findings of the second Science for DRM 2020 book. It focuses in particular on Chap. 6 of the book which provides recommendations to four groups of society that can contribute to reducing disaster risk by converting research findings into concrete applications in practical disaster risk management: policymakers, practitioners (including here civil protection groups), scientists and citizens.

5.3.4 Global Conflict Risk Index

In the context of the EU conflict Early Warning System (EWS),[10] the Global Conflict Risk Index (GCRI) (Halkia et al. 2020) provides quantitative information about the risk of violent conflict in non-EU countries. Such risk information, complemented with qualitative input, forms the basis for subsequent prioritization by the EU services. Therefore, as a tool, it positions itself in the very beginning of the EU conflict EWS cycle which identifies, prioritizes and assesses countries at risk of violent conflict, in line with the EU's global strategy for foreign and security policy, and more specifically in the context of conflict prevention and mediation support.

The GCRI methodology assesses the risk of violent conflict from a quantitative point of view, by using statistical modelling based on open source information. The input data relate to a number of quantitative indicators reflecting historical structural conditions within each country, correlated to the likely occurrence of violent conflict. This reliance on quantitative data only makes it a neutral and evidence-based starting point for the EU conflict risk assessment, thus keeping any initial political bias to a minimum. The output of the underlying statistical regression models is expressed and communicated to the appropriate EU services on a yearly basis, in terms of the probability and the intensity of violent conflict within each country in the next one to four years. Based on these results, an initial risk category is assigned by the competent EU service to each country, which then drives the follow-up discussions and final decision on the actual assessed risk.

---

[9] https://drmkc.jrc.ec.europa.eu/knowledge/Gaps-Explorer.

[10] https://eeas.europa.eu/headquarters/headquarters-Homepage/426/conflict-prevention-peace-building-and-mediation_en.

Due to confidentiality purposes, both the results of the GCRI index and the final assessment of the risk of violent conflict in non-EU countries are internal to the EU services involved in the overall EU conflict EWS process and therefore not publicly available.

5.3.5 *Natech Risk Assessment and Mapping*

Natural-hazard impact at industrial facilities can trigger so-called Natech accidents with releases of hazardous substances (Krausmann et al. 2017). A prominent example of a Natech accident was the Fukushima nuclear power plant accident after the Great East Japan earthquake and tsunami in 2011. Climate change and growing human development will increase Natech risk in the future. Natech risk is a bridge between the technological and natural risk worlds. While DRR frameworks in general do not really address technological risks, the DRMKC, with its all-hazards and multi-stakeholder approach, also embraces technological risks and hence Natech.

For governments and operators to identify Natech risks and hotspots in their territory, Natech risk maps are needed but are scarce. Following calls by governments to fill this gap, the JRC developed the rapid Natech risk assessment and mapping system RAPID-N[11] which is a free Web-based service.

RAPID-N allows the assessment and mapping of Natech risks with minimum data input. It incorporates a unified methodology which estimates on-site natural-hazard parameters, determines damage probabilities of industrial plant equipment under natural-hazard loading and assesses the probability and severity of possibly triggered Natech events (Girgin and Krausmann 2013). The output of RAPID-N simulations can be used for land use and emergency planning before natural hazards hit an area, but also for damage assessment and early warning after an event. Figure 5.3 shows an example of a RAPID-N output map with its impact areas.

The JRC organizes training courses on the use of RAPID-N and performs Natech risk assessment case studies jointly with interested countries.

5.3.6 *Natech Risk Management Performance Indicators*

Natech risk exists anywhere in the world where hazardous industry is located in areas prone to natural hazards. Unfortunately, there is no base for comparing risk trends over time, rendering comparison of Natech risk management performance in countries impossible. The JRC has developed simple indicators to provide a baseline for countries to gauge their progress in Natech risk management, but also for donors to channel their funding to where it is most needed. The indicators can be used as a

[11] https://rapidn.jrc.ec.europa.eu/.

Fig. 5.3 Example output of a regional RAPID-N risk assessment. The circles show impact endpoint distances around facilities affected by an earthquake

simple self assessment of countries over time, but also for comparing the status of different countries at the same point in time (Krausmann et al. 2019).

5.4 Conclusions

The emergence of new and the intensification of existing risks requires a joint effort to enhance society's resilience to these risks. The DRMKC aims to support this process. Following an integrated disaster risk management approach, where the whole is greater than the sum of its parts, the DRMKC gathers different European Commission services, EU countries and the disaster risk community in an attempt to manage risks in a more coordinated way. This leads to an enhanced effectiveness of better coordinated policies based on knowledge transfer across sectors.

The driving force of DRMKC is collaboration. Relying on the expertise of the wide DRM community, the DRMKC proposes and coordinates actions to consolidate collective knowledge across scientific disciplines and economic sectors to co-develop inovative solutions for existing needs. It also reinforces the existing links across DRM- and DRR-related policies. These are the ingredients of a more resilient society.

The DRMKC proposes opportunities for co-developing collective knowledge through a number of activities presented in this chapter along with other complemantary instruments available on the online platform such as:

(i) the quarterly DRMKC newsletter[12] prepared by the Community of Practice of DRMKC which gathers 12 Directorates General and institutions of the European Commision,

[12] https://drmkc.jrc.ec.europa.eu/overview/newsletter.

(ii) the scientific outputs[13] which pool information from scientific results and expertise, to boost transfer of research outputs to end-users and facilitate the task of making sense of existing knowledge.
(iii) and the Learning Corner[14] which provides learning opportunities about methodological and conceptual approaches, guidelines and good practices in the field of DRM.

While knowledge is abundant, gaps do, of course, still exist. The DRMKC contributes to mapping these gaps, fostering in that way a more focused and needs-driven development. The DRMKC also makes a conscious effort, in line with the Sendai Framework for DRR, to fully embrace all types of risks which may so far not have been considering sufficiently or at all in DRR frameworks, e.g. technological accident risk, CBRNe, cybersecurity, etc.

At the core of the Knowledge Network Science Pillar, DRMKC will facilitate the engagement of all DRM stakeholders interested in contributing to its development. Starting from the mentioned existing components, the aim is to expand the Science Pillar into an open community where each and everyone can suggest additional components, trigger new scientific debates and, collectively, shape its future development.

References

Aitsi-Selmi A, Murray V, Wannous C, Dickinson C, Johnston D, Kawasaki A, Stevance AS, Yeung T (2016) Reflections on a science and technology agenda for 21st century disaster risk reduction: based on the scientific content of the 2016 UNISDR science and technology conference on the implementation of the sendai framework for disaster risk reduction 2015–2030. Int J Disaster Risk Sci 7(1):1–29. https://doi.org/10.1007/s13753-016-0081-x

Antofie E, Marrin Ferrer M, Luoni S, Eklund G (2020) Decision making improvement for disaster risk management (DRM) through technological support: DRMKC Risk Data Hub, GRRASP (geospatial risk and resilience assessment platform) and RAPID N (rapid natech risk analysis and mapping system) 3rd DRMKC RDH Workshop, 16th of Oct 2019. Publications Office, LU. https://data.europa.eu. https://doi.org/10.2760/44907

Casajus Valles A, Marin Ferrer M, Poljansek K, Clark I (eds) (2020) Science for disaster risk management 2020: acting today, protecting tomorrow. EUR 30183 EN, Publications Office of the European Union, Luxembourg. ISBN 978-92-76-18181-1. https://doi.org/10.2760/438998

Coppola E, Jones R, Sörensson AA, Pinto I (2021) Summary for policymakers of the working group I contribution to the IPCC Sixth Assessment Report. Application/xml. NERC EDS Centre for Environmental Data Analysis. https://doi.org/10.5285/35A7EE81A50C4B95AB59F9BD128F9B63

Commission Implementing Decision (EU) 2021/1956 of 10 November 2021 on the establishment and organisation of the Union Civil Protection Knowledge Network (notified under document C(2021) 7939) (Text with EEA relevance) C/2021/7939 OJ L 399, 11.11.2021, pp 1–7. http://data.europa.eu/eli/dec_impl/2021/1956/oj

[13] https://drmkc.jrc.ec.europa.eu/knowledge/scientific-output.

[14] https://drmkc.jrc.ec.europa.eu/knowledge/learning-corner.

Communication to the commission C(2016)6626, Data, Information and Knowledge Management at the European Commission. https://ec.europa.eu/transparency/documents-register/detail?ref=C(2016)6626&lang=en

Decision No 1313/2013/EU of the European Parliament and of the Council of 17 December 2013 on a Union Civil Protection Mechanism Text with EEA relevance, OJ L 347, 20.12.2013, pp 924–947. http://data.europa.eu/eli/dec/2013/1313/oj

Decision (EU) 2019/420 of the European Parliament and of the Council of 13 March 2019 amending Decision No 1313/2013/EU on a Union Civil Protection Mechanism PE/90/2018/REV/1 OJ L 77I , 20.3.2019, pp 1–15. http://data.europa.eu/eli/dec/2019/420/oj

European commission, directorate-general for european civil protection and humanitarian aid operations (ECHO) (2021) Overview of natural and man-made disaster risks the European Union may face : 2020 edition. Publications Office. https://data.europa.eu/doi/10.2795/1521

Girgin S, Krausmann E (2013) RAPID-N: rapid natech risk assessment and mapping framework. J Loss Prev Process Ind 26(6):949–960. https://doi.org/10.1016/j.jlp.2013.10.004

Halkia M, Ferri S, Schellens MK, Papazoglou M, Thomakos D (2020) The global conflict risk index: a quantitative tool for policy support on conflict prevention. Progress Disaster Sci 6(April):100069. https://doi.org/10.1016/j.pdisas.2020.100069

IASC and the European Commission (2022) INFORM REPORT 2022; Shared evidence for managing crises and disasters, EUR 31081 EN, Publications Office of the European Union, Luxembourg. ISBN 978-92-76-52775-6. https://doi.org/10.2760/08333

Krausmann E, Cruz AM, Salzano E (2017) Natech risk assessment and management: reducing the risk of natural-hazard impact on hazardous installations. Elsevier, Amsterdam, Boston, Massachusetts

Krausmann E, Girgin S, Necci A (2019) Natural hazard impacts on industry and critical infrastructure: natech risk drivers and risk management performance indicators. Int J Disaster Risk Reduction 40(Nov):101163. https://doi.org/10.1016/j.ijdrr.2019.101163

Poljansek K, Marin Ferrer M, De Groeve T, Clark I (eds) (2017) Science for disaster risk management 2017: knowing better and losing less. EUR 28034 EN, Publications Office of the European Union, Luxembourg. ISBN 978-92-79-60678-6. https://doi.org/10.2788/688605

Poljansek K, Casajus Valles A, Marin Ferrer M (eds) (2019) Recommendations for national risk assessment for disaster risk management in EU: approaches for identifying, analyzing and evaluating risks, Version 0, EUR 29557 EN, Publications Office of the European Union, Luxembourg. ISBN 978-92-79-98366-5. https://doi.org/10.2760/084707

Poljansek K, Casajus Valles A, Marin Ferrer M (eds.) (2021) Recommendations for national risk assessment for disaster risk management in EU: where science and policy meet, Version 1, EUR 30596 EN, Publications Office of the European Union, Luxembourg. ISBN 978-92-76-30256-8. https://doi.org/10.2760/80545

Spirito L, Castello P, Mortara B, Wilikens M (2017) Knowledge management for policy. European Commission, JRC110151

World Meteorological Organization (2021) WMO releases Atlas of impacts of weather, climate and water extremes 1970–2019

Chapter 6
Infrastructure Resilience: A Framework for Assessment, Management and Governance

Craig Davis, Bilal M. Ayyub, Sue McNeil, Kiyoshi Kobayashi, Hirokazu Tatano, Masamitsu Onishi, Yoshikazu Takahashi, Riki Honda, Toshio Koike, and John van de Lindt

Abstract An infrastructure resilience framework is introduced presenting processes, tools and outcomes for system assessment, management and governance. This can be used for analysis or design of systems prior to or following actual disruptions. For this framework, infrastructure is defined as the basic physical and organizational structures and facilities (e.g., building clusters and lifeline systems) needed for the operation of a society or enterprise. The framework emphasizes infrastructure systems and how they support community resilience through eight key elements. These elements provide the foundation for system assessments, management and governance of resilient infrastructure systems. Elements underpinning assessments identify the framework processes for evaluating existing or proposed systems and provide a strong basis for management and governance. The elements primarily underpinning management and governance identify the processes and tools needed to provide the outcomes to ensure resilient infrastructure systems; they also draw from and provide feedback for improving the other elements. Resilience management

C. Davis (✉)
C.A. Davis Engineering, Santa Clarita, CA, USA
e-mail: cadavisengr@yahoo.com

B. M. Ayyub
Department of Civil and Environmental Engineering, University of Maryland, Maryland, USA

S. McNeil
Civil and Environmental Engineering, University of Delaware, Delaware, USA

K. Kobayashi · H. Tatano · M. Onishi · Y. Takahashi
Kyoto University, Kyoto, Japan

R. Honda
University of Tokyo, Tokyo, Japan

T. Koike
International Center for Water Hazard and Risk Management, Public Works Research Institute, Tsukuba, Japan

J. van de Lindt
Department of Civil and Environmental Engineering, Colorado State University, Fort Collins, CO, USA

H. Tatano and A. Collins (eds.), *Proceedings of the 4th Global Summit of Research Institutes for Disaster Risk Reduction*, Disaster and Risk Research: GADRI Book Series,
https://doi.org/10.1007/978-981-19-5566-2_6

and governance utilizes the knowledge and results of system assessments and incorporates other tools to engineer, manage and govern resilient systems that properly support the resilience of the communities they serve.

Keywords Infrastructure · Resilience · Lifeline infrastructure system · Building · Framework

6.1 Introduction

Large-scale earthquakes, floods and other natural hazards continue to threaten communities all over the world and affect their economies and social well-being. As such there is a dire need to decrease community risks and increase their resilience (UNISDR 2015). The Third UN World Conference on Disaster Risk Reduction was held in Sendai City, Miyagi Prefecture, Japan in March 2015. The "Sendai Declaration", among other things, called for the strengthening of disaster risk reduction to mitigate loss of lives and assets worldwide. In other words, a call to strengthen critical infrastructure. Infrastructure systems are essential for supporting communities, changing the equity imbalance, and increasing resilience (United Nations 2016). It is therefore essential to develop means for increasing infrastructure system resilience against severe hazards.

This paper fulfills the need for an international framework providing a standard way to assess and implement resilient infrastructure systems. Having a common framework will improve the global understanding of resilient infrastructure systems and aid in focusing institutions toward common goals for research, development, and implementation. The Japan Society of Civil Engineers and American Society of Civil Engineers Infrastructure Resilience Division are taking an international leadership role to develop and implement this framework for all to know how to work toward creating resilient infrastructure systems in support of developing community resilience. The proposed infrastructure resilience framework identifies processes, tools and outcomes for system assessment, and management and governance useful for analyzing and designing systems prior to or following actual disruptions.

The purpose for creating an infrastructure system resilience framework is to identify methodologies, processes, and procedures through which the resilience of all infrastructure systems can be consistently evaluated and implemented. The primary objective of 'infrastructure resilience' is to limit damage and losses and restore infrastructure services to support community resilience following an event. The proposed framework allows for standardized tools to be developed through coordinated research efforts in support of and to operationalize infrastructure resilience.

There are many different resilience frameworks relating to infrastructure and communities in existence, several of which are summarized in Appendix I. They tend to fit within three categories:

1. Very high level frameworks intending to create a conceptual structure for global, national, regional, or community-wide resilience, which are supported by infrastructure systems.
2. Narrowly focused frameworks intending to create a conceptual structure for limited aspects of an overall process needed to evaluate infrastructure resilience, such as formulating a functionality curve, or establishing performance objectives.
3. Some frameworks cover specific infrastructure systems such as water, hospitals, electric power, particular phases like recovery, or address certain hazards such as an earthquake or flood. By not addressing the combination of research, development, and implementation needed or not incorporating system services, or only addressing resilience relative to specific indicators, hazards or phases and not including true system functionality and operability, these frameworks have limited applicability.

A resilience framework, covering all infrastructure systems and hazards, incorporating the ability to evaluate system functionality and operability, or just indices, for the complete set of elements needed to create resilient infrastructure systems has not previously been fully conceptualized. This paper describes such a framework that focuses on research, development and implementing infrastructure resilience with an understanding of its role within the broader context of overall community resilience. It has an emphasis on infrastructure systems and is intended to be applicable to all possible hazards. The critical inter-relations between assessing infrastructure resilience and the need to manage and govern the systems with an understanding of the economics or financing of resilience as well as potential social and economic losses, all in support of community resilience, are brought together within a single framework.

6.2 Infrastructure Resilience Framework

Figure 6.1 diagrams the proposed infrastructure resilience framework identifying processes, tools and outcomes for system assessment, management and governance useful for analyzing and designing systems prior to or following actual disruptions. For this framework infrastructure is defined as the physical (built) and organizational structures and facilities (e.g., building clusters and lifeline systems) needed for the operation of a society or enterprise.

Resilience is defined as the ability to prepare for and adapt to changing conditions and withstand and recover rapidly from disruptions. Resilience includes the ability to withstand and recover from deliberate attacks, accidents, or naturally occurring threats or incidents (PPD 2013). Ayyub (2014a) provides an intentional definition as "the resilience of a system is the persistence of its functions and performances under uncertainty in the face of disturbances" to enable its measurement through metrics. Davis (2019a) defines a resilient infrastructure system as one that is "managed to provide safe and reliable services to customers/users, cope with chronic stressors,

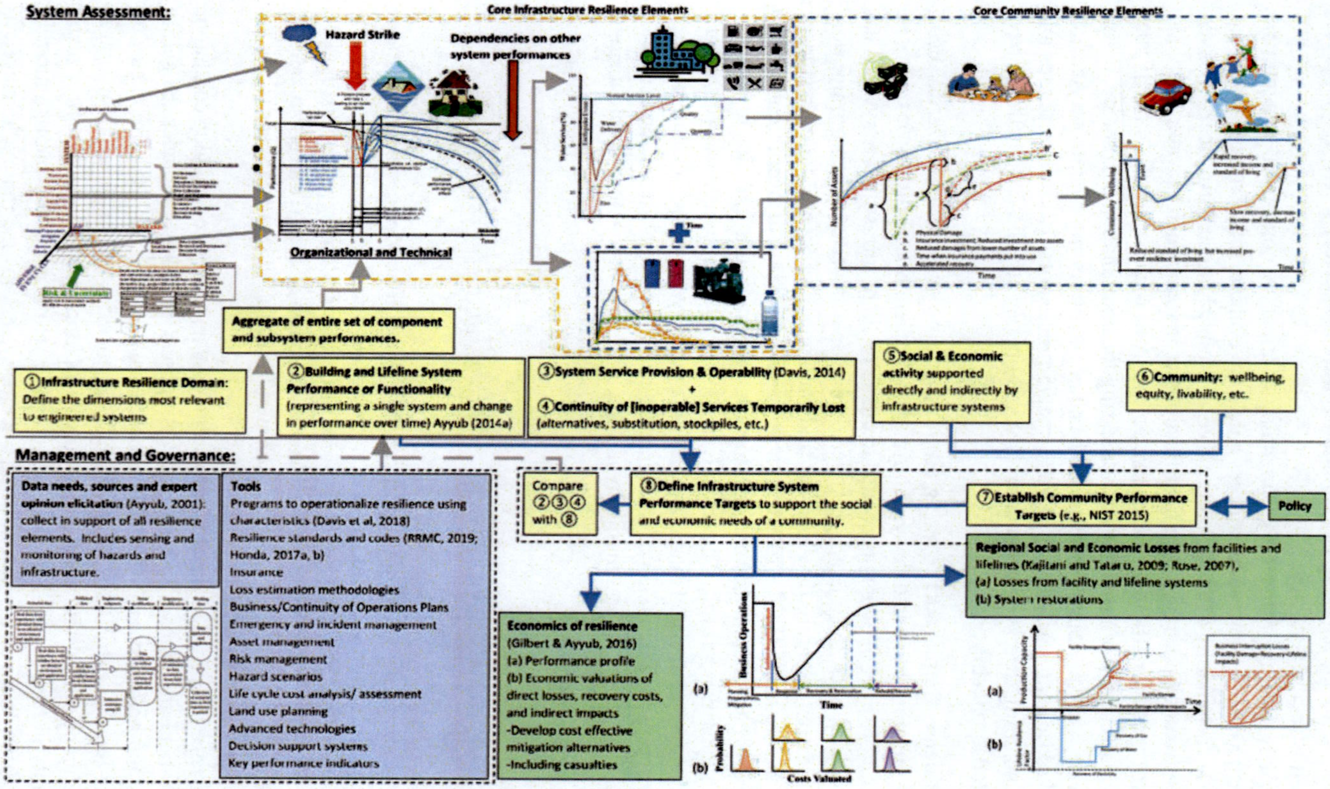

Fig. 6.1 Infrastructure resilience framework. Gray and blue arrows, respectively, show flow for system assessment and management and governance processes. Element numbers are in circles

and accommodate hazard-related impacts with ability to continue providing services or limit service outage times tolerable for community recovery efforts" (modified from Davis et al. (2018)). Bruneau et al. (2003) notes that resilience can be achieved by enhancing the ability of a communities infrastructure (e.g., lifelines, structures) to perform during and after an earthquake [or other hazard strike], as well as through emergency response and strategies that effectively cope with and contain losses and recovery strategies that enable communities to return to levels of pre-disaster functioning (or other acceptable levels) as rapidly as possible. Resilience is the aggregate of many dimensions. Infrastructure resilience focuses on a subset of dimensions related to the built systems (buildings, their clusters, and lifeline systems) important for community resilience. The built systems support social institutions so they can provide services needed for community resilience. Social institutions include family, economic, government, health, education, community service, religious, cultural, and media organizations (NIST 2015).

The framework emphasizes infrastructure systems and how they support community resilience. Appendix II describes the relationship between built, social, and natural environmental systems and the role of infrastructure within the broader context of resilience. Figure 6.1 shows elements numbered 1 to 8 making up the processes for the infrastructure resilience framework. Each of these elements as well as their connection to other elements is described below.

Elements 1 to 6 collectively describe the system assessment. Elements 7 and 8 are important to managing and governing infrastructure resilience. Elements 1 to 4 represent the core infrastructure system resilience. Elements 4 to 6 represent the core community resilience elements. Element 4 is a key crossover element linking the resilience of infrastructure systems with that of the community.

As shown in Fig. 6.1, proper assessment of these elements results in social and economic outcomes for the infrastructure systems in terms of policy, economics of resilience, and communities in relation to regional social and economic losses.

Tools useful for managing and governing infrastructure resilience are identified in Fig. 6.1. These include the data collection (Ayyub 2001, 2002), and understanding its range of uncertainty, in support of all resilience elements. Data collection and sharing improves adaptive capacity, decision-making, and situational awareness. Other resilience management and governance tools include developing lifeline system resilience programs, resilience-based performance standards and codes, loss estimation as previously described, business continuity, emergency management, asset management, risk management (Ayyub 2014b), hazard scenarios, life cycle cost assessment, decision support systems, key performance indicators, and land use planning, to name a few. These and other tools can be developed further in support of the infrastructure resilience framework.

Figure 6.1 is broken into two sections horizontally across the middle. The upper portion is designated as 'System Assessment' and the lower portion as 'Management and Governance'. These sections are inter-related as shown by the arrows in Fig. 6.1 linking elements on either side of the line. The 'System Assessment' identifies the framework processes for evaluating existing or proposed systems. The 'Management and Governance' identifies the processes and tools, including results from the 'System Assessment', needed for proper resilience management and/or governance to provide the outcomes to ensure resilient infrastructure systems. Resilience management and governance utilizes the knowledge and results of system assessments and incorporates other tools to engineer, manage, and govern resilient systems that properly support the resilience of the communities they serve.

The processes in Fig. 6.1 can be applied as an evaluation for a hypothetical event or scenario, or for an actual event. In both cases there is the opportunity for learning and making improvements based on the results. The framework is applicable at the spatial scale of the infrastructure system and communities they serve.

6.2.1 Element 1: Infrastructure Resilience Domain

Element 1 in Fig. 6.1 and enlarged in Fig. 6.2 defines the dimensions of infrastructure resilience. Resilience is a multifaceted notion and the aggregate of many dimensions. The ASCE Infrastructure Resilience Division (IRD) defined this infrastructure resilience domain as shown in Fig. 6.2 (Davis et al. 2018). The domain is structured around three primary dimensions: Civil Infrastructure and Lifeline Systems (shown on vertical axis), Hazards (shown on horizontal axis to right), and the Adverse Event Cycle (shown on the horizontal axis pointing out of the plane of paper). These primary dimensions make up a matrix of resilience cubes as schematics, with each cube having a System-Hazard-Event cycle locus. Figure 6.2 explicitly outlines the cube for Communication System-Flood-Preparation. Each cube identifies the dimensions needed to be addressed to achieve true infrastructure resilience; for each system, all hazards of importance, and every step in the event cycle need to be addressed.

Figure 6.2 identifies the infrastructure systems on the vertical axis, which are broken down into two primary categories: lifeline systems and building clusters. The traditional lifelines are itemized on the vertical axis as communication, electric power, water, wastewater, inundation protection, natural gas, liquid fuels, solid waste management, and transportation systems. Building clusters are defined as "a set of buildings—and supporting infrastructure systems—that serve a common function such as housing, healthcare, retail. Clusters are not necessarily geographically co-located and may be distributed throughout the community" (NIST 2015). Building clusters are systems of buildings working together to provide specific types of services to social institutions. All engineering disciplines are needed in support of these systems.

Infrastructure systems are large and complex geographically distributed networks made up of numerous interlinked components that were built over long periods of time, under a wide variety of conditions, using many different materials and construction techniques. The systems include all subsystems, regardless of who owns or operates them, from source to delivery, from collection to disposal, and so on. Some parts of each system are more fragile than others. The vulnerabilities of each system are dependent upon the geospatial layout of the system and distribution of vulnerable components relative to the geographic distribution of the hazards to which

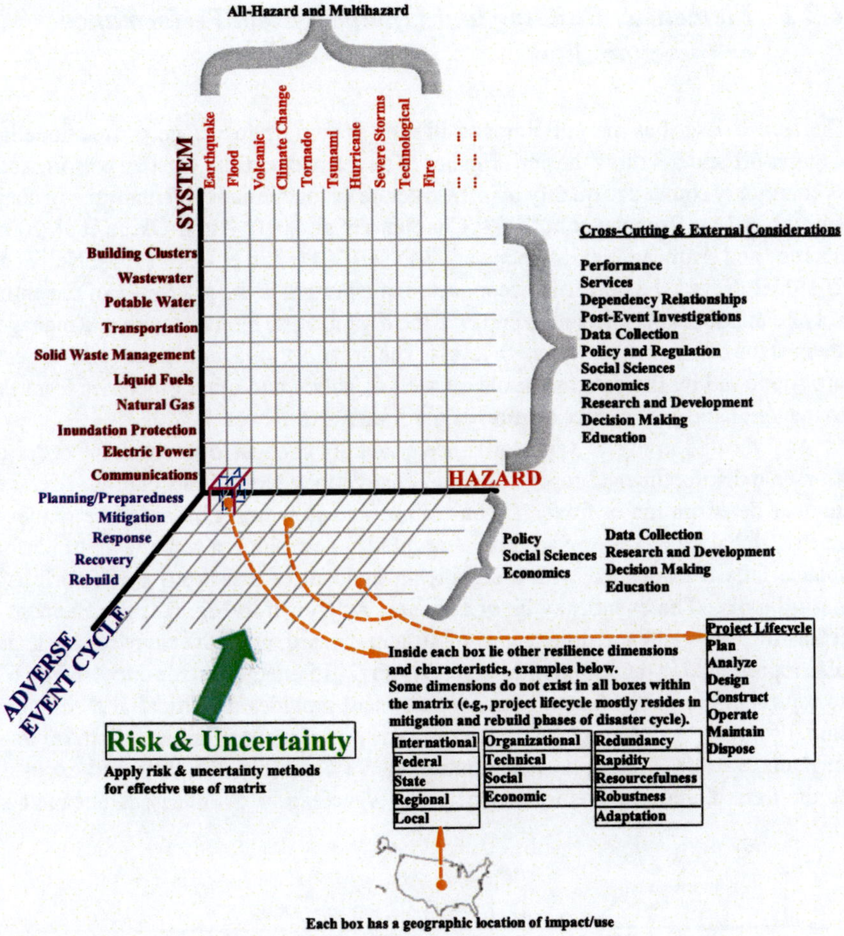

Fig. 6.2 ASCE infrastructure resilience domain. Modified from Davis et al. (2018)

the systems are exposed. These large systems traverse many geologic formations, geographies and hazards (e.g., rock, soil, landslides, rivers, reclaimed land, etc.). Severe hazards can impact large areas with varying intensity resulting in significant variation of effects (e.g., flooding, wind, or landslides in some locations but not in others). The infrastructure vulnerabilities and hazards they are exposed to often have time dependent variables.

Davis et al. (2018) provides more detailed explanation of the infrastructure resilience domain depicted in Fig. 6.2.

6.2.2 *Element 2: Building and Lifeline System Performance or Functionality*

Element 2 describes the building and lifeline system performance or functionality. Figures 6.1 and 6.3 show the performance of a system over time. System performance is commonly equated to quality of infrastructure or functionality by many researchers (Ayyub 2014a; Bruneau et al. 2003; Cimellaro et al. 2010; Attoh-Okine et al. 2009; Tierney and Bruneau 2007; Croope and McNeil 2011; Liu et al. 2017; McNeil et al. 2019). However, the performance measure of a system is the aggregate of the entire set of component performances and how the organization interacts with and manages them at the component and system levels. The organizational and managerial aspects are supported by the data and tools, as well as policy and other outcomes from the managerial and governance portion of the framework.

As indicated in Fig. 6.3, performance gradually changes over time, for better or worse, and sometimes abruptly. Gradual change may come from chronic stressors such as deterioration or financial constraints, and slow onset hazards like sea level rise or global temperature variation. Abrupt change results in some level of performance failure associated with a significant incident or disastrous event such as a hazard strike. The event may be of any size. As shown in Fig. 6.1, the Element 1 infrastructure resilience dimensions of system, hazard, and adverse event cycle are also represented in the Element 2 curve. Ayyub (2015) suggested the curve in Fig. 6.3 to represent the performance of any system and provides analytical and computational frameworks. Abrupt loss of performance is a result of a hazard strike, level of damage to the system as a function of its vulnerability, fragility, and exposure to the hazard, and impacts to the operating organization. Similarly, Bruneau et al.

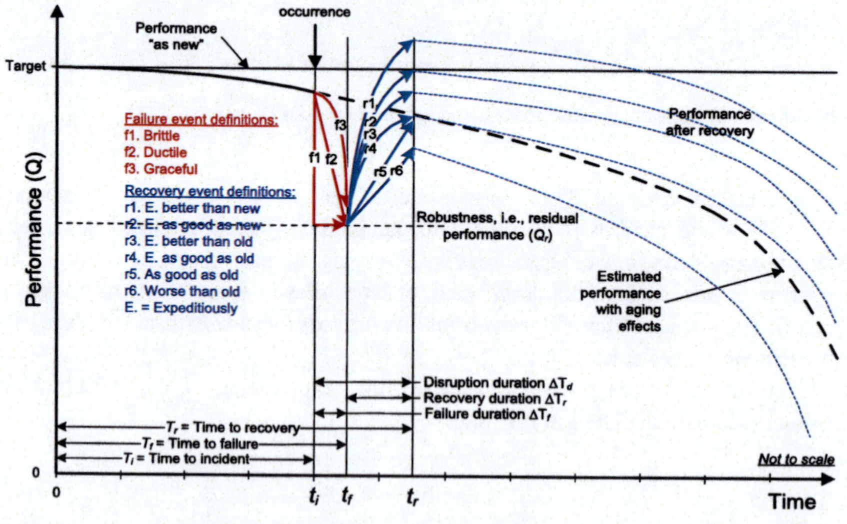

Fig. 6.3 Infrastructure (building or lifeline) system performance over time (Ayyub 2015)

(2003) use the curve to represent robustness, the capacity to minimize the initial losses, which is a function of the systems' and its components pre-event condition. The time to recovery following an abrupt performance loss is a result of planning, preparedness, response, recovery, and rebuild capabilities and capacity, and is closely associated with emergency and crisis management competencies. The system failure may be brittle, ductile, or graceful. Recovery may be to a performance level worse than old, as good as old, or expeditiously (meaning accelerated recovery using additional resources) as good as old, better than old, as good as new, or better than new. In extreme cases recovery may include complete abandonment and rebuild as was the case for all systems in some towns in China following the 2008 Wenchuan earthquake and in Japan following the Great East Japan Earthquake and Tsunami (Cabinet Office Japan 2019; Brook and Chi Yin 2017). The curves in Fig. 6.3 capture the technical and organizational dimensions of systems requiring the physical systems and the organizations that manage them to withstand hazard-related impacts and recover quickly from them. Socio-technical systems emerge from interplay of the technical and organizational dimensions combined with how they interact with social dimensions of resilience (Appendix II). Davis et al. (2018) summarize organizational, technical, social, and economic characteristics needed for resilient lifeline systems. Van der Merwe et al. (2018) present a framework for assessing resilience of socio-technical systems. The technical and organizational aspects are equally important for creating resilient infrastructure systems.

The performance of each infrastructure system is dependent upon the performance of other systems. Loss of performance may result for direct system damages as a result of a hazard strike, or from loss of dependent services even if the system was not damaged (e.g., loss of electric power may impact water system performance due to inability to pump).

Element 2 also considers the potential for secondary damage resulting from cascading effects of the primary damage inflicted by an initial hazard strike. For example, if an earthquake or flood were to cause a dam to break, then the massive flood wave from the reservoir causes severe secondary damage to the infrastructure in the inundation zone; or a radiation release from a nuclear power plant contaminating the regional environment.

6.2.3 Element 3: System Service Provision and Operability

Element 3 identifies the system service provision and operability (Davis 2021). Every infrastructure system provides services to the community. For hospital and health care building clusters, the services are related to health care, for water systems the services are related to water usage, and similarly for other infrastructure systems. The services provided by each system can be defined in terms of the basic service categories (Davis 2021) essential for supporting community resilience. The ability to continue providing these service categories, or rapidly restore them if they are lost after a hazard strike, is related to the Element 2 system performance or functionality, which is dependent upon the technical and organizational dimensions. Post-event performance or functionality is measured relative to pre-event normal functionality, which is assumed to be 100% (a system or building may have less than 100% functionality before an event if it is not working completely proper). Loss of functionality following an event results in a measure less than 100% until it is completely restored to pre-event levels. In this context, as an example, loss of power to a hospital is considered a reduction in functionality below 100%, even if the power is temporarily supplied through emergency generation; the hospital building system is not considered 100% functional until the electric power is restored. Improved functionality following restoration and mitigation after an event can result in a functionality measure greater than 100%. A highly resilient system can be operable and achieve the provision of the core services to the community prior to achieving full functionality restoration. The system is considered to have 100% operability when all the core community services are able to be met, which for a highly resilient system will occur prior to returning to 100% functionality. Similar to functionality, post-event operability may be less than 100%. There are conditions where lifeline infrastructure system services can be provided following an event, but the building or customers may not be able to receive them for various reasons; for example, houses destroyed by a tornado cannot receive water or power services. Service restoration is a function of the dynamics of how the system is repaired to incorporate redundancy and isolation capabilities through existing system layout or from adaptations and resourcefulness including human ingenuity. These, combined with the robustness of the components to a hazard strike, define the absorptive capacity of a system. Using the same example for power loss to a hospital, the building system is considered 100% operable when using the generator to provide power if it can still achieve the provision of all core community services prior to the electric power service being restored.

Figure 6.4 presents a more complicated example of the Los Angeles Water System following the 1994 Northridge Earthquake. The water system service categories, treated as performance dimensions per Fig. 6.3, are defined as (Davis 2014):

Water Delivery: This service is fully achieved when the system can distribute water to customers, but the water delivered may not meet quality standards (requires public notification), pre-event volumes (requires water rationing), fire flow requirements (impacting firefighting capabilities).

Water Quality: This service is fully achieved when water quality at customer connections meets pre-event standards. Potable water meets health standards (public notices for water use are removed), including minimum pressure requirements to ensure contaminants do not leach into the system.

Water Quantity: This service is fully achieved when water flow to customers meets pre-event volumes (water rationing removed).

Fire Protection: This service is fully achieved when the system can provide pressure and flow of a suitable magnitude and duration to fight fires.

Figure 6.4 plots the service category loss and restoration along with functionality. Functionality is calculated using (Davis 2013). Operability is plotted as the fulfillment of all the service categories. As seen in Fig. 6.4, full operability was returned in 12 days after the earthquake, but full functionality was not restored for another 6 years

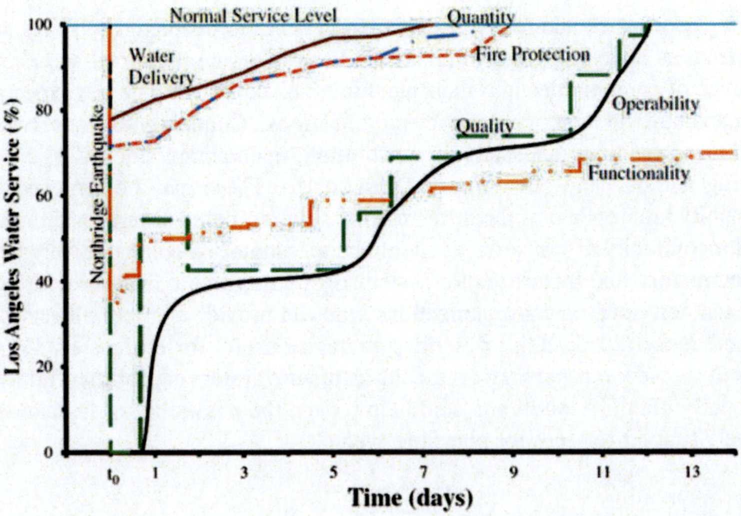

Fig. 6.4 Los Angeles water system service restorations following the 1994 northridge earthquake. Operability is not a service; it is a milestone delineating collective restoration of water delivery, quality, quantity, and fire protection services (Davis 2014)

(Davis 2014). At 12 days after the event all customers were able to resume normal use of water services even though the water system was not functioning as it was prior to the event. This is a result of organizational and technical resilience characteristics, primarily having a well-trained and equipped staff possessing the ability to rapidly mobilize to the hazard strike and adapt to the situation combined with a highly redundant system with isolation capabilities. This example illustrates how operability is a measure of an infrastructure system supporting community resilience and functionality is a measure of the infrastructure system resilience. Davis (2019b) and Davis (2021) describe the inter-relation between functionality and operability in greater detail.

6.2.4 Element 4: Continuity of Services

Element 4 describes the continuity of inoperable services that is, the services which are temporarily lost. Understanding the lost services is supported by Element 3. Resilient communities have ability to manage without normal infrastructure services during the time they are lost. This requires the social units within communities to understand and integrate the technical and organizational resilience of infrastructure systems into their planning. It is unlikely that large complex infrastructure systems can prevent some loss of services from significant hazard strikes. Resilient infrastructure systems can accommodate service losses and restore them to the community when they are needed relative to the size and likelihood of events and uncertainty of system performance and recovery. This requires the infrastructure systems and the organizations operating them to understand and integrate the social and economic resilience of communities into their planning (i.e., community needs, expectations, and adaptability to temporary service modifications). Continuity of lost services can be achieved by using alternatives, substitution, maintaining stockpiles, curtailing or going without, and other means (Rose 2016). These may be achieved by the community independent of the infrastructure systems, but to ensure resilience, planning for continuity of lost services should be accomplished collaboratively between the community and infrastructure system operators in coordination with government and non-government organizations who can provide aid until all services are returned. Rose (2016, 2009) describes strategies useful for businesses. Coordination will provide transparency on the infrastructure system capabilities, allow users to properly plan for conditions, and help reduce the possibility of inadequate and untimely support services for plausible events.

6.2.5 Element 5: Social and Economic Activity

Element 5 describes the social and economic activity of a community. Economics is identified as a social system in Appendix II but is highlighted in Element 5 as a critical and necessary system to track along with the other social systems. As related to infrastructure resilience, this element pertains to those social and economic aspects supported directly and indirectly by the infrastructure systems. Infrastructure systems, as previously defined, are needed for the operation of society and enterprise; without infrastructure systems social and economic functionalities are diminished. This element is supported by Elements 3 and 4. The increased ability to maintain services and rapidly restore any lost services from Element 3, and provision of continuity during service losses in Element 4, improve the resilience of social and economic systems. Additionally, the planning, preparedness, mitigation, emergency response, recovery, and rebuilding capabilities of social and economic institutions (including government and businesses), along with financial mechanisms such as insurance, loans, and public assistance is essential to assessing community resilience. Understanding Element 5 is vital to knowing how to establish community performance targets in Element 7.

Regional social and economic losses include all direct and higher order (indirect) potential impacts to communities resulting from infrastructure system service losses. These are described in Gilbert et al. (2015), Rose (2016, 2009, 2004), Gilbert and Ayuub (2016), Kajitani and Tatano (2018) and summarized here to include, but not limited to: direct and higher order economic losses to property and infrastructure, persons, households, and businesses, including business interruption and unemployment; non-economic losses to/from people (casualties, physical or mental health impacts, etc.), key governmental services, social networks and systems including loss of social capital, cultural heritage and artifacts, standard of living, and the environment (pollution, greenhouse gas emissions, loss of environmental capital).

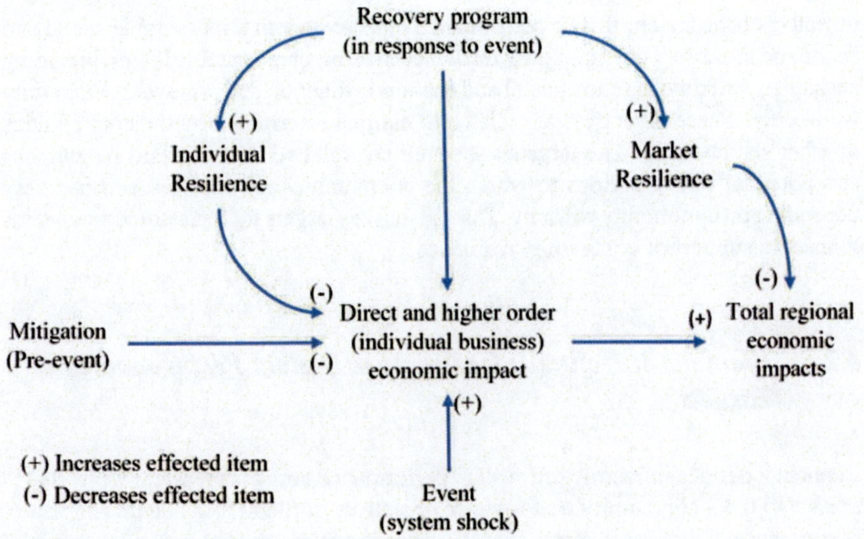

Fig. 6.5 Economic resilience and its offsetting role on event losses. Modified from Mysiak and Markandya (2015); originally formulated based in information from Rose (2007)

The distribution of losses is important for evaluating equity (United Nations 2016). Figure 6.5 shows how individual and market resilience influence the impacts from an event on individual businesses and regional economy.

6.2.6 Element 6: Community

Element 6 focuses on understanding the contribution of the environment, economic, health, equality, cultural, security, education, and political components of a community in maintaining itself and fulfilling the various needs of residents (Kusel and Fortmann 1991; What Works Wellbeing 2017). This is supported by Element 5 through the very definitions of infrastructure and resilience. The amount of infrastructure service losses and their durations can have a significant impact on communities and their well-being and livability. Understanding the local components making up Element 6 community and its well-being is vital to knowing how to establish community performance targets in Element 7.

6.2.7 Element 7: Establish Community Performance Targets

Element 7 establishes community performance targets. Identifying target community performance objectives is essential to determining target infrastructure performance objectives for Element 8. The performance objectives define measurable social and economic indicators of community resilience in terms of acceptable losses, including casualties, and time to return social and economic function and pre-event community well-being. These objectives are related to metrics determined by the communities or other stakeholders. The targets are developed relative to the hazard magnitudes and potential return periods to which the communities are exposed and can vary depending on community capacity. This establishes targets for infrastructure systems to meet in support of community resilience.

6.2.8 Element 8: Define Infrastructure System Performance Targets

Element 8 defines infrastructure system performance targets consistent with what is needed to meet community performance objectives in Element 7. Identifying target performance objectives in terms of lost services and time to restore them is essential to creating resilient infrastructure systems. Without target objectives it is difficult to design resilience into the systems and impossible to predict if the systems can provide the anticipated support for community resilience relative to the hazard exposure. Mieler et al. (2015) and the International Infrastructure Management Manual (INGENIUM and IPWEA 2015) describe processes for linking community resilience goals to specific performance targets for the built environment. NIST (2015) describes a practical approach for identifying infrastructure system recovery targets considering social goals. This element includes the need to modify existing codes and standards to be properly aligned and allow components and systems to be designed to meet performance targets needed to achieve community resilience (NIST 2016; RRMC 2019). This element also incorporates concepts of immediate occupancy (NIST 2018), Functional Recovery (FEMA/NIST 2021) and anti-catastrophe design by adding features to enhance performance beyond safety, recoverability and serviceability for the design-basis event and ultimately improve infrastructure and community resilience (Honda et al. 2017a, b; NIST 2017). Element 8 is the link to developing outcomes for economics of resilience and regional social and economic losses. This includes establishing reasonable and affordable performance objectives.

6.2.9 Feedback

As shown in Fig. 6.1, the results of system performance from Elements 2, 3, and 4 are compared with target infrastructure system performance objectives from Element 8. If the performance is better than the target performance objectives, then the system is considered to meet the expected resilience. If the performance is less than the target performance objectives then modifications may be needed, possibly using advanced technologies, and the system reassessed as shown by the dashed arrow in Fig. 6.1. Modifications may include pre-event mitigations to the infrastructure system (e.g., hardening components or adding redundancy and isolation) or improvements to the management and governance structures (e.g., modifying regulations or implementing management processes such as emergency exercises).

The social, economic and community performances from Elements 5 and 6 along with the community performance targets from Element 7 may also need to be re-evaluated but this is not directly shown in Fig. 6.1. This comparison should be undertaken collaboratively by engineers, sociologists, economists, regional planers, and other experts and is essential for ensuring infrastructure systems perform resiliently in support of the community needs.

As identified in Appendix II, the infrastructure organization and the people within are integrally linked to the community resilience Elements 5 and 6. As a result, infrastructure resilience in not independent of the overall community resilience and should account for performance and service losses from all dimensions/sectors (e.g., within the built, natural, social) through social learning, as shown in Fig. 6.6, when assessing if modifications are needed. In this way, the proposed infrastructure resilience framework can capture all the influential antecedent and post-event factors inherent to community resilience. Social learning occurs when beneficial actions are formulated into institutional policy (e.g., pre-event mitigation, planning and preparedness improvements) for handling future events (Cutter et al. 2008) and for managing post-event activities (e.g., shelter, evacuation, containing cascading hazards, reconstruction).

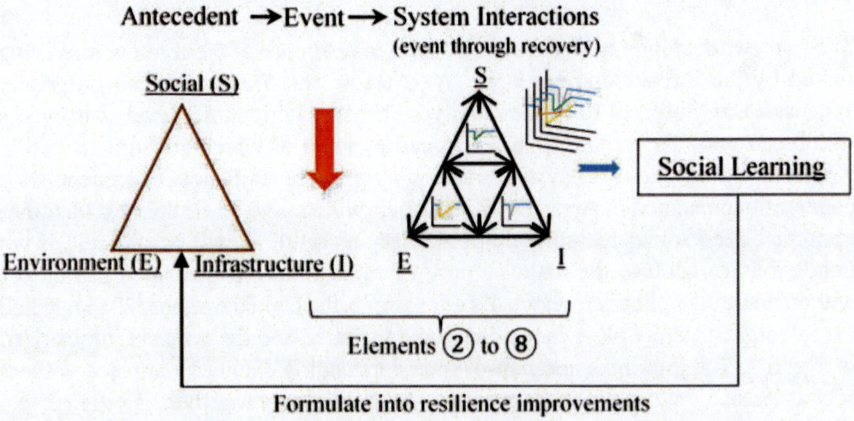

Fig. 6.6 Feedback from social learning through interaction of social, environment, and infrastructure systems within the broader community. See Appendix II describing interactions between social, environment, and built infrastructure systems

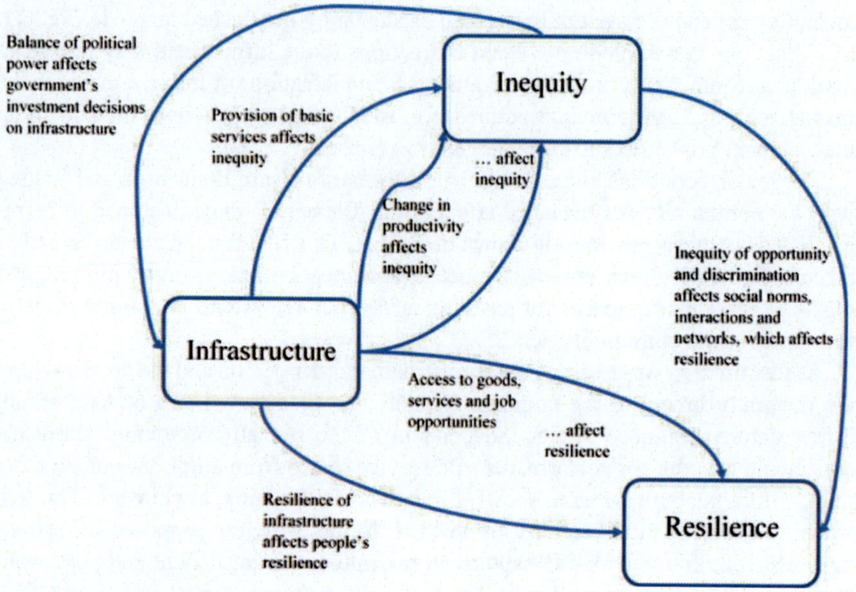

Fig. 6.7 Evidence map of the infrastructure—inequality—resilience nexus. Modified from United Nations (2016). *Source* From the authors of United Nations (2016) elaborations based on inputs by experts and literature review

6.3 Multidisciplinary

The framework incorporates the targeted level of resilience of the communities being served by the infrastructure systems. As a result, this framework can be greatly enhanced when utilized in conjunction with broader community level frameworks which can properly capture Elements 5 and 6, some of which are summarized in Appendix I. The entirety of systems making up the core resilience components for a community presented in Appendix II, and their wide range of disciplines, identifies important need for increased multidisciplinary research. It will be difficult, if not impossible, to achieve the main purpose of the framework presented in Fig. 6.1 without multidisciplinary research. As an example, the United Nations (2016) studied the role infrastructure plays in resilience and inequity, and the nexus is summarized in Fig. 6.7. The interlinks and improved understanding cannot be attained without such extensive multidisciplinary teams working diligently together.

6.4 Conclusions

An infrastructure resilience framework for assessment, management and governance was presented to provide a standard way for researching, evaluating, and creating resilient infrastructure systems which may be impacted by any hazard. The critical inter-relations between assessing infrastructure resilience and the need to manage and govern the systems with an understanding of the economics of resilience as well as potential social and economic losses, all in support of community resilience, are brought together within this single framework. The framework has eight elements and identifies needed tools and key outcomes and incorporates processes to evaluate, measure, and document the resilience level for all credible hazards, incorporate system interdependencies, define system performance levels consistent with the communities served, develop organizational policy, employ socio-technical learning and improvement processes, understand potential regional losses, identify the economics of resilience actions, further develop programs and procedures, and monitor, collect and constructively apply data to prioritize and implement resilience actions. The development of this framework is an important step toward fostering the ability to engineer resilient infrastructure systems that properly support the resilience of the communities they serve.

6.5 Appendix I: Review of Existing Resilience Frameworks

There are many resilience frameworks in the literature covering aspects from how to model specific infrastructure and performance elements in Fig. 6.1 to the entire community public health, social-economic activity, and well-being. This appendix is not an exhaustive attempt to review all frameworks and models (Rose 2009) summarizes additional frameworks), but instead summarizes a subset relevant to the infrastructure resilience framework presented herein.

Bruneau et al. (2003) introduced a conceptual framework to quantitatively assess and enhance the resilience of communities to earthquakes. This conceptual framework intended to coordinate research efforts toward quantifying measures to improve community resilience applicable to all systems but initially focused on electric power and water lifeline systems, acute-care hospitals, and organizations responsible for emergency response which are crucial to overall community resilience. The Bruneau et al. (2003) research was based on earthquake hazards, but identified resilience properties of robustness, redundancy, resourcefulness, and rapidity and the technical, organizational, social, and economic dimensions which are applicable to all potential hazard strikes (e.g., Davis et al. (2018)).

Cimellaro et al. (2010) established a framework for analytical quantification of disaster resilience, focusing on the earthquake hazard. Following Bruneau et al. (2003), this framework essentially quantifies the functionality curve in terms of robustness and rapidity and describes how to include resourcefulness and redundancy. Cimellaro et al. (2010) apply their method to a hospital and set of hospital buildings, i.e., hospital building clusters. Similarly, Ayyub (2014a) established a framework for quantifying the functionality curve. Rose (2016) presents a framework linking community level resilience goals to specific design targets for infrastructure systems and components creating indicators for system robustness but not for recovery. Balaei et al. (2018) developed a simplified framework for measuring water system resilience using technical, organizational, social, and economic indicators, but the results do not quantify the amount of system damage, services lost or restoration time.

Cutter et al. (2008) presented a place-based framework for assessing disaster resilience at the local or community level (disaster resilience of place, DROP, model). The model initiates with antecedent conditions of social, environmental and built systems. It then considers how the antecedent conditions interact with a hazard event, creating immediate effects. The immediate effects are modified by coping responses in the community. These accumulate to the total hazard impact, which is then compared to absorptive capacity of the community. If absorptive capacity is not exceeded, then degree of recovery is high. If absorptive capacity is exceeded, then degree of recovery is low, but can be increased with adaptive capability. Social learning can also increase the degree of recovery to future events. This framework is scalable to systems working within the size of a community.

The PEOPLES resilience framework has seven holistic, quantitative resilience dimensions and hierarchical lead indicators that stand for the state of functionality of systems in communities (Renschler et al. 2010). The seven dimensions are Population and Demographics, Environmental/Ecosystem Services, Organized Governmental Services, Physical Infrastructure, Lifestyle and Community Competence, Economic Development, and Social-Cultural Capital. It is a performance-based management framework applicable at scales ranging from individual, local, regional, and national to global.

NIST (2015) developed a community resilience planning guide for buildings and infrastructure systems. This guide was developed to help communities address the resilience challenges through a practical approach considering social goals and their dependencies on the built environment. The built environment is made up of building clusters and lifeline infrastructure systems. Using a six-step planning process communities can develop customized resilience plans by bringing together all relevant stakeholders, establishing community level performance goals, and developing and implementing plans to become more resilient. The guide outlines how to characterize a community's social institutions and the built environment supporting them. Desired performance goals are then established for the built environment to meet the community's social needs for three event levels, each successively more extreme. The built environment performance goals target times when the system is needed to be at 30, 60, and 90% functional. Estimating anticipated performance for the event levels allows for a gap analysis, which is used to identify and prioritize projects, then develop and implement a resilience plan.

The Rockefeller 100 resilient cities program created a City Resilience Framework (Rockefeller Foundation and ARUP 2014). The framework falls within four categories: the health and well-being of individuals (people); infrastructure and environment (place); economy and society (organization); and leadership and strategy (knowledge). Within these four categories lie 12 indicators. The indicators are complemented by qualities that distinguish a resilient city from one that is simply livable, sustainable or prosperous. The three layers of the City Resilience Framework, categories, indicators and qualities, help articulate city resilience. The framework helps facilitate a common understanding of resilience among city stakeholders, identifies critical gaps, where action and investment is needed to improve resilience, and where deeper analysis or understanding is required. The framework enables cities to carry out an objective assessment of their resilience and measure progress against an initial baseline city resilience index (Rockefeller Foundation and ARUP 2015).

The Sendai Framework for Disaster Risk Reduction 2015–2030 (SFDRR) was adopted by the United Nations in 2015. The SFDRR is the successor to a series of United Nations programs aimed at reducing risks and increasing resilience on a global scale at the national and regional levels. The Hyogo Framework for Action 2005–2015: Building the Resilience of Nations and Communities to Disasters preceded the SFDRR and was successor to the 1994 International Strategy for Disaster Reduction, the 1994 Yokohama Strategy for a Safer World, and the 1989 international framework for action (UNISDR 2015).

The SFDRR has four priorities for actions: (1) Understanding disaster risk; (2) Strengthening disaster risk governance to manage disaster risk; (3) Investing in disaster risk reduction for resilience; and (4) Enhancing disaster preparedness for effective response, and to "Build Back Better" in recovery, rehabilitation and reconstruction. It's goal is to "Prevent new and reduce existing disaster risk through the implementation of integrated and inclusive economic, structural, legal, social, health, cultural, educational, environmental, technological, political and institutional measures that prevent and reduce hazard exposure and vulnerability to disaster, increase preparedness for response and recovery, and thus strengthen resilience." The SFDRR has 13 guiding principles and seven targets, including "Substantially reduce disaster damage to critical infrastructure and disruption of basic services, among them health and educational facilities, including through developing their resilience by 2030." The SFDRR is very broad and far reaching which has specific focuses on resilience and infrastructure systems.

An analytical framework on risk and resilience (CEB 2018) was created to provide the United Nations system a common tool to integrate the three UN pillars (peace and security, human rights, and development) with the 2030 Agenda for Sustainable Development (United Nations 2015). The 2030 Agenda for Sustainable Development identifies 17 sustainable development goals (SDGs) with 169 targets and provides an integrated framework for action aimed at markedly improving the lives of the people, protecting natural resources, and fostering peaceful, just and inclusive societies. The framework on risk and resilience contains three elements summarized here as: systems thinking, general relationship between risk and resilience, and a prevention lens. The systems thinking used to describe the fundamental relationships among risks and other sustainable development issues at the global, regional, national and subnational levels. The general relationship between risk and resilience, called an equation by CEB (2018) identifies the inverse relationship between risk and resilience in that commonly actions to increase resilience also reduce risks to a community which can be used to organize efforts across the pillars to define collective positive outcomes. A prevention lens fosters a proactive approach to increase resilience and decrease risks.

The SDG's and related frameworks are critically linked with infrastructure resilience. Waage et al. (2015) pose a framework for organizing and understanding the relationship between the 17 SDGs. They create three concentric rings, labeling the outer ring as natural environment (3 SDG's), center ring as infrastructure (7 SDGs), and inner ring as well-being (6 SDGs). One SDG is excluded from this framework because it is an overarching goal. The infrastructure goals address essential functions of modern societies necessary to deliver the well-being goals. In their framework, infrastructure goals represent a domain for global development goal setting with particularly strong effects on well-being and environment goals and relevance to global health and environmental impact. Infrastructure goals draw on common natural resources to provide the services needed for the well-being goals; upon which infrastructure systems should not sacrifice environmental goals in favor of well-being, but instead draw a sustainable and resilient balance. SDG 9 states "build resilient infrastructure, …" without which, a community can be neither sustainable nor resilient and the well-being goals cannot be maintained.

Infrastructure affects resilience through its effect on access of people to goods, services and job opportunities, which influence the ability of people to adapt to shocks. The quality, design, distribution, inter-relation and operation of infrastructure also affect the resilience of the infrastructure itself, which has an effect of people's resilience to economic, social and environmental shocks. Disasters impact the socio-economic fabric of communities when they damage the infrastructure and its functionality (United Nations 2016).

The Composite of Post-Event Well-being (COPEWELL) framework and system dynamics model (Links et al. 2018) focuses on the functioning and well-being of communities over time after a disaster. The framework incorporates factors thought to contribute to community functioning over time after an event, including natural systems, engineered systems, and countermeasures, which together reflect prevention and mitigation factors; population vulnerability, inequality, and deprivation; pre-event preparedness activities; social cohesion as a surrogate for emergent collective behaviors during and after the event; characteristics of the event itself; external resources after the event; formal emergency management activities, both planned and improvised, during and after the event; and community activities, both planned and emergent. It intends to model post-event community well-being at the county level throughout the USA through use of readily available indicators. As such, it does not specifically incorporate all the infrastructure systems within a community (e.g., wastewater and energy systems and not specifically included but may be indirectly accounted for).

The National Academies of Sciences, Engineering, and Medicine (NASEM) (2019) describes a community resilience framework for the Gulf Resilience Program (GRP). They recommend that each community in the Gulf Research Program community resilience initiative, the GRP should develop and employ a community resilience framework that includes: (1) community engagement to engender buy-in around resilience priorities, goals, and leadership; (2) resilience across multiple community capitals (i.e., natural, built, social, financial, human, and political); (3) measures and ways to track progress that are useful to decision makers; and (4) investments in resilience that result in multiple benefits. The proposed measurements are generally of higher-level indicator types.

The ASME-ITE Risk Analysis and Management for Critical Asset Protection (RAMPCAP) Plus methodology (Brashear and Jones 2010) standard deals with security as well as natural hazards, sets requirements for all hazards risk and resilience analysis and management of water and wastewater systems, and prescribes methods to address the requirements set forth. This has been used to develop standards for specific systems such as water and wastewater (American Water Works Association 2010). The threats addressed include terrorism, natural hazards, product or waste-stream contamination, supply chain hazards, and proximity hazards. The American Water Works Association (2010) identifies manmade hazards or accidents, hurricanes, earthquakes, tornadoes, and floods, and further recommends wildfire, ice storms, extreme cold weather, avalanche, tsunami, landslide, mud slide, and other hazards be included "if the probability of occurrence and the consequences are as high or higher than the four basic natural hazards in the specific location." The procedure attempts to establish a common base for system risk and resilience assessment based only on revenue loss and economic activity, making it difficult for drinking water, wastewater, and stormwater systems to understand how they are supporting community resilience. The Environmental Protection Agency developed the VSAT Web Tool (United States Environmental Protection Agency 2019) to assist the RAMCAP process, but also only makes assessments at the component level. There is limited or no direction on how to assess the combined impacts to system service losses and restoration from large events effecting multiple priority assets (e.g., earthquakes can damage 100's of components) and no guidance on performance levels needed by the systems. This essentially can result in an (*n*-1) approach looking at service losses and restoration one component at a time, or alternatively assumes all vulnerable components are damaged in a single event. The RAMCAP (Brashear and Jones 2010) approach is most useful for hazards impacting one or a few independent components of linear systems.

The Civil Restoration with Interdependent Social Infrastructure Systems (CRISIS) decision support model (Loggins et al. 2019) optimizes repair of damaged civil infrastructures based on stakeholder-determined priorities for the restoration of social infrastructure services that depend on the damaged civil systems following an extreme event. It is envisioned as an education and training tool that can serve as a common platform to bring together public officials, emergency managers, infrastructure service providers, healthcare providers, social service providers, local businesses, and other community leaders to plan and prepare for extreme events before they occur.

Table 6.1 summarizes the relationships between the current framework presented herein and those summarized above. In Table 6.1, the column identifying type describes the general context of existing frameworks as very high level, narrowly focused (e.g., a specific element in Fig. 6.1), or specific to a hazard or particular infrastructure system. The relationship column identifies a relative ranking. All the city and community social, economic, and well-being frameworks are useful for understanding and enhancing the present framework through Elements 4, 5, 6, and 7.

6.6 Appendix II: The Social, Built Infrastructure, and Natural Environment Systems Relationships

This appendix provides a broad perspective on the role infrastructure systems have in creating resilience. This perspective initiates with a set of definitions, then formulates the relationships between social, built infrastructure and natural environment systems.

6.6.1 *Definitions*

Social System: the patterned series of inter-relationships existing between individuals, groups, and institutions and forming a coherent whole (https://www.merriam-webster.com/dictionary/social%20system).

Social institutions: family, economic, government, health, education, community service, religious, cultural, and media organizations (NIST 2015).

A social group is two or more people who influence each other.

Infrastructure Systems: the basic physical (built) and organizational structures and facilities (e.g., building clusters and lifeline infrastructure systems) needed for the operation of a society or enterprise.

Natural systems: Climate, earth, life, universe (climatology, geology, biology, astronomy).

Ecology: the branch of biology that deals with the relations of organisms to one another and to their physical surroundings.

Community: places (such as towns, cities, or counties), designated by geographical boundaries (NIST 2015).

Socio-ecological systems: human–environment interactions.

Socio-technical systems: human-technical interactions.

6.6.2 *System Relationships*

From the definitions, built infrastructure systems are needed for the operation of a society or enterprise. They also draw resources from the natural environment. The environment has been impacted so significantly by human activity that it no-longer can be considered as behaving independent of anthropological activity. This

Table 6.1 Relationship between the present and existing resilience frameworks

Framework	Type	Relationship	Comments
Bruneau et al. (2003)	High level focus on earthquakes	Consistent	Present framework advances earlier work
Cimellaro et al. (2010)	Specific to buildings and hospital functionality	Falls within present	Covers specific elements of present framework
Ayyub (2014a)	Specific to functionality	Falls within present	Covers specific elements of present framework
Mieler et al. (2015)	Specific to robustness not recovery	Falls within present	Covers specific elements of present framework
Balaei et al. (2018)	Specific to water	Falls within present	Covers specific elements of present framework
DROP (2018)	High level	Present falls within at community scale; Falls within present at infrastructure scale	Existing framework can be idealized at multiple scales. Present mirrors and expands on existing
NIST (2015)	High level	Compatible	Present framework encourages to go beyond a simple review of existing and include more advanced approaches
PEOPLES (Renschler et al. 2010)	High level	Compatible	Present framework complements and enhances existing physical infrastructure dimension
Rockefeller Foundation and ARUP (2014)	High level	Compatible	Present framework complements and enhances existing
NAESM (2019)	High level	Compatible	Present framework complements and enhances existing built capital dimension
COPEWELL (Links et al. 2018)	High level	Compatible	Existing framework can be informed by present and possibly enhanced by it. Existing can be used to enhance Element 6

(continued)

Table 6.1 (continued)

Framework	Type	Relationship	Comments
SFDRR (UNISDR 2015)	High level	Compatible	Present framework is necessary for accomplishing the SFDRR targets[1]
Waage et al. (2015)	High level specific to sustainable development goals (SDG)	Compatible	Present framework supports existing and is necessary to achieve the SDG's, especially SDG 9
CEB (2018)	High level	Compatible	Present framework supports existing and is necessary to achieve the SDG's
RAMCAP (Brashear and Jones 2010; American Water Works Association 2010)	High level, (American Water Works Association 2010; United States Environmental Protection Agency 2019) are specific to water and wastewater systems	Falls within present	Existing has limitations to specific components and limits resilience assessment to revenue loss and economic activity
CRISIS (Loggins et al. 2019)	Specific to restoration phase	Falls within present	Present framework supports existing. Existing has practical application of some elements

[1]Present framework directly supports the following SFDRR targets
- Substantially reduce disaster damage to critical infrastructure and disruption of basic services
- Reduce direct disaster economic loss
- Substantially reduce global disaster mortality
- Substantially reduce number of affected people globally

includes inner space (only outer space may be considered semi-independent of human activity). Thus, as shown in Fig. 6.8, these systems must be considered as part of a socio-ecological system.

Communities are made up of systems of systems. They reside within the larger concept of socio-ecological systems. Their resilience is highly dependent upon the three primary sets of systems shown in Fig. 6.8: social, natural, and built infrastructure.

Figure 6.8 shows the relationships between the primary systems (social, natural environment and built infrastructure) making up socio-ecological systems. Each of the systems making up the broader social, natural environment, and built infrastructure are shown in boxes at the corners of a triangle. The triangle identifies how within the socio-ecological system, all the systems within the social, natural, and

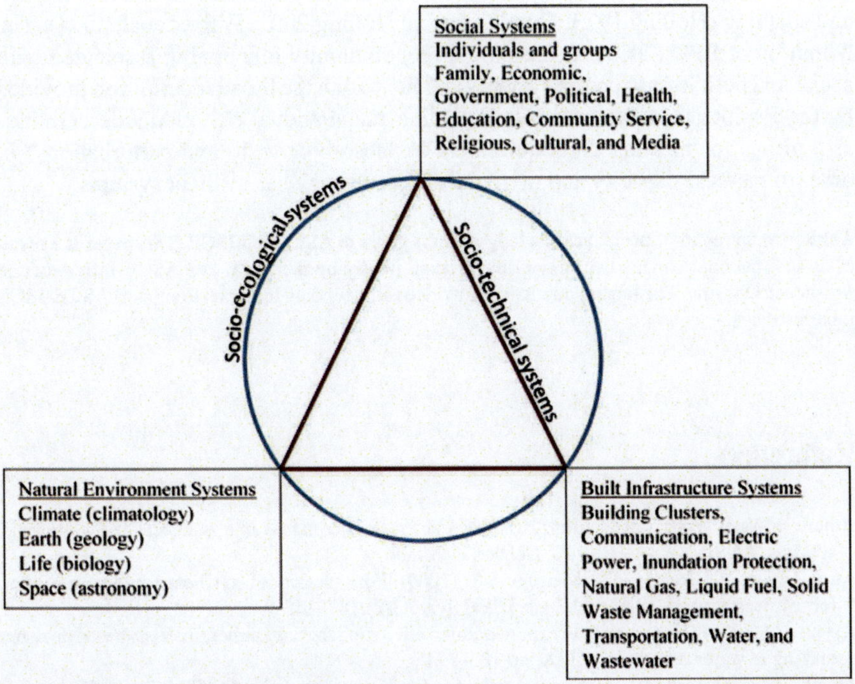

Fig. 6.8 Systems influencing community resilience

infrastructure domains interact, directly or indirectly. The interactions change based on geographic location.

Built infrastructure can only be operated by the organizational (human) interaction with the technical built components, making infrastructure systems a socio-technical system functioning within the broader socio-ecological system as shown on the right leg of the triangle in Fig. 6.8.

Hazards are essentially outcomes from the social, natural, and infrastructure domains. As a result, hazards come from within the broader socio-ecological system. They may be influenced by a single domain (e.g., volcanic eruption from the natural; terror attach from the social; nuclear release from the built), or a combination of domains (e.g., heat islands created from global warming trends in urban areas using paved infrastructure). This means that protecting communities from hazards requires the proper actions to (1) prevent the hazards, or reduce their potential for development, when they are created by anthropological activity, and (2) monitor and understand the physical and chemical basis for natural occurring hazards.

The Resilience Alliance (https://www.resalliance.org/resilience) defines resilience in terms of socio-ecologic systems as the capacity of a social-ecological system to absorb or withstand perturbations and other stressors such that the system remains within the same regime, essentially maintaining its structure and functions. It describes the degree to which the system is capable of self-organization, learning and adapting (Holling 1973; Gunderson and Holling 2002; Walker et al. 2004). The definition of PPD (2013) focuses more on community functioning associated with social and built infrastructure systems and fits within the broader definition provided by the Resilience Alliance, while including preparational risk reduction activities as a means for meeting the requirement to "absorb or withstand perturbations" in order to increase capacity and drive the ability to engineer resilient systems.

Acknowledgements Special acknowledgement is given to ASCE and JSCE. This paper is a result of an international cooperative project undertaken jointly by the JSCE and ASCE Infrastructure Resilience Division. The Japan Foundation provided some funding for travel to support ASCE-JSCE joint meetings.

References

American Water Works Association (2010) Risk and resilience management of water and wastewater systems. ANSI/AWWA J100–10, Denver, Colorado

Attoh-Okine N, Cooper AT, Mensah SA (2009) Formulation of resilience index of urban infrastructure using belief functions. IEEE Syst J 3(2):147–153

Ayyub BM (2001) Uncertainties in expert-opinion elicitation for risk studies. In: Risk-based decision making in water resources, vol IX, pp 103–121

Ayyub BM (2002) Elicitation of expert opinions for uncertainty and risks. CRC Press, FL

Ayyub BM (2014b) Risk analysis in engineering and economics, 2nd edn. Chapman and Hall/CRC Press, FL

Ayyub BM (2014a) Systems resilience for multihazard environments: definition, metrics, and valuation for decision making. Risk Analysis J 34(2):340–55. https://doi.org/10.1111/risa.12093

Ayyub BM (2015) Practical resilience metrics for planning, design, and decision making. ASCE-ASME J Risk Uncertainty Eng Syst, Part A: Civil Eng. https://doi.org/10.1061/AJRUA6.0000826

Balaei B, Wilkinson S, Potangaroa R, Hassani N, Alavi-Shoshtari M (2018) Developing a framework for measuring water supply resilience. ASCE Nat Hazards Rev 19(4). https://doi.org/10.1061/(ASCE)NH.1527-6996.0000292

Brashear J, Jones J (2010) Risk analysis and management for critical asset protection (RAMCAP Plus). In: Part 2. cross-cutting themes and technologies, in risk modeling and vulnerability assessment, wiley handbook of science and technology for homeland security. https://doi.org/10.1002/9780470087923.hhs003

Brook D, Chi Yin S (2017) Here a city shall be wrought. Harpers Maga. https://pulitzercenter.org/reporting/here-city-shall-be-wrought. Accessed 5 Nov 2019

Bruneau M, Chang SE, Eguchi RT, Lee GC, O'Rourke TD, Reinhorn AM, Shinozuka M, Tierney K, Wallace WA, Winterfeldt D (2003) A framework to quantitatively assess and enhance the seismic resilience of communities. Earthq Spectra 19:733–752

Cabinet Office Japan (2019) Disaster management in Japan. http://www.bousai.go.jp/kyoiku/fukko/shisai/higashinihon_en.html. Accessed 4 Nov 2019

CEB (United Nations System Chief Executives Board for Coordination) (2018) Adopting an analytical framework on risk and resilience: a proposal for more proactive, coordinated and effective United Nations action. In: High level committee on programs, prepared by a task team led by the world food programme, Annex III, CEB/2017/6, endorsed November 2017, published 1/31/18. https://www.unsceb.org/content/analytical-framework-risk-and-resilience. Accessed 7 Apr 2019

Cimellaro GP, Reinhorn AM, Bruneau M (2010) Seismic resilience of a hospital system. Struct Infrastruct Eng 6(1–2):127–144; Civil contingencies secretariat of the cabinet office, London (2003)
Croope S, McNeil S (2011) Improving resilience of critical infrastructure systems post-disaster for recovery and mitigation. Transportation research record no 2234, pp 3–13
Cutter SL, Barnes L, Berry M, Burton C, Evens E, Tate E, Webb J (2008) A place-based model for understanding community resilience to natural disasters. Glob Environ Change 18:598–606
Davis CA (2013) Quantifying post-earthquake potable water system services. In: 6th China-Japan-US trilateral symposium lifeline earthquake engineering, Chengdu, China, May 28-June 1, 2013, TCLEE monograph, vol 38. ASCE, Reston VA, pp 19–26
Davis CA (2014) Water system service categories, post-earthquake interaction, and restoration strategies. Earthq Spectra 30(4):1487–1509
Davis CA (2019a) A proposed performance based seismic design process for lifeline systems. In: ISSMGE, 7th international conference on earthquake geotechnical engineering. Rome, IT
Davis CA (2019b) Infrastructure system resilience: functionality and operability. In: 2nd international conference on natural hazards and infrastructure, paper no 200. Chania, Greece
Davis CA(2021) understanding functionality and operability for infrastructure system resilience. Nat Hazards Rev 22(1):06020005. https://doi.org/10.1061/(ASCE)NH.1527-6996.0000431
Davis CA, Mostafavi A, Wang H (2018) Establishing characteristics to operationalize resilience for lifeline systems. Nat Hazards Rev 19(4)
FEMA/NIST (2021) Recommended options for improving the built environment for post-earthquake reoccupancy and functional recovery time, federal emergency management agency, national institute of standards and technology, FEMA-P2090/NIST SP-1254
Gilbert S, Ayyub BM (2016) Models for the economics of resilience. ASCE-ASME J. Risk Uncertainty Eng Syst, Part A: Civ Eng. https://doi.org/10.1061/AJRUA6.0000867
Gilbert SW, Butry DT, Helgeson JF, Chapman RE (2015) Community resilience economic decision guide for buildings and infrastructure systems, NIST SP 1197. National Institute of Standards and Technology, Gaithersburg, MD. http://dx.doi.org/https://doi.org/10.6028/NIST.SP.1197
Gunderson LH, Holling CS (eds) (2002) Panarchy: understanding transformations in systems of humans and nature. Island Press, Washington DC
Holling CS (1973) Resilience and stability of ecological systems. Annu Rev Ecol Syst 4:1–23
Honda R, Akiyama M, Kataoka S, Murono Y, Nozu A, Takahashi Y (2017a) Seismic design method to consider “anti-catastrophe” concept. In: Proceeding 16th world conference on earthquake engineering. Santiago, Chile
Honda R, Akiyama M, Nozu A, Takahashi Y, Kataoka S, Murano Y (2017b) Seismic design for ‘anti-catastrophe’—a study on the implementation as design codes. J JSCE 5:346–353; Originally published in the J Japan Soc Civ Eng Ser A1 72(4):I_459–I_472
INGENIUM and IPWEA (2015) International infrastructure management manual, version 5.0. In: Association of local government engineering NZ Inc (INGENIUM) and the institute of public works engineering of Australia (IPWEA)
Kajitani Y, Tatano H (2018) Applicability of a spatial computable general equilibrium model to assess the short-term economic impact of natural disasters. Econ Syst Res 30(3):289–312
Kusel J, Fortmann LP (1991) What is community well-being? In: Kusel J, Fortman L (eds) Well-being in forest-dependent communities, vol 1. Forest and Rangeland Resources Assessment Program and California Department of Forestry and Fire Protection, Berkley, California, pp1–45
Links JM, Schwartz BS, Lin S, Kanarek N, Mitrani-Reiser J, Kirk Sell T, Watson CR, Ward D, Slemp C, Burhans R, Gill K, Igusa T, Zhao X, Aguirre B, Trainor J, Nigg J, Inglesby T, Carbone E, Kendra JM (2018) COPEWELL: a conceptual framework and system dynamics model for predicting community functioning and resilience after disasters. Disaster Med Public Health Preparedness, Soc Disaster Med Public Health Inc 12(1):127–137. https://doi.org/10.1017/dmp.2017.39

Liu Y, Mc Neil S, Herning G (2017) Integrating resilience concepts with pavement management: a case study in Delaware. In: Proceedings world conference on pavement and asset management (WCPAM2017). Italy

Loggins R, Little RG, Mitchell J, Sharkey T, Wallace WA (2019) CRISIS: modeling the restoration of interdependent civil and social infrastructure systems following an extreme event. Nat Hazards Rev 20(3)

Mieler MB, Stojadinovic RB, Comerio M, Mahin S (2015) A framework for linking community-resilience goals to specific performance targets for the built environment. Earthquake Spectra 31(3):1267–1283

Mysiak J, Markandya A (2015) Economic costs of droughts. In: XEROCHORE working package 2 Brief. European commission. Available at https://www.researchgate.net/publication/267934571_ECONOMIC_COSTS_OF_DROUGHTS. Accessed 21 Apr 2019

National Academies of Sciences, Engineering, and Medicine (NASEM) (2019) Building and measuring community resilience: actions for communities and the gulf research program. The National Academies Press, Washington, DC. https://doi.org/10.17226/25383

Mc Neil S, Liu Y, Ramirez-Villamizar AS (2019) Infrastructure resilience: from concept to performance to decisions. In: Proceedings 7th international conference "bituminous mixtures and pavements. Thessaloniki, Greece

NIST (National Institute of Standards and Technology) (2015) Community resilience planning guide for buildings and infrastructure systems, vol 1190. NIST Special Publication 1190. http://dx.doi.org/https://doi.org/10.6028/NIST.SP.1190v1

NIST (National Institute of Standards and Technology) (2016) Critical assessment of lifeline system performance: understanding societal needs in disaster recovery. In: NIST GCR 16-917-39, Prepared by applied technology council for the national institute of standards and technology. Gaithersburg, MD

NIST (2017) Implementation guidelines for executive order 13717: establishing a federal earthquake risk management standard [ICSSC Recommended Practice (RP) 9]. U.S. Department of Commerce, National Institute of Standards and Technology

NIST (2018) Research needs to support immediate occupancy building performance objective following natural hazards events [NIST Special Publication 1224]. U.S. Department of Commerce, National Institute of Standards and Technology

Presidential Policy Directive (PPD) (2013) Critical infrastructure security and resilience. PPD-21, Released 12 Feb 2013. Available at: https://obamawhitehouse.archives.gov/the-press-office/2013/02/12/presidential-policy-directive-critical-infrastructure-security-and-resil. Accessed 25 June 2013

Renschler CS, Fraizer AE, Arendt LA, Cimellaro GP, Reinhorn AM, Michel Bruneau M (2010) A framework for defining and measuring resilience at the community scale: the PEOPLES resilience framework. U.S. Department of Commerce National Institute of Standards and Technology, Office of Applied Economics Engineering Laboratory NIST GCR 10–930. Gaithersburg, MD

Rockefeller Foundation and ARUP (2014) City resilience framework. https://www.rockefellerfoundation.org/report/city-resilience-framework/. Accessed 10 March 2019

Rockefeller Foundation and ARUP (2015) City resilience index. https://assets.rockefellerfoundation.org/app/uploads/20160201132303/CRI-Revised-Booklet1.pdf. Accessed 10 March 2019

Rose A (2004) Economic principles, issues, and research priorities in hazard loss estimation. In: Okuyama Y, Chang SE (eds) Modeling spatial and economic impacts of disasters. Advances in spatial science. Springer, Berlin, Heidelberg

Rose A (2007) Economic resilience to natural and man-made disasters: multidisciplinary origins and contextual dimensions. Environ Hazards 7(4):383

Rose A (2009) Economic resilience to disasters. In: Community and regional resilience institute, CARRI research report 8, published articles & papers, paper 75. http://research.create.usc.edu/published_papers/75

Rose A (2016) Benefit-cost analysis of economic resilience actions. Oxford Research Encyclopedia of Natural Hazard Science, Oxford university Press, USA

Risk and Resilience Measurement Committee (RRMC) (2019) Resilience-based performance: next generation guidelines for buildings and lifeline standards. In: ASCE infrastructure resilience division, IRP3. American Society of Civil Engineers

Tierney K, Bruneau M (2007) Conceptualized and measuring resilience. TR News 250, 14–17, ⟨http://onlinepubs.trb.org/onlinepubs/trnews/trnews250_p14-17.pdf⟩. 8 Apr 2015

UNISDR (United Nations Office for Disaster Risk Reduction) (2015) Sendai framework for disaster risk reduction 2015–2030. In: 1st Ed., United Nations, Switzerland. https://www.unisdr.org/files/43291_sendaiframeworkfordrren.pdf. Accessed 7 Apr 2019

United Nations (2015) Transforming our world: the 2030 agenda for sustainable development, A/RES/70/1. https://sustainabledevelopment.un.org/post2015/transformingourworld/publication. Accessed 7 Apr 2019

United Nations (2016) The infrastructure—inequality—resilience nexus. In: Chapter 2 in the global sustainability development report, 2016 edn. Department of Economic and Social Affairs, New York

United States Environmental Protection Agency (2019) Vulnerability Assessment Tool, version 2.0. https://vsat.epa.gov/vsat/. Accessed 17 Nov 2019

Van der Merwe SE, Biggs R, Preiser R (2018) A framework for conceptualizing and assessing the resilience of essential services produced by socio-technical systems. Ecol Soc 23(2):12. https://doi.org/10.5751/ES-09623-230212

Waage J, Yap C, Bell S, Levy C, Mace G, Pegram T, Unterhalter E, Dasandi N, Hudson D, Kock R, Mayhew S, Marx C, Poole N (2015) Governing the UN sustainable development goals: interactions, infrastructures, and institutions. Lancet Glob Health, Comment 3(5):PE251–E252. https://doi.org/10.1016/S2214-109X(15)70112-9, https://www.thelancet.com/journals/langlo/article/PIIS2214-109X(15)70112-9/fulltext. Accessed 2 Apr 2019

Walker B, Holling CS, Carpenter SR, Kinzig A (2004) Adaptability and transformability in social-ecological systems. Ecol Soc 9:5

What Works Wellbeing (2017) What is community wellbeing? Conceptual review, pp 67. whatworkswellbeing.org, Accessed 22 Apr 2019

Chapter 7
Harnessing Scientific Knowledge and Technological Innovation for Disaster Risk Reduction (DRR) in Sub-Saharan Africa-Case of Social Media

Desmond Manatsa and Lucy Sakala

Abstract The biggest challenge to disaster risk reduction (DRR) in Sub-Saharan Africa is being presented by climate change due to the continent's acute vulnerability inherent in multiple stresses and low adaptive capacity. It is common knowledge that the task of managing disaster risks and disaster events is heavily dependent on scientific knowledge and evidence-based technique which doubles up as a pillar for investment in DRR and used as a basis for formulating disaster risk management techniques. In this way, the application of science and technology has the potential to substantially reduce losses of lives and property. The current Intergovernmental Panel on Climate Change special report (AR6) on managing the risk of extreme events and disasters thoroughly interrogates this aspect and provides the essential evidence base. Equally important is to note that climate change appreciation and discourse demands high scientific standards. As such, its acceptance and readiness to move to reduce its impacts through DRR techniques requires much more systematic and proven sets of data, methods and tools and much stronger scientific and technical support capacities than is the case at present in Sub-Saharan Africa. The Sendai Framework for DRR 2015–2030 (SFDRR) sent clear signals to the mandate of science, technology, and innovation in driving government activities that is based on developing and sharing the knowledge and solutions needed to improve the resilience of communities. As such, there is need to harness the new global scientific challenges to DRR if Sub-Saharan Africa is to remain afloat in the DRR framework. Fortunately, the current science and technology advances have reached levels where a global village has

D. Manatsa (✉) · L. Sakala
Faculty of Science and Engineering, Bindura University of Science Education, Bindura, Zimbabwe
e-mail: dmanatsa@gmail.com

D. Manatsa
Department of Geography, Kwa Kwa Campus, Free State University, Phuthaditjhaba, South Africa

D. Manatsa · L. Sakala
Africa Alliance for Disaster Risk Institutes (AADRI), Bindura University of Science Education, Bindura, Zimbabwe

H. Tatano and A. Collins (eds.), *Proceedings of the 4th Global Summit of Research Institutes for Disaster Risk Reduction*, Disaster and Risk Research: GADRI Book Series,
https://doi.org/10.1007/978-981-19-5566-2_7

been created which makes available more than ever before, immeasurable amounts of information relevant to DRR through a simple mouse click. In this way, more actors from different disciplines and policy areas can harness the emerging scientific challenges for the enhancement of DRR in Sub-Saharan Africa. This paper is predominantly concerned with how disaster risks and losses in Sub-Saharan Africa can be further reduced through harnessing science and technology innovation in the case of Social Media.

7.1 Introduction

Sub-Saharan Africa is notorious of being the only continent whose share of reported disasters in the world total has increased over the past decades, especially those related to climate change. This is despite the continuous advances in science and technology that includes innovations at global scale whose exploitation in the developed world has led to a significant reduction in hazard-related risks. Since these advances in science and technology innovations are now easily accessible at the click of a mouse button, in the ever increasingly consolidated global village, we are now faced with an unacceptable scenario where Sub-Saharan Africa is progressively failing to reverse the upward trend in climate change-related disaster damage. In other words, the continent can be conveniently described as 'swimming in an ever-improving pool of DRR-related science technology innovations but losing more in terms of life and property'. Obviously, there are gaps between the available advances in science and technology innovation on the one hand and how this can be harnessed into making Sub-Saharan Africa more resistant. It appears the continent's capacity to harness these advances in order to reduce hazard-related risk has evidently not been exploited to their full potential. Proper adoption of these advances by the continent should lead to better understanding and improved disaster management which ultimately enhances risk reduction and adaptation. This is true, especially to Sub-Saharan Africa's biggest threat which is climate change as her economic activities are already being highly influenced by climate variability in various aspects. As such, disaster impacts have become an impediment to sustainable development in Sub-Saharan Africa and hence is retarding advances toward achieving the SDGs. It then becomes imperative to harness the unfolding scientific knowledge that is being exhibited through advances in innovation and technology for the continent's improvement in DRR but in this case considering the role of Social Media and Data analytics.

Sub-Saharan Africa's disaster profile is closely connected with the vulnerability of its population and economies that are exacerbated by minimal coping capacities. Though the region is endowed with rich natural resources, it has a large population that is not supported by economic and social development. It faces all potential hazards in the world that are climate change instigated and hence increasing with global warming. These climate-related hazards are characterized by high frequency and intensity which inflicts large disaster casualties, material, and economic losses of the highest proportion relative to global disaster total. The resulting disasters are predominantly hydrometeorological, with droughts on the top followed by floods, with the most vulnerable places being along the major river systems and poorly cited flood-prone urban areas. Climate change is already aggravating the magnitude and frequency of these extreme weather events. To make matters worse, most of the countries have limited resources to invest in DRR, while exhibiting minimal fiscal space to fund relief and recovery efforts after a major disaster. This was evidenced by the effects of Cyclone Eline of 2000 in southern Sub-Saharan Africa where the related impacts are still vivid even up to today and that of the recent Cyclone Idai in March 2019 whose full recovery from the devastation remains far beyond the region's capacity.

The impacts of these disasters precipitate from the fact that in Sub-Saharan Africa; though DRR policies and institutional mechanisms do exist at various degrees of completeness, their effectiveness is limited, hence the need for a strategic approach that is informed by new scientific knowledge and the exploitation of the current advances in technology innovation. Their adoption and sharing can facilitate its increased use for improved DRR which translates to enhanced climate change adaptation. This paper provides basic knowledge on how to harness a selected approach of scientific knowledge and technological innovation for the enhancement of DRR that is contextualized through the Sendai Framework for DRR: 2015–2030 in Chap. 2. Chapter 3 looks at the high vulnerability of Sub-Saharan Africa to hydrometeorological hazards before suggesting the adoption of social media and big data analytics to reduce these vulnerabilities in Chap. 4, while Chap. 5 concludes the paper.

7.2 Sendai Framework for DRR 2015–2030 on Technology Innovation for Sub-Saharan Africa

Due the seemingly unending cycle of unprecedented climate change induced natural disasters, it has never been more imperative to connect sociological and scientific research and technological innovation to policy and practice (Smajgl and Ward 2013). In fact, the need for increased application of innovation and technology for disaster risk reduction (DRR) has never been greater in order to foster new development and implementation of more effective evidence-based approaches. As such, the increased frequency and intensity of extreme climate events requires accelerated DRR efforts which are supported by innovative solutions to minimize the damage and wide-ranging secondary impacts of future climate change-related events. In this respect, the Sendai Framework for Disaster Risk Reduction 2015–2030 (SFDRR) was designed to encourage an increased focus on science and technology as a key tool to confront these immerging global challenges. Its major thrust was to predominantly shift the focus from managing disasters as prescribed in its successor instrument of the Hyogo Framework for Action (HFA) 2005–2015 to institutionalize innovative ways in reducing risks as the global roadmap for lessening human and economic loss as a direct result of disasters (Carabine 2015).

In this change of approach, an improved appreciation of risk in all its dimensions of environment, hazards, exposure, and vulnerability was emphasized. These had to take account of underlying disaster risks such as inequality, climate change, urbanization, the increase in population density, and environmental degradation. This implies that governance should ensure that DRR is factored into planning and development at all levels across all sectors and all facets of the disaster cycle. Hence, this new paradigm shift to risk-sensitive development becomes a vehicle through which sound foundations in knowledge and understanding of DRR contribute to the attainment of the sustainable development goals (STGs). The principal objective being to avail the full potential of science and technological innovative in strengthening individual countries' disaster management plans while empowering local communities and individuals in disaster situations. Strengthening DRR decision-making at community, local, national, and international level, it is necessary to provide a robust and accessible scientific evidence base. Understanding these attributes as priorities at various levels assists decision-makers and scientists in developing the implementation plan to consider how science, technology, and innovation can be an enabling factor for DRR.

However, while adopting 'Understanding Disaster Risk' as its main priority for action, the Sendai Framework is anchored on the advances in science and technology innovation as the primary driver to prevent and reduce disaster risk. Hence, SFDRR encourages better access and support for innovation and technology as well as increased investment in DRR to develop new innovations that are both cost-effective and beneficial when applied in all disaster management phases. In other words, scientific and technical work on DRR should be prioritized in development through fronting science-based risk knowledge and technology innovation. In fact, implementing SFDRR should not be a daunting task as different stakeholders have recognized and reaffirmed the pivotal role that is played by science and technology innovation in several major global and regional forum. Interestingly, Sub-Saharan Africa was the first region to officially adopt a position on a post-2015 framework on DRR at its 5th Session of Sub-Saharan Africa Regional Platform and 4th High Level Meeting on DRR. This was subsequently endorsed at the African Union Summit in January 2015 giving SFDRR continent-wide scope recognition and impact. Even the 3rd UN World Conference on DRR was characterized high-level African participation with the continent unprecedentedly speaking with one voice to highlight the related specific needs of the region. Though not exclusively related to DRR, Zimbabwe has readily embraced this paradigm shift and has taken upon itself to ensure that science and technology innovation plays a pivotal role as the panacea to its current economic problems through the newly rebranded ministry which is now effectively referred to as the Ministry of Higher and Tertiary Education, Science, and Technology innovation.

Nevertheless, while this SFDRR emphasis has taken deep roots in the developed world that includes Asia since its inception in 2015, Sub-Saharan Africa, though had initially embraced SFDRR with much pomp and hive, still lags far behind in the implementation aspect. Despite this reality and in the face of all the successes that are demonstrated by the developed world, Sub-Saharan Africa currently fails to excel herself in terms of the adoption of science and technology for DRR. As such, the 2030 global goals on DRR set by the United Nations will remain a mirage if DRR indicators such as disaster casualties and property losses and their proportions including poverty levels relative to the rest of the world do not decrease in Sub-Saharan Africa. Therefore, in order to achieve the global targets, there is an urgent need to strengthen the overall science and technology-based DRR capacity in Sub-Saharan Africa. It is on record that an implementation plan of action underpinned by scientific evidence has the potential to save lives, more accurately target investment, and contribute to greater resilience over the coming decades (ODI and CDKN 2014; Izumi et al. 2019). In this context, communities and governments need to be creative and scout for novel means to build resilience using some of the latest developments in science and technology. Adoption and accelerating in this noble drive may not be farfetched for Sub-Saharan Africa as most of the individual country priorities have proved to be clustered and clear, reveling the existence of enhanced local demand for greater science in DRR decision-making and solutions.

7.3 Current Vulnerabilities of Sub-Saharan Africa to Hydrometeological Hazards

Due to the general limited capacity to finance the building of resilience to disasters, it is a fact that the magnitude of disaster impact is inordinately very high in developing countries. Sub-Saharan Africa, where two-third of least developed countries globally are located, is prone to recurrent disasters, be it manmade or natural. Here, the population densities are relatively high and characterized by low social development that is compounded by poverty and accelerated unplanned urbanization processes. This makes the sub-region one of the most disaster-prone sub-continents in the world. Although it is synonymous with wars and bad governance, here, disasters are largely caused and exacerbated by natural hazards and climate change. In this regard, hydrometeorological extremes events top the list of being responsible for more than 95 percent of the disasters. Of these, drought and floods occur most frequently with drought coming out as the most common disaster, accounting for the largest proportion of casualties and people affected on the continent every year (Manatsa et al. 2010). The mega drought of Ethiopia of 1983 which was unprecedented in world media coverage due to its impact with about 300,000 deaths and the devastating Southern Africa droughts of 1991 that had the largest area coverage affected by a drought in the regional modern history are just but examples of the intensities of Sub-Saharan African droughts when they strike at times.

The decade 2005–2015 shows that while the global and other continent's trends of deaths from natural disasters are decreasing that of Sub-Saharan Africa is increasing. This period saw Sub-Saharan Africa experiencing an average of 157 disasters per year where around 10,000 people lost their lives annually with a total of 180 million people being affected. The Horn of Africa and southern parts of Sub-Saharan Africa endured mostly El Niño induced droughts, while floods occurred in other parts of the sub-continent, especially those due to tropical cyclones that mostly bring along heavy tolls in Mozambique and Zimbabwe. A case in point is the recent Tropical Cyclone Eline of 2000, which by killing more than 100 000 people, broke the region's record of disaster impacts by any previous tropical cyclone. But, with the advent of enhanced climate change, the risk associated with these weather extremes has been increasing and hence culminating into a cycle of hydrometeorological disasters Royal Society (2014), i.e., a series of floods and droughts a few years apart or even back to back within the same season. For example, in 2013 and 2014, Cape Town, South Africa had one of its highest annual rainfalls ever which made its six main reservoirs that serve the city to be unprecedentedly almost 100% full. Then, the prolonged drought set in immediately after. These reservoir levels fell to 71 percent in 2015 to 60% in 2016 and reached 38% in 2017 and were so critically empty in 2018 that a state of emergency in order to conserve water was declared. Dire was the situation that it was even suggested that icebergs be towed to the city to supply freshwater. In Bulawayo, one of Zimbabwe's largest cities, the post-2014 drought caused months of severe water shortages which resulted in 72 h of water rationing a week and decommissioning of more than 10 dams which supplied water to the city. The 2019/20 season saw a severe drought that killed a significant number of wildlife and livestock, especially cattle and elephants over southern Africa, but come January and February 2020, numerous severe floods were recorded throughout the region.

While these recurrent droughts pose major challenges of how to grow and provide adequate supply of food to feed the growing urban population at a time of regional water scarcity, flash floods and tropical cyclone-induced floods destroy infrastructure and hence push back development by several decades. With the projected increase in weather and climate extremes, there is need use science and local innovation to design methods and structures that do not only reduce the impact on the communities livelihoods but which can withstand the realized disaster situations.

The list of positive results achieved through the adoption of science and technology innovation in DRR the world over is now limitless, hence demonstrating the achievement of great strides which buttress why SFDRR had to prioritize science and technology innovation as the primary driver of DRR. That being the case, the future for Sub-Saharan Africa's DRR is not all that bleak. Asia and the Americas, for example, which also had a fair share of world disasters were able to significantly reverse the upward trend of disaster impacts through embracing and acting on possibilities presented by SFDRR. This means that Sub-Saharan Africa also has the potential to significantly reduce the casualties and losses due to disasters if science and technology innovation is adopted in earnest.

7.4 Adoption of Social Media and Big Data Analytics by Sub-Saharan Africa

The world has recently been inundated with an exponential increase in the amount of information and data that are openly available, be it emanating from satellites and drones collecting data from above, or from crowd sourced information and social media from local communities. This has inevitably resulted the accumulation of incredibly huge amounts of idle data. Consequently, there is currently a major international drive toward 'open data' and 'open science' so as to easily assess this unexploited data with great potential for relevant and useful information for DRR. As such, when analyzed holistically, this vast data can readily provide valuable insight for understanding the risks and establishing a common operating environment for DRR.

In Sub-Saharan Africa, activities related to science and technology innovation are mostly undertaken by national publicly funded institutions, such as national research institutes, universities, and government agencies which inevitably are endowed with relatively easy access to the Internet. By its nature, science research and applied science have a predominant international character that is made possible through international publications and peer reviewing, extensive sharing of data and results, foreign training, and heightened interaction among individual scientists. These processes are not only critical to shaping and achieving high-quality research and to providing solid foundations for public policy and action, including for DRR but also contribute to the rapid dissemination and application of related information. Hence, the post-Sendai process for Sub-Saharan Africa should take advantage of this created environment to link up with the developed world where the implementation of the SFDRR has proved to be effective. This provides possibilities to develop a roadmap for capacity building by demonstrating advanced technology and best practices from the developed world. In fact, there is a recognized need for international science partnerships to provide more than assessments of local scientific information, particularly where different kinds of knowledge can contribute solutions. This includes sharing of new trends and the best practices of using big data and social media and scouting for the best methods to promote crowdsourcing data collected from social media for enriching contents of local DRR. Here, only, the following Scientific and Technology Innovation approaches that can be adopted by Sub-Saharan Africa are described:

7.4.1 Social Media Approach Using Big Data Analytics and Data Mining

The social network service/system hereafter simply referred to as 'social media', is an online space for making connections with others, whereby users create a profile and connect through file-sharing, emails, messages, or comments. The development of computers, smart phones, and tablets allows for the proliferation of social media use. It thus avails vast opportunities with unlimited spatial extend to educate people, especially the youth who happen to be the biggest users, on knowledge of hazards. It also allows for collection of disaster data, gives voice to people, especially during an emergency, and provides information on logistic and humanitarian needs. As such, social media effectively replaces the traditional way of disaster information dissemination through radio and television. It has great potential, especially in terms of developing early warning systems that identify risk at various spatial and temporal scales. On the other hand, big data analytics is the process of collecting, organizing, and analyzing these large sets of data (called big data) emanating from users so as to discover patterns and other useful information which can be effectively used in DRR.

The birth of big data analytics and data mining from social media has inevitably brought with it crowdsourcing which is a process that takes advantage of data and information contained within the wide spread mobile and smartphone users, to obtain real time and local information from a wider public. The advantage of integrating this technology is that it allows for pooling of the public's contribution on the location of evacuees, individual geo-referenced needs of disaster affected communities which includes the identification of locations which are at risk of future disasters, etc. Since analyzing disaster risk information from macro-level data limits the analysis of the impact at the local level, crowdsourcing complements the analysis by allowing incorporation of higher resolution data points. Hence, when this local data are integrated into DRR analysis and plans, it can stimulate further local actions based on local risks. This process therefore also doubles up by enabling participating communities to effectively learn about their risks, thus becoming an avenue of disaster risk communication. As a result, such citizen science that enhances participatory mapping is gaining recognition within the DRR community.

The provision of these approaches is not complex as it predominantly relies on existing infrastructure where the two-way communications and interlinks provided through social media are the used channel to collect inputs and disseminate information. As an additional measure to traditional sensor-based data, social media provides non-structural, real time, and direct responses to disaster managers in formats that are highly diverse including either texts, pictures, videos, or live stream. Its simplicity in approach has made it become one of the most convenient and popular channels to report situations and emergencies in the developed world. In this respect, since social media is now the most common means of effective communication in Sub-Saharan Africa, there is no apparent reason why it cannot be considered as an effective tool for Sub-Saharan Africa as well. In fact, big data analytics and data mining of social media, if adequately adopted in the sub-region, can become the most conducive and effective technique for effective emergency management during different phases of disaster management processes. Hence, enhancing digital preparedness by social media for DRR and emergency response should be universally adopted and so that it becomes a new trend of evidence-based disaster management which can offer continuous and local information to test preparedness and reaction conducted by the public sector the world over.

Except for Sub-Saharan Africa, several case studies have demonstrated how the concept of big data and social media usages has become extremely handy and used with ease in preparedness, response, recovery, and mitigation phases to reduce the hazard, risk, and vulnerability of the affected communities. This emerging approach can be used in all sectors to facilitate information communication or knowledge transfer during disaster situations among local communities, non-government organizations, private sector, government stakeholders, and volunteering organizations. Thus, social media offers direct channels to reach end users to take action in reducing possible adverse impacts by, for example, tropical cyclones and floods. In this way, users can provide specific and location-based alerts and receive instant updates, thus harnessing the requisite provision to arrive at instantaneous decisions during disaster situations. As such, big data analytics thus provides a novel instrument that conforms to the speed with which disasters occur and thus aids in enhancing the required swiftness in DRR decision-making. Hence, this approach is a typical example of how innovative applications demonstrate the use of the state of the art information and telecommunication technology to enhance coverage and efficiency of information facilitation to enhance DRR.

Consequently, the use of big data analytics from social media in DRR has transformed the way people perceive and respond to disaster information and even to decision-making processes by government or other disaster management-related organizations. Since the public is usually the first responders when disasters strike, they gather social capital, either directly or through social media, in the form of the mobilization of skills, leadership, networks, and support systems. For example, after Cyclone Idai of March 2019, in southern Africa, people posted texts, photos, and personal experiences via social media, which led to the pouring in of resources in a very short time which became a cost-effective way for soliciting for donations. Similar generous donations from communities were also realized after the Haiti earthquake in January 2010, and in the aftermath of the Great East Japan Earthquake and Tsunami of March 2011, the availed correct information through mass participation was used much more prominently. The creation of social cohesion and promotion of therapeutic initiatives can be enhanced through the use of social media by the public to inform each other, their families, and friends that they are safe after an extreme event. Eventually, data obtained through social media can then be collected and analyzed by researchers for education and decision-making purposes. However, some negative effects of social media use have been observed in relation to disaster information. For example, immediately after the 2010 Chile earthquake, when information from government/authorities was somehow constrained, rumors circulated about an impending earthquake and tsunami. As a result, scientists had trouble with the authorities as the information they gave were used and quoted inappropriately, which lead to chaos and insecurity among people.

Big data and crowdsourcing data does not only bring benefits to DRR but also enhances emergency preparedness. The information thus generated is therefore a vital strategy of DRR, operation of emergency relief, and sustainable development of the economy and livelihood. As such, information and alerts of natural disasters or hazards on big data and social media would provide the best approach to effectively integrated partnership between public and private sector to mitigate the adverse effects. Eventually, these technologies can be extended to become beneficial for adaptation and can assist in making local communities to become more resilient to disasters.

7.4.2 *Challenges to the Effective Use of Social Media in DRR for Sub-Saharan Africa*

However, simple as it may seem to adopt these technological innovations of social media and big data analytics in DRR, the region must also show some commitments to demonstrate willingness of embracing this dimension toward fully implementing SFDRR. Information technology is now data intensive in the global village, with the vast amount of new data for use to enhance DRR activities for Africa being predominantly generated in the developed world. New global scientific knowledge has not only become increasingly complex, but extremely huge amounts are increasingly being produced and made freely available at a mouse click more than at any other time in history. For example, the Earth observation is now producing more than 10 TB of free and open data per day. This is also true for the climate change models whose outputs are being continuously updated and availed to the general public through open-source avenues. Hence, new demands which require novel approaches for data handling and processing are now presenting themselves of which Sub-Saharan Africa has limited capacity to handle.

While the risk of DRR actors cutting corners to use the available data from different disciplines of which they mostly lack expertise to analyze let alone interpret has tremendously increased, the major challenge that the region is facing in general is the limited capacity to access this free data. The cost of bandwidth is beyond the majority of the regional countries, and hence, the data are not ready available where it is needed most. It is logical that for easy access to data, the cost of bandwidth needs to be reasonable. The UN Broadband Commission recommends that an affordable cost of a gigabyte of data should not exceed 2 percent of the average monthly income. But, for Sub-Saharan Africa, the average cost of a gigabyte of data is well in excess of 9 percent of a monthly income for most countries, compared to around 1.5 percent for Asia and even less for the Americas and Europe. Even though apart from the bandwidth price, Internet speeds across Africa which largely constrict data downloads are still far below the global minimum standard. This presents multiple challenges for Internet-sourced data use which is very critical for the flow of social media information. Such a scenario militates against the both accessibility and reliability and hence provision of social media data for use in disaster management activities. The opportunity here is for individual sub-regional countries to find innovative ways to not only ensure that the data are easily accessed but to package the data so that it can still convey the required information and at a lower cost. Deliberate policies can also be enacted to facilitate easy access to data. In any case, it is a fact that the benefit accrued for countries when it comes to reducing casualties, and loss and damage to property from disaster impacts by simply using up-to-date online data are immeasurable. With the changing nature of the science-policy interface, science advisory services the world over, it is inevitable that appropriate initiatives from Sub-Saharan Africa evolve to meet these challenges.

7.5 Summary and Conclusion

Science and technological innovation for DRR has always existed in one form or another in all countries. However of recent, innovations in methods, tools, and analyses have made significant leaps in finding solutions to DRR challenges, as more data are becoming widely generated and made accessible (Royal Society 2014). The resulting knowledge, related to the role of science and technological innovation in providing evidence for policy, is gaining prominence worldwide. This has resulted in growing demand for multidisciplinary enquiry to address the complex and inter-related problems of climate change, disasters, and sustainable development (Hellmuth et al. 2011). In this respect, it is quite evident that DRR is becoming one of the key areas in which science and innovations in technology can make an exceptional impact to development and reduce disaster impacts, especially through sourcing relevant data to enable action from social media.

Adopting SFDRR has now been proven as a panacea in reversing the upward trend in disaster impacts. With the necessary support, this framework can be easily implemented by taking a cue from the developed world's successful stories for replication in Sub-Saharan Africa. This thrust, while offering highly adoptable applications to local contexts, neither requires large budgets nor demands inaccessible technology. Only that it demands adoption that conforms to local and indigenous knowledge so as to be effective and derive widespread acceptance. In this respect, there is no justification for Sub-Saharan Africa to be the only continent which should continuously break her own record of disaster impacts.

Consequently, increased disasters due to the changing climate should act as a wake-up call to Sub-Saharan countries. It has potential to open a vast array of development opportunities to visionary and innovative countries. Proper communication of this scientific developmental evidence to policymakers is imperative so as to ensure relevant policy support. The SFDRR signals a clear mandate to the science, technology, and innovation community to work together with governments in developing and sharing the knowledge and solutions needed to improve the resilience of communities, save lives, and reduce disaster losses. Hence, the international science community should act as a key stakeholder to help Sub-Saharan African countries effectively implement SFDRR. As such with the correct framing, climate change proofing is able to take Sub-Saharan Africa on new development pathways that will avoid some of the complacency of the West and excesses of the East through the incorporation of science and technological innovation in DRR planning for the future.

References

Carabine E (2015) Revitalising evidence-based policy for the Sendai framework for disaster risk reduction 2015–2030: lessons from existing international science partnerships. PLOS Currents Disasters. http://currents.plos.org/disasters/article/policy-forthe-sendai-framework-for-disaster-risk-reduction-2015-2030-lessons-from-existing-international-science-partnerships-2/

Hellmuth ME, Mason S, Vaughan C, van Aalst M, Choularton R (eds) (2011) A better climate for disaster risk management. International Research Institute for Climate and Society (IRI), Columbia University, New York, USA

Izumi T, Shaw R, Ishiwatari M, Djalante R, Komino T (2019) 30 innovations for disaster risk reduction by IRIDeS, Keio University, the University of Tokyo, UNU-IAS, CWS Japan, Japan, p 80

Manatsa D, Mukwada G, Siziva E, Chinyanganya T (2010) Analysis of multidimensional aspects of agricultural droughts in Zimbabwe using the standardised precipitation index (SPI). Theoret Appl Climatol. https://doi.org/10.1007/s00704-010-0262-2

ODI and CDKN (2014) The future framework for disaster risk reduction: a guide for decision-makers, 2nd edn. London, UK, Overseas Development Institute and Climate and Development Knowledge Network

Royal Society (2014) Resilience to extreme weather. The Royal Society Policy Centre Report 02/14. The Royal Society: London, UK

Smajgl A, Ward J (2013) A framework to bridge science and policy in complex decision making arenas. Futures 52:52–58

Part III
Group Discussion Sessions

Chapter 8
Concept Notes for Group Discussion Sessions

Shoko Arakaki, Masahiro Chigira, Rodrigo Cienfuegos, Andrew Collins, Ana Maria Cruz, Craig Davis, Tom De Groeve, Haruo Hayashi, Srikantha Herath, Yoshiki Ikeda, Michel Jaboyedoff, Wilma James, Sameh Kantoush, Kenji Kawaike, Hiroshi Kawase, Toshio Koike, Paul Kovacs, Elisabeth Krausmann, Wei-Sen Li, Yuki Matsushi, Irwan Meilano, Robin Eve Miller, James Mori, Khalid Mosalam, Mahua Mukherjee, Virginia Murray, Soichi Nakajima, Eiichi Nakakita, Kazuyoshi Nishijima, Maria Camila Suarez Paba, Indrajit Pal, Ram Sateesh Pasupuleti, Jean-Paul Pinelli, Ortwin Renn, Charles Scawthorn, Subhajyoti Samaddar, Rajib Shaw, Takahiro Sayama, Jörgen Sparf, Kaoru Takara, Tetsuya Takemi, Hirokazu Tatano, Annisa Triyanti, and Irina Zodrow

The concepts notes were developed to facilitate discussion among various stakeholders gathered from around the world to share information on their ongoing research activities and current research projects, find collaborative activity partners, showcase research achievements, and challenge further improvements to disaster risk reduction and resilience to disasters. Discussions, in particular, evaluated research activities carried out by members of GADRI and discuss ways to further promote and contribute to the 2016 Science and Technology Roadmap for the implementation of the Sendai Framework Agenda.

Through the group discussion sessions, various avenues were explored on the governing nature of empirical research on disaster risk reduction and resilience to disasters conducted by various research institutions nationally and globally, and

S. Arakaki · M. Chigira · R. Cienfuegos · A. M. Cruz · C. Davis · T. De Groeve · H. Hayashi · S. Herath · Y. Ikeda · M. Jaboyedoff · W. James (✉) · S. Kantoush · K. Kawaike · H. Kawase · T. Koike · P. Kovacs · E. Krausmann · W.-S. Li · Y. Matsushi · I. Meilano · R. E. Miller · J. Mori · K. Mosalam · M. Mukherjee · V. Murray · S. Nakajima · E. Nakakita · K. Nishijima · M. C. S. Paba · I. Pal · R. S. Pasupuleti · J.-P. Pinelli · O. Renn · C. Scawthorn · S. Samaddar · R. Shaw · T. Sayama · J. Sparf · K. Takara · T. Takemi · H. Tatano · A. Triyanti · I. Zodrow
Kyoto University, Kyoto, Japan
e-mail: james.wilmatheonesta.6r@kyoto-u.ac.jp

A. Collins
Northumbria University, New Castle, UK

H. Tatano and A. Collins (eds.), *Proceedings of the 4th Global Summit of Research Institutes for Disaster Risk Reduction*, Disaster and Risk Research: GADRI Book Series,
https://doi.org/10.1007/978-981-19-5566-2_8

especially on how evidence-based research results are utilized or implemented in mitigating disasters by policy makers. Emphasis was placed on integrated disaster research with a wider focus on collaborative research activities and how these activities are directly or indirectly contributing to the 2016 Science and Technology Roadmap and the Sendai Framework Agenda.

Topics explored at the group discussion session included the following:

Group Discussion Session I:

I-A: Water and Weather Related
I-B: Earthquake and Volcano Disaster Related
I-C: Geohazard Related
I-D: NaTech and Cross-Cutting Issues Related
I-E: Social and Human Sciences including Health Related

Group Discussion Session II:

II-A: GADRI Contributions to the 2016 Science & Technology Roadmap-how best could GADRI promote the SFDRR 2030 agenda?
II-B: SDGs, climate change and adaptation-what engagement mechanisms and research linkages are needed to influence research directions among policy makers, governments, localities, media and other groups?
II-C: Nation's Synthesis: Concept Design and Implementation, and the SFDRR Agenda 2030-what systemic synthesis can be built?
II-D: Research Funding-where is the funding for disaster risk reduction activities coming from and is this invested to the right causes and the areas that are most needing support?
II-E: Data Working Group-what are the real strengths and weaknesses of data and risk management?

At the end of the group discussion sessions, the participants provided:

- Inputs to the contextualization of the 2016 Science & Technology Roadmap.
- These inputs were submitted at the UN Global Platform 2019 held in Geneva, Switzerland in May 2019.
- Learned from the accumulated research knowledge of each research institute.
- Shared information on ongoing project activities and achievements.
- Explored opportunities for collaborative and empirical research activities.

Day 1—Wednesday, 13 March 2019

Session I-A—Water and Weather-Related Disasters

Chair: Dr. Tetsuya Takemi

Co-chairs: Dr. Sameh Kantoush, Dr. Kenji Kawaike, Prof. Wei-Sen Li, Prof. Ortwin Renn, Prof. Paul Kovacs, Prof. Mahua Mukherjee

This session will focus on water and weather-related disaster from scientific, engineering, and implementation perspectives. The water and weather-related disasters will be caused by extreme meteorological phenomena at various spatial scales. Under changing climate, these extreme phenomena are anticipated to intensify and hence to become more disastrous. Therefore, to mitigate and prevent water and weather-related disasters, we need not only to understand the impacts of extreme weather under current meteorological conditions but also to assess the impacts of changing climate. The understanding of the risk is decomposed into the understanding of physical mechanisms for the responses of disaster-spawning extreme weather to global warming, changes in statistics of extreme hazard events, and responses of built environment to the extreme events. The session will discuss the state-of-the-art concepts, methodologies, and approaches from scientific, engineering, and implementation viewpoints. Past experiences of the GADRI institutes on evidence-based scientific and engineering approaches and best practices/implementations will be shared during the session. In order to provide basic ideas for Group 2 "SDGs and climate change and adaptation" in Discussion Session II, this session will focus on the past research and implementation experiences and share those experiences to the participants. Furthermore, by recognizing rapid urbanization trend in developing countries and re-shaping of urban districts in developed countries, sustainable development of urban weather and climate will be discussed as well.

The purpose of the session is to highlight the interactions between natural events (such as earthquakes, floods, and extreme weather) and human actions. In the modern understanding of disasters, they cannot be labeled as natural versus societal, but they represent co-production of natural forces and human activities. Many human activities release, promote or intensify catastrophic impacts, others are designed to reduce, mitigate and absorb natural forces that could lead to major damages. It is therefore mandatory to understand technological, social and natural disaster as a systemic interplay of interrelated forces. Natural hazards are part of a larger system of nature-human relationships that necessitate new integrated risk management approaches and concepts. In urban areas, heat island coupled with heat wave is an impending urban climate phenomena. Areas receiving fairly good to very good rainfall are showing strain in water resource both in terms of inundation as well as drought. Thus, importance of managing water including storm water is of paramount importance for today's time, especially for urban areas. Furthermore, importance of Eco-DRR to mitigate climate and water-related risks should be recognized.

Key topics to be discussed:

- The interplay of natural disasters and human settlements
- The effectiveness and efficiency of risk management for a combination of natural and social/technological hazards
- Resilience of social adaptation strategies for natural and socio-technical hazards
- Human behavior in disaster situations
- Trend analysis of rainfall pattern with climate change and global warming
- Urban climate map for hot-spot identification
- Water scarcity -Global examples
- Monitoring system
- Water resource management- best case examples
- Blue- Green Infrastructure

Key questions to be addressed:

- How can public authorities cope with combinations of natural, technological, and social disasters?
- How can preparedness for interacting disasters be improved?
- What kinds of interventions are suitable for multi-level disasters caused by a combination of natural, social, and technological hazards?
- How to integrate Eco-DRR?

Results and recommendations

This group focused on water and weather-rated disasters from the scientific, engineering and implementation perspectives with emphasis on the state-of-the-art concepts, methodologies, and approaches. Evidenced-based research results, best practices, and approaches were shared and discussed during the session. Highlights showed the interaction between natural events and human-related events. Discussions also focused on understanding risks and dealing with emerging risks in each region.

The group emphasized the importance of technical development and communication for weather forecasting or other water related warnings; and the important of communicating with local agencies and local governmental official or other stakeholders. The group highlighted particular points that were of concern in particular based on past research implementation and experiences: (1) understanding of risk and transfer of science and engineering knowledge on hazards and risks from one region to another; (2) dealing with emerging risks in each region and interrelationships between academic, decision-makers, and stakeholders; and (3) filling the gaps between academic sector and decision-makers.

The group pointed out that the scientific, engineering knowledge and human society need to forge cooperation between designers, protected infrastructure, and analysis on behavior responses in crisis situations by understanding cultural diversity around the world. In addition, education is vital not only for researchers but also for facilitators in developing countries. Non-structural measures should be incorporated into disaster mitigating measure.

It was also noted that integration of technological and social sciences from the beginning of the analysis on design in structural and non-structural measures are needed. Once again emphasis was placed on the importance of education, for example, education for mass media and local society.

Discussions focused on how to understand risk in general sense. Increase capacity for disaster management is important to reduce existing risk and to prevent risk creation and to build resilience in recovery. Risk creation may emerge from the trade-off between the human interferences and upcoming challenges to the socio-economic factors at longer time-scales. Not to create new risks under urbanization and climate change is an important issue.

Climate change at a local level is not well known by decision makers and general public especially in developing countries. For example, the impact of extreme weather on agricultural and its climate impact should be conveyed to the general public and decision makers. Specific to the risk from water and weather-related phenomena, sharing weather climate data among countries is important. Business continuity plan must be well recognized based on the experiences of damaging tsunami as an example.

It is recommended that GADRI as a platform should prioritize how non-structural measures are incorporated in disaster risk reduction and that GADRI can be a platform to transfer scientific and engineering knowledge from one region to the other by understanding cultural and societal diversity in each region.

Session I-B: Earthquake and Volcano Disasters

Chair: Prof. James Mori, Prof. Yoshiki Ikeda, Prof. Hiroshi Kawase

Co-chairs: Prof. Khalid Mosalam, Mr. Soichi Nakajima, Dr. Craig Davis

We plan to review recent research topics in Seismology, Earthquake Engineering, and Volcanology at GADRI institutes. Efforts include research activities, applied applications that implement research results, and field investigations of recent significant events. We want to highlight areas where current work is addressing gaps there were identified at the previous GADRI summits. Also, for future research efforts, we hope to emphasize topics where important results are anticipated and identify areas where more multi-disciplinary work can be effective. Important points of these discussions should be summarized in terms of GADRI contributions to the 2016 Science & Technology Roadmap.

Key topics to be discussed:

- Progress at individual institutes for research in Seismology, Earthquake Engineering and Volcanology. How has this work contributed to gaps and issues that were identified at the last GADRI summit?
- Future research plans,
- Short-term research topics that can lead to signify results in a few years.
- Long-term research goals that need to be initiated and continued.
- Collaborative work with other fields where multi-disciplinary efforts need to be improved.
- Organizational issues that should be addressed by GADRI to improve cooperative work between institutes.
- Priorities for GADRI contributions to the 2016 Science & Technology Roadmap.

Key questions to be addressed:

- What are the remaining research gaps that need to be filled to improve our understanding of earthquake, tsunami, and volcano hazards and risks?
- What are the best organizational models for research, development of practical applications, and education efforts?
- What are the important tasks for scientists and engineers in public awareness and education efforts?
- How should the concepts of uncertainty be communicated to the public?
- What are the roles of scientists and engineers for improving laws and regulations related to these hazards and risks?
- What improvements are needed for response and recovery efforts following

- Large earthquake, tsunami and volcano disasters?

Results and recommendations

1. Understanding disaster risks

Short Term

- Promote open sharing of data
- Use modern AI and machine language for analyzing big data.
  This can encourage open data sharing.
- Encourage use of satellite data for reconnaissance. Encourages open data sharing
- Develop methods of site characterization (seismic, landslide, and liquefaction)
- Promote success stories, such as volcano predictions.
- Promote interdisciplinary studies, and especially monitoring, with landslides, environmental, ocean studies, and induced earthquake. Monitoring can be piggy-backed with other applications.
- Carry out detailed post-earthquake surveys
- Invest in monitoring efforts.
  - Building monitoring
  - Microzonation in cities
  - Basic regional seismic networks for developing countries
  - Undersea tsunami monitoring
- Monitoring can be costly and depends on country. Maintenance is always an important issue.

Long-term

- Non-engineered structures in developing countries is an important problem.
  - need inventory of buildings
  - develop retrofit methods
  - carry out historical studies for past earthquakes
- Need to transfer appropriate engineering technology from developed countries. Is base isolate on appropriate technique in developing countries.

2. Strengthening Governance
   Communication is most important !

Short term

- Promote effective "evidence-based" scientific and engineering reports.
  Should be consensus based and concise. Need good written communication from technical people.
- GADRI should
  - encourage teaching students to communicate with govt. officials

- Encourage teaching policy makers to communicate with technical people. Need effective methods, such as field trips and good audio-visuals.
- encourage teaching technical people to communicate with public
- encourage more interdisciplinary communication among different technical
- facilitate social scientists to teach technical people how to better communicate

All of above should be continued in Long Term

- Present good examples of effective communication, such as USGS earthquake hazard maps and City of Los Angeles retrofit program.

Long Term

- GADRI should encourage comprehensive studies of real examples of response and recovery following significant disasters. These would be very useful evaluations to be used in the next crisis.

3. Investing in Risk Reduction

Short term

- Slow bad practices, simple building, and engineering practices can provide effective low cost solutions.
- Education and training very important
- Distribution of information to many people about simple hazard and risk information. Very culturally dependent. Appropriate information in Japan is different from Malaysia.
- Provide information about retrofit practices, again depends on country
- Develop series of GADRI MOOC

Long term

- Have to deal with the difficult problem of low probability but high risk hazards. Events that happen once every thousand years or more, but when happen damage will be very severe.
- Need to set priorities for implementing scientific results

4. Enhancing Disaster Preparedness

Short and Long Term

- Should encourage setting priorities and choosing efficient solutions. Cannot do everything so priorities need to be set. GADRI institutes can provide reliable science-based information on hazards and risks to provide information for setting priorities.
- Promote early warning systems for earthquakes and volcanoes. Very dependent on resources and expertise of country

Session I-C: Geohazard

Chair: Dr. Yuki Matsushi and Prof. Masahiro Chigira, Prof. Michel Jaboyedoff

Co-chairs: Dr. Rodrigo Cienfuegos, Dr. Indrajit Pal

This group focuses on strategy for geohazard mitigation in the next decade, regarding especially to the priority 2 in the Sendai Framework for Disaster Risk Reduction (SFDRR): "strengthening disaster risk governance to manage disaster risk." In the last summit in 2017 (the 3rd GSRIDRR), we confirmed the direction of our research effort for the various types of geohazards (landslides, rockfall, liquefaction, etc.), which was intended to review our current status for understanding disaster risk (priority 1 in the SFDRR). On another front of this directive research action, we do need to pay simultaneous effort for disaster risk reduction, overviewing the global situation under an enhanced condition of geo-disaster triggering by the changing climate and earthquakes. In this discussion session (Group 3 at Day 1 of the 4th GSRIDRR), we would like to discuss how we can utilize our scientific and engineering knowledge to make social implementation for risk assessment and mitigation of geo-disaster, by filling gaps between, research community, policy/decision-makers, practitioners, and local stakeholders.

(1) The session will seek input from participants on the following topics:
    Strategy for mainstreaming the geo-disaster mitigation at diverse levels.
(2) How can we make human-oriented, truly useful hazard maps?
(3) Road to make an informative database for policy/decision-makers and practitioners.
(4) Required input for assessment, governance, and management of geo-disaster risks.
(5) How can we generalize our research knowledge to make social implementation over diverse local conditions?
(6) Establishing tight linkages between geo-disaster risk monitoring and inputs from metrological, seismological, geological, and geomorphological big data.

The purpose of this discussion session is to deepen our recognition for the current status and near-future tasks to utilize our scientific and engineering knowledge for risk assessment and mitigation of geo-disasters. The discussion will focus also on the strategy for filling the gaps between research community, policy/decision-makers, practitioners, and local stakeholders relating to the governance and management of geo-hazards and/or geo-risks. In this discussion group, we will try finally to get a scope toward social implementation of recent scientific and engineering advances in the next decade.

Key topics to be discussed:

The session will seek input from participants on the following topics:

- Strategy for mainstreaming the geo-disaster mitigation at diverse levels.
- How can we make human-oriented, truly useful hazard maps?
- Road to make an informative database for policy/decision-makers and practitioners.
- Required input for assessment, governance, and management of geo-disaster risks.

- How can we generalize our research knowledge to make social implementation over diverse local conditions?
- Establishing tight linkages between geo-disaster risk monitoring and inputs from metrological, seismological, geological, and geomorphological big data.

Key questions to be addressed:

- Can we achieve the geo-disaster mitigation by solving the issues listed above?
- What kinds of research needs do arise?
- What is the role of GADRI in the way to social implementation?

Results and recommendations

- Discussion focused on improvement and assessment of vulnerabilities of geohazard;
- The need to push for global coordination of projects.
- Need to establish early warning systems and finding flexible solutions to improve community participatory tools.
- Another point of importance is that geo hazard, and especially landslides and its perception by people is that it is localized.
- A way to solve the problem with the available means:
  - First, it requires multi-level communication and exchange of dialogue between the politician and all stakeholders.
  - It is necessary to improve the dialogue between scientists and other stakeholders.
  - People need to be informed about the real risks and see how they accept the situation. It requires introduction of participatory approaches to teach about existing risks and how to live with risks.
- A suggestion for GADRI:
- To bring in the good and bad experiences about geo hazard mapping.
- Geo hazard mapping is really important for the disaster risk reduction strategy implementation. By comparison and extracting main processes, it is possible to see how best to implement disaster risk reduction and foresee the accuracy of different situations.
- Set-up recommendations to implement disaster risk reduction by means of sharing knowledge and improvement of assessment of the vulnerability of geo hazard which is up to now has proven to be very precise.
- To adapt communication to transmit the good knowledge and explore good practices for human-induced landslides.
- To disseminated exchanges around GADRI by creating an exchange platform for GADRI remembers. Through the platform share experiences and especially mapping and DRR experiences, and their implementation. These include good and bad experiences and also sharing of data, recommendations, etc.

Also refer to the report under "Report on the 4th Global Summit of GADRI—Sect. 5.3

Session I-D-a: Social Dimension of Risk

Chair: Prof. Andrew Collins and Dr. Subhajyoti Samaddar

Co-chair: Prof. Yuichi Ono, Dr. Masamitsu Onishi, Dr. Ram Sateesh Pasupuleti,

Disasters are, or become, socially and culturally constructed. Two dimensions of the social construction of risks are (i) that hazards that become risk through socio-economic, political, and cultural activities of people and (ii) people's creation, reproduction, or manufacturing of disaster risks to meet their own needs, growth, aspirations, survival, and livelihood securities. As such, risk is always a subjectively interpreted process. In such circumstances, both real and perceived risk perceptions are crucial to dealing with disasters. Risk perception influences willingness or intention to respond and prepare against disasters, whilst there are usually limitations in the measurement of actual risk for events that may be variously more random or hitherto not yet experienced. The implementation of DRR strategies in society therefore should understand as much as possible of the social dimension of risk to embrace an understanding of how risk is built, perceived, shared, and communicated. In recent decades, voluminous studies have been carried out on understanding the social dimension of risks. A question for this session is to consider whether these are more diagnostic in nature, perhaps limited in understanding the scientific basis of risk perception, underlying factors of risk development, and so on. It is suggested that there remains a big gap in the scientific findings that can be used for effective risk communication and risk governance. It is proposed that integrating a revitalized social construction of risk into our research is critical for effective implementation of disaster risk reduction. The focus would then be placed not only on how risks are perceived and constructed (or deconstructed) but also how they are communicated, shared among stakeholders and based on countermeasures planned and initiated. This session will provide opportunity for participants, particularly researchers from different disciplines, to share their opinions and observations on better integration and utilization of knowledge on the social dimension of disaster risks in relation to disaster risk communication and governance and in that sense what is needed for more effective implementation DRR strategies.

Purpose and objective of this session are:

- To identify how research on the social construction of risk can be used for effective DRR?
- To consider how the understanding of risk can be better circulated and discussed among stakeholders to reach a shared recognition of the social dimensions of risk.
- To work out how the social dimensions of understanding risk can help to design the countermeasures of DRR.
- To reflect and report on future research directions relating to the social dimension of disaster risk.

Key topics to be discussed:

- Understanding risk for improved preparedness.
- Social construction of risk, risk perception, and risk communication: How risk is perceived, amplified, constructed, deconstructed, and shared among stakeholders.
- Risk governance: what it is and how conflict associated with risk can become cooperation and consent for disaster risk reduction.
- Concern assessment and scoping: Underlying factors influencing risk, risk perception, and social implementation of DRR.
- Understanding social construction of risk as a process that can be used for more effective implementation of DRR strategies.

Key questions to be addressed:

- To what extent are hazards, risks, and disasters within society currently well known?
- How can the social construction of risk be effective for DRR?
- How can an improved understanding of risk be communicated around varying DRR stakeholders?
- In what way does understanding of the social dimensions of risk help drive DRR?

Results and recommendations:

The discussions focused on the context of risk and living with the risks. To define risk, risk itself is a social construct due to the nature of unequal distribution of resources, skills, and abilities, and also due to the fact that victims are from different genders, differently abled individuals, social setups, and so forth.

When it is about the social dimension, it is thought about even before it addresses the society. It starts with 'I'. The concept of I, is accountable for we, you, and ours, and theirs, and it continues to grow. Further, the accountability of I is very important. How does one look at the risk? How is it not only considered in one's own context rather it can be looked in a holistic manner.

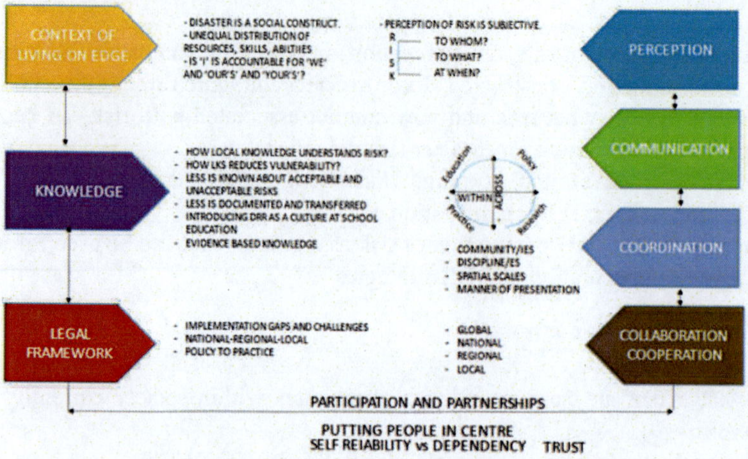

Copy of PPT presentation by Dr. Ram Sateesh Pasupuleti

Also refer to the report under "Report on the 4th Global Summit of GADRI—Sect. 5.4.

Session I-D-b: Health and Disaster Risk Management

Chair: Prof. Virginia Murray

Co-chair: Prof. Andrew Collins, Ms. Robin Eva Miller

Health is a key element of the Sendai Framework and the other landmark 2015 agreements including the SDGs and Paris climate change agreements. Emphasis is placed on population vulnerability as expressed through health status at all levels including those that are marginalized, young, old, disabled, and those that have non-communicable diseases, are migrants, of any indigenous origin, or otherwise identified. Under the Sendai Framework agreement, building resilient health systems through integration of all-hazards disaster risk management within health care and well-being is already increasingly recognized as essential. Capacity development of health professionals at all levels in understanding disaster risk reduction and management will therefore have to become a priority. Meanwhile, engaging disaster risk means recognizing the role of multi-sectoral practices and therefore multidisciplinary health and well-being research for disaster prevention and response strategies well-beyond the formalized health sector. An all-hazards approach demands an inclusive risk management system, of disaster preparedness for effective response to "Build Back Better" in recovery, rehabilitation, and reconstruction as continuous quality improvement processes of health systems and modes of human well-being. Research can deliver evidence informed health and disaster risk and management approaches—these should be addressed in this session. Discussions along these lines are likely to produce further comment as a resource to facilitate how these goals can be addressed. This session should examine the gaps in research that need to be addressed to provide more effective guidance for health emergency and disaster risk management. With the number of events increasing significantly over the last years in terms of hazards being identified through their health impacts, and emerging knowledge on the long-term health and well-being recovery of the populations affected. However, strategic actions on health emergency and disaster risk management, through a WHO related thematic platform[1] and research network[2] hopes to get this core aspect of the 2030 agenda to advance much further. The health area in its broader definition will also need to be clearly represented within the GADRI.

Purpose and objectives of this session are:

- To identify pathways to better representations and mechanism for health and well-being related research in wider disaster risk reduction scholarship.
- To identify existing and likely further health and disaster risk management linkages, representations within the GADRI membership.
- To develop additional aspects of the topic, including where this would mean broadening the role of GADRI in health and disaster risk management cooperative enterprises.

Key topics to be discussed:

As above:

- Capacity building and research for health and disaster risk management: thinking about the basics.
- Understanding institutionalized and non-institutionalized parameters of health and disaster risk management.
- Future effectiveness of health and disaster risk management research and enterprise in contexts of rapid global and local change.
- New frontiers for health and disaster risk management, and aspects requiring action from GADRI based on all of the above.

Key questions to be addressed:

- What do different participants consider to be central concerns for health and well-being in disaster risk management?
- How might non-health professionals, and health professionals, progress health through improvements in disaster risk management?
- How healthy is healthy enough in terms of the limits of disaster risk management?,i.e., what is an acceptable risk to health and what new scope is there for health sensitized disaster management?

---

[1] **WHO Thematic Platform for Health Emergency and Disaster Risk Management**—*WHO Thematic Platform for Health Emergency and Disaster Risk Management Terms of Reference pdf, 245 kb.*

[2] **WHO Thematic Platform for Health EDRM Research Network**—*WHO Thematic Platform on Health Emergency & Disaster Risk Management Research Network pdf, 83 kb.*

- What is the future for health and well-being orientated DRR with GADRI?

Results and recommendations

It is very difficult to isolate health related risks from the disaster as there will be consequences to health after a disaster. Health and safety are an integral part of DRR.

Various factors that contribute to health risks are:

- Direct and indirect risks
- The disciplinary orientation gaps in understanding risks related to health and DRR, and the unequal resource allocation.
- Market-driven risk
- Lack of basic hygiene conditions –
- Accessibility is another important point—how communities have access to medical needs during the emergency situations.
- Accountability is where and how the supply chain processes and to ensure necessary legal frameworks and protocols are followed.
- Affordability—this is one of the hardest of the factors in developing countries—sometimes it is not possible to have affordable medical treatment due to high cost.
- Advocacy of right practices is how to disseminate information to follow the right approach. Advocacy has to go into the bottom level through building partnerships across various stakeholders.

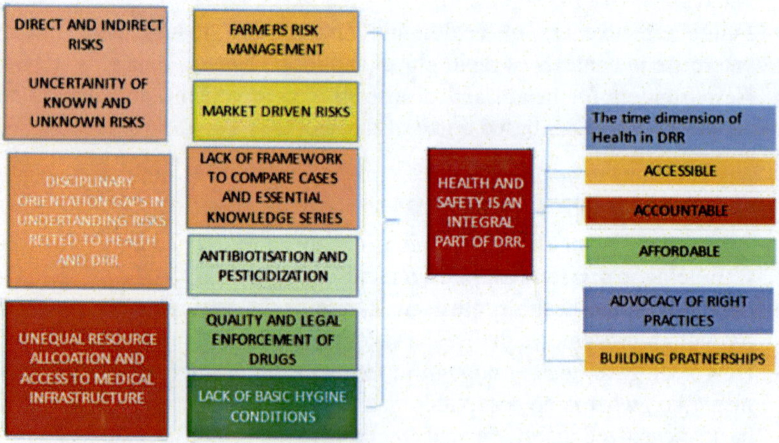

Copy of slide from PPT presentation by Dr. Robin Miller

Also refer to the report under "Report on the 4th Global Summit of GADRI—Sect. 5.4.

Session I-E: NaTech and Its Cross-cutting Issues

Chair: Prof. Ana Maria Cruz, Dr. Elisabeth Krausmann

Co-chairs: Dr. Maria Camila Suarez Paba

Natechs refer to natural hazard-triggered technological accidents. The Natech concept and Natech related research is relatively new. The aim of this discussion session is to share information on and evaluate the status of Natech risk reduction activities in general and Natech research specifically, carried out by member institutes of GADRI. Furthermore, the session will discuss how Natech research can be promoted and better aligned with the Sendai Framework's 2016 Science and Technology Roadmap, including delineating specific actions and priority areas for future research. As with other group discussion sessions, we hope to inform the outcomes of the discussion through the Global Platform 2019.

The Sendai Framework for Disaster Risk Reduction 2015–2030 highlights the need to better understand different hazards, and for the first time specifically addresses technological and Natech (natural hazard triggered technological) hazards. These complex hazards have the potential for catastrophic consequences. Examples of major past Natechs include many fires, explosions and oil spills at refineries in Sendai and Ichihara cities caused by the Great East Japan earthquake and tsunami in March 2011; oil and gas spills during Hurricane Katrina in 2005; and multiple fires and toxic gas releases caused by the Kocaeli earthquake in Turkey in 1999; among others.

There are many challenges for Natech risk reduction due to the need for multidisciplinary efforts which require bringing together diverse stakeholders from different government agencies, industry, and potentially affected communities that do not normally work together or interact. In this session, we will discuss advances in Natech research and current challenges and gaps.

The aim of this discussion session is to share information on and evaluate the status of Natech risk reduction activities in general, and Natech research specifically, carried out by member institutes of GADRI. Furthermore, the session will discuss how Natech research can be promoted, and better aligned with the Sendai Framework's 2016 Science and Technology Roadmap, including delineating specific actions and priority areas for future research. As with other group discussion sessions, we hope to inform the outcomes of the discussion through the Global Platform 2019.

Key topics to be discussed:

In line with the overall objectives of the discussion sessions, in this session we would like to discuss the Natech concept, and share information about the status of Natech risk management at the country level as well as internationally. Then, we would like to review the research achievements in this field, and identify the research challenges and gaps. We would like to focus the discussion by going through the stages of risk assessment and management, risk communication, risk perception, and risk governance, thereby also discussing the science-policy interface. Furthermore, we will take a second look at the research achievements, challenges, and gaps by considering the contributions from the view point of the disaster risk management cycle.

This discussion session is highly multi-disciplinary, given the nature of Natechs risks having the potential to affect large areas with ripple effects that cut across national boundaries due to cascading effects and supply chain impacts. Thus, Natech risk reduction necessarily involves the consideration of cross-cutting issues, including socio-economic factors, governance issues, and climate change, and among others.

Key questions to be addressed:

- What are the key achievements, and what are the main Natech research challenges and gaps?
- Are evidence-based research results utilized or implemented for Natech disaster risk reduction?
- What are the main factors that hamper effective Natech risk reduction?
- What contributions can this group make to the 2016 Science & Technology Roadmap in this field; and what specific actions and priority areas should be promoted?
- How can GADRI promote Natech risk management and research in general, and through the SFDRR 2030 agenda?

Results and recommendations

An enriched discussion was supported with ideas from diverse perspectives, all of which pointed at promoting and expanding the visibility of Natechs at the academia/research, industry, and governmental levels. In addition, the need for a multi-stakeholder interaction was raised, which calls for Natech considerations to create links between industry and researchers, government, and citizens.

Based on the above, the following ideas were highlighted:

More research is needed to support and develop Natech accident modeling and simulation, this could be addressed by considering:

- Cascading events, risk reduction.
- Natech issues by bringing complex system analysis models
- Dynamic risk assessment
- Bayesian networks and systemic risks

Furthermore, other research fields need to be strengthened, these include:

- Natech-economic impacts
- Risk communication, evacuation planning, and evacuation behavior differences
- Multi-stakeholder interactions for complex scenarios

- Risk governance issues
- Social impacts of Natechs from the physical and psychological point of view
- Considerations of societal conditions (poor or rich) on affected people
- Cross-cutting issues-systemic risk approach

Since there is limited awareness and understanding of Natech scenarios among all stakeholders, it is important to bring industry to the discussion to:

- Raise awareness to adopt prevention measures
- Encourage Natech related information disclosure (anonymously)
- Build trust with academia and researchers.
- Open the door for collaborative research and funding (involve universities)
- Promote a systemic collection of Natech data

All the above ideas grounded in the need for a paradigm shift to expand traditional technological risk management to a holistic systems perspective that contributes to societal resilience at the same time that promotes risk awareness and strengthens multi-stakeholder interactions.

On the other hand, Natech research can be promoted, and better aligned with the Sendai Framework's 2016 Science and Technology Roadmap, including delineating specific actions and priority areas for future research for the next 3 years by considering:

- Complex risks in land use control and planning
- Training and education of industrial facility staff, workers, government officials, etc., through concerted actions (First efforts to be carried out in Colombia)
- Academically, bringing in Natech to broaden the scope of disaster risk management
- Participatory disaster risk reduction strategies
- Natech Narrative Knowledge (storytelling based on residents' voices)
- Imagimulation of Natech scenarios (Imagination + Simulation) to analyze what could happen.
- Performance indicators for Natech risk reduction.
- Development of Natech framework for resilience (testing and validation)
- Contribute to the Natech accident database: eNatech at the JRC

Day 2—Thursday, 14 March 2019

Session II-A: GADRI Contributions to the 2016 Science & Technology Roadmap - how best could GADRI promote SFDRR 2030 Agenda?

Chair: Prof. Hirokazu Tatano, Ms. Shoko Arakaki, Ms. Irina Zodrow

Co-chairs: Prof. Renn Ortwin, Prof. Paul Kovacs, Prof. Irwan Meilano, Ms Annisa Triyanti

The science and technology community, as well as other stakeholders, came together at the Science and Technology Conference on the implementation of the Sendai Framework for Disaster Risk Reduction 2015–2030 in January 2016 in Geneva and adopted the 'Science and Technology Roadmap to Support the Implementation of the Sendai Framework for Disaster Risk Reduction 2015–2030'.

In order to reflect and align with global and regional trends and developments in the area of disaster risk reduction, resilience and sustainable development since 2016, the S&T community is updating and "contextualizing" the S&T roadmap. The revision is led by the UNDRR Global Science and Technology Advisory Group (G-STAG) and the Integrated Research for Disaster Risk (IRDR) program of the International Science Council (ISC).

Among others, the revision includes better reflection of the full spectrum of Sendai hazards (natural, man-made, environmental, biological, and technological) and their cascading, interlinked nature, and increasing recognition of the need for a coherent implementation of all 2030 Agenda agreements (Sendai, SDGs, Paris Climate Agreement, Financing for Development, New Urban Agenda, and Agenda for Humanity), in particular at the local and community level. The revision will also aim to add clear action plans to the S&T roadmap and encourage specific commitments and specific contributions from the S&T community to support its implementation.

An initial draft of the revised roadmap has been shared with the S&T community for feedback in February 2019. The 4th GSRIDRR organized by GADRI will be another milestone for the finalization of the contextualization. In particular, it is critical for major players such as GADRI, to actively contribute to the implementation of disaster risk reduction activities and initiatives and reflect risk-sensitive development and resilience in its vision and strategies.

Building on the discussions at GSRIDRR, the final revised S&T roadmap will be presented at the Science and Policy Forum, held at Global Platform meeting from 13 to 14 May during the Global Platform 2019 (GP19) in Geneva. The Forum is expected to then adopt the "contextualized" S&T roadmap, record commitments and contributions by the S&T community and agree on a regular review process.

The purpose and objective of this session is to review the draft revised S&T roadmap by the participant, with specific focus on the current context versus 2016 and how to overcome potential barriers and make use of opportunities for enhanced engagement by the S&T community.

Key topics to be discussed:

The session aims at providing occasion for DRR S&T community members, especially GADRI members, to discuss

- The 'Science and Technology Roadmap to Support the Implementation of the Sendai Framework for Disaster Risk Reduction 2015–2030': What is it for? and what is missing
- Status of implementation and capitalization on current and existing scientific and technical knowledge;
- Current status of the "contextualization" of the S&T roadmap

Key questions to be addressed:

- What are the key concerns of the S&T communities to be reflected in the revised S&T Roadmap?
- How and by whom will the implementation process be monitored and evaluated?
- What is the role of GADRI can play in the implementation of the S&T roadmap?
- What specific contributions can the GADRI members and broader global science and technology community make to the implementation of the roadmap or parts of it?

Results and recommendations

What are the key concerns of the S&T communities to be reflected to the S&T roadmap of UNISDR?

- The availability of data to assess activities and initiatives of the science and technology community. It will require partnerships with other actors within the science technology community and beyond, including with diverse and relevant societal actors and private sectors.
- The degree of inclusiveness. There should be an inclusive process, engaging all stakeholders and use both system and action orientation to increase evidence-based policy making.
- The alignment between Sendai Framework and other frameworks such as SDGs and Paris Agreement. That is where the synergy is needed as the science and technology community has a role in ensuring there is coherence among the processes.
- Emphasis on the importance of the local knowledge and the mechanisms of utilization. For example, using local knowledge is a validation of what is currently being done in the science and technology community. These are needed capture the real problems and design solutions.
- It is important to investigate how and by whom will the implementation process of this roadmap be monitored and evaluated? And how to evaluate impact?

What is the role of GADRI in the S&T roadmap implementation?

- GADRI could contribute in terms of stocktaking of other processes like the global platform and regional platforms to monitoring updates at the implementation of roadmap. GADRI has been developing so much with the new alliances, regional alliances and it's a very good opportunity to harness the network to implement the S&T roadmap.

  How and by whom will the implementation process be monitored and evaluated?

- There is a need of monitoring mechanism. That can be a product like report, document, etc. One of the things that is most important is also make sure that this monitoring mechanism realize the multidisciplinary approaches of the products and also the process is well and to link with the current platform like Sendai Monitoring and voluntary commitments.

Session II-B: SDGs, Climate Change and Adaptation - what engagement mechanisms and research linkages are needed to influence research directions among policy makers, governments, localities, media and other groups?

Chair: Dr. Tetsuya Takemi, Prof. Yuichi Ono

Co-chairs: Prof. Soichi Nakajima, Dr. Kazuyoshi Nishijima, Prof. Eiichi Nakakita, Prof. Srikantha Herath

The water and weather-related disasters will be caused by extreme meteorological phenomena at various spatial scales. Under changing climate, these extreme phenomena are anticipated to intensify and hence to become more disastrous. Therefore, to mitigate and prevent water and weather-related disasters, we need not only to understand the impacts of extreme weather under current meteorological conditions but also to assess the impacts of changing climate. In the Group 1 discussion in discussion session I, we understand the state-of-the-art concepts, methodologies, and approaches from scientific, engineering, and implementation viewpoints from the past research and practical experiences. Based on the discussion in discussion session I, this session will discuss SDGs and climate change and adaptation. Under the IPCC framework, a number of the GADRI institutes are conducting studies on climate changes and their impact assessments from various points of view. Therefore, during this session we would like to share the current activities on studies related to climate changes and their impact assessments from the GADRI institutes. Some experiences on the linkages among the research community, the policy makers, and the practitioners will be shared. From the common understanding on these issues among the participants, we will discuss how the GADRI community influences or direct policy makers for achieving SDGs and climate change adaption.

The purpose of this session is to introduce current research activities on climate change assessment and adaptation by the participating institutes, share the current knowledges from those studies, and discuss what engagement mechanisms and research linkages are needed to influence research directions among policy makers, governments, localities, media, and so on. Climate change is a global concern for the natural and human environments. Among the SDGs, climate change adaptation is listed as "Goal 13: Take urgent action to combat climate change and its impacts". For adapting climate change, precise assessment of climate change on natural disaster, water resources, and human and eco-systems is important. Thus, firstly the current understanding on climate change assessment from scientific and engineering viewpoints will be shared and discussed. Secondly, the adaptation strategy for climate change will be discussed. Experiences on the linkages among the research community, the policy makers, and the practitioners will be shared. From the common understanding on these issues among the participants, we will discuss how the GADRI community influences or direct policy makers for achieving SDGs and climate change adaption.

Key topics to be discussed:

- Current understanding on the studies on impact assessment of climate change
- How climate change affects natural disasters, water resources and management, eco-systems, and human environment?
- How are the GADRI institutes involved in climate change studies?
- How to move forward to contribute to adaptation strategies for climate change?
- How to link climate change assessment studies to setting adaptation strategies in order to contribute to achieve the SDGs?

Key questions to be addressed:

- How the GADRI community influence or direct policy makers for achieving SDGs and climate change adaption?
- How to direct academic researches to policy making processes?

Results and recommendations

The group discussion recognized the importance of evidence-based decision-making concerning SDGs and climate change adaptation to come up with better mechanisms to integrate research outcomes into policy making so that research outcomes could adequately influence policy makers, governments, localities, media, and other stakeholders. The group discussion led to identify the importance of better communication mechanisms to narrow gaps between research outcomes of climate science, including climate change adaptation and policy making. While IPCC reports provide strong basic findings at the global level, actual policy making happens at the national and regional levels and GADRI has roles there. The group recognized the need for a guideline linking the Sendai Framework for Disaster Risk Reduction, Sustainable Development Goals, and Paris Agreement at the global level as there is a great deal of confusion at the national levels, including government and academia. One proposal was to identify and prioritize some of the most significant indicators from them to monitor as there are too many indicators to follow. Finally, the group discussed how GADRI and its researchers could improve the situations and identified some key issues to be reported at the next GADRI Summits.

Here are the results of three items specifically discussed in this group.

1. How researchers in GADRI could better communicate with policymakers on climate change adaptation and SDGs.

   Climate change issues become scapegoats for some of the hydro-meteorological and large-scale disasters, instead of recognizing the malpractices and inactions by governments and policy makers. The group discussed on how this trend could be reversed with a phased approach from both a short-term and a long-term perspective. The group also discussed how to better communicate with climate impacts and SDGs from the viewpoints of communication method, education system, communication engagement at organization/institution level, and implementation planning.

2. How can researchers/GADRI develop an inventory/atlas/portal of climate change adaptation measures around the world and make them (the adaptation measures) usable/implementable in other locations.

   The group proposed that these points could be resolved by developing a database and interactive web-based platform and disseminating adequate information to certain sectors.
3. How to differentiate/relate climate change adaptation (CCA) from/to disaster risks reduction (DRR)

   The group agreed that climate change adaptation and disaster risk reduction are related and inseparable as disaster risk is a function of hazard, vulnerability, exposure, and capacity. Hazard used to be purely natural, but now is changing due to anthropogenic effects over climate change. With regard to linkages with SDGs, climate change adaptation and disaster risk reduction, a few recommendations from the group were made as follows: (1) there should be a guideline linking SFDRR, SDGs and CCA at the country level, and to reinforce the guidelines, the information should be provided by the UN as a blue print in compliance with experts including the GADRI members; (2) it is necessary to focus or prioritize some of the significant indicators accounting (present and future in terms of SDGs) for risks to enable policy makers to focus on the implementation more practically and effectively.

   Lastly, group discussions identified following points for GADRI members to consider and report the progress back to the next GADRI Summits.

   The group agreed that climate change adaptation and disaster risk reduction are related and inseparable as disaster risk is a function of hazard, vulnerability, exposure, and capacity. Hazard used to be purely natural, but now is changing due to anthropogenic effects over climate change. With regard to linkages with SDGs, climate change adaptation and disaster risk reduction, a few recommendations from the group were made as follows: (1) there should be a guideline linking SFDRR, SDGs and CCA at the country level, and to reinforce the guidelines, the information should be provided by the UN as a blue print in compliance with experts including the GADRI members; (2) it is necessary to focus or prioritize some of the significant indicators accounting (present and future in terms of SDGs) for risks to enable policy makers to focus on the implementation more practically and effectively.

   Lastly, group discussions identified following points for GADRI members to consider and report the progress back to the next GADRI Summits.

Session II-C:

Knowledge Management and Science Synthesis (https://www.scj.go.jp/ja/member/iinkai/kagakubosai/pdf/kagakubosai-siryo2404-4a.pdf)

Nation's Synthesis: Online Synthesis System and Pre-Synthesis

Chair: Prof. Toshio Koike, Dr. Tom De Groeve, Dr. Elisabeth Krausmann

Co-chairs: Prof. Haruo Hayashi, Dr. Indrajit Pal, Prof. Rodrigo Cienfuegos.

In conformity with global agreements such as the Sustainable Development Goals and the Paris Agreement on Climate Change, the (https://www.undrr.org/publication/sendai-framework-disaster-risk-reduction-2015-2030) Sendai Framework for Disaster Risk Reduction, and strengthened the foundation for relationships between disaster risk reduction and management.

Science and technology within society is recognized as the boundary to reinforce resilience to disasters and sustainable development. Support from global stakeholders such as GADRI (http://gadri.net/), IRDR (https://www.irdrinternational.org/) and the Scientific and Technological Advisory Group of UNDRR (https://www.undrr.org/implementing-sendai-framework-partners-and-stakeholders/science-and-technology-action-group) are imperative for scientists and practitioners to work closely to achieve the desired outcomes in disaster prevention and management especially in local contexts.

In this context, the UNISDR Science and Technology Conference on the Implementation of the Sendai Framework in January 2016, adopted the Science Technology Roadmap (https://www.preventionweb.net/files/45270_unisdrscienceandtechnologyroadmap.pdf) for implementation of Sendai Framework. The Global Forum on Science and Technology for Disaster Resilience was in Tokyo in November 2017 where "The Tokyo Statement (2017)" agreed to:

(1) formulate guidelines for supporting national platforms for DRR by making the best use of science and technology, and
(2) produce a synthesis system on science and technology for disaster risk reduction.

The group on knowledge management and science synthesis will consolidate efforts by the science and technology communities to share their inputs at the forthcoming GP 2019 (https://www.scj.go.jp/ja/member/iinkai/kagakubosai/pdf/kagakubosai-siryo2404-4a.pdf):

(1) To make a design of "Online Synthesis System (OSS)", which is an information infrastructure for the science & technology community and stakeholders from local to national levels involved in DRR to share information on the status of the current DRR efforts implemented based on science and technology and good practices and lessons from domestic and overseas cases.
(2) To undertake a preliminary synthesis based on the Science Technology Road Map of 2016 and the Recommendations from the Global Resilience Forum 2017 by involving a wide science & technology community and establish a basis for scientific knowledge and information source group for OSS.
(3) To strengthen the science technology partnership for its effective contribution to the Global Platform (GP), and develop plan for Science Technology conference as pre-event of GP.

(Report of the 4th Global Summit of Disaster Research Institutes) Many challenges are ahead of us to realize the implementation of practical DRR measures by incorporating scientific and technological knowledge into policy development and public behavior. We should promote and strengthen development of a system in which scientists and practitioners in each country can assist with national platforms for disaster risk reduction, with international support through enhanced inter- and transdisciplinary cooperation. We also need to coordinate ongoing scientific and technological research activities at national, regional, and global levels to establish a support system for disaster risk reduction through comprehensive, effective, and sustainable transdisciplinary cooperation. During the past ten years, several scientific challenges have been done.

One of the most typical one is "Disaster Reduction Hyperbase (DRH)". DRH is a web-based database to disseminate appropriate disaster risk reduction technology and knowledge, so called "implementation technology" consisting of implementation-oriented technology, process technology and transferable indigenous knowledge. The disaster risk management knowledge center also published a report on the practical use of scientific knowledge in disaster risk management actions in Europe and the scientific evidence base and its practical use in various areas of disaster risk management.

Each national platform should play an active role as a hub organization where the science and technology community meet with other relevant stakeholders such as policymakers, practitioners, private enterprises, and citizens' groups to discuss issues from multiple perspectives, nurture transdisciplinary collaboration between society and science and technology, and thereby increase each nation's disaster preparedness, coping capacity, and resilience.

This group reviews the efforts that have been/are being made to formulate science and technology for DRR and shape evidence-based decision-making for DRR in the national platform and proposes a new approach to be taken by all stakeholders based on a strong commitment of the science and technology community (https://rikaindia.com/wp-content/uploads/2019/07/Science-Tech-in-DRR.pdf).

Key topics to be discussed:

- Current status of science and technology in DRR policy-making and community of practice.
- Roles of periodic synthesis to promote the use of science and technology in policy-making and community of practice.
- Contents and functions to be required to a synthesis system on science and technology for disaster risk reduction.
- Strategies and implementation planning for realizing synthesis of scientific evidence in a timely and accessible manner in society.

Key questions to be addressed:

- Why is it that we cannot fully exploit advanced science and technology to solve disaster-related problems and that losses are still increasing?
- How is science and technology used to shape social and political decision-making in the context of hazards and disasters?
- Toward development of a synthesis system:
- Why do we need the system?
- What are included in the system?

- To whom are the system addressed?
- How to organize an implementation group to develop the system?
- What are the outcomes from the system?

Results and recommendations

Data Issues:

- Gaps in unified online systems for data sharing,
- Guidelines and institutional framework for data sharing and data use need to be established.
- Local government (s) do not efficiently translate/utilize the available data
- Last mile connectivity with the simplified information flow (which could be useful for local community).

Science and Policy Coordination:

- Need to identify the policy needs and specific gaps in implementation
- Promoting dialogue between the scientific community and policy-makers.
- Risk communication needs to be strengthen for policy formulation.

Interdisciplinary:

- Lack of incentives for researchers to foster interdisciplinary work.
- National level funding mechanisms need much encouragement for interdisciplinary work.
- Need to join natural disasters with technological accidents → Natech.

Short-term Plan (3-year plan):

- Prepare and submit questionnaire to GADRI members (after GADRI BoD endorsement) to make situation assessment among the different institutes/countries.
- Make strategy for promoting nations' synthesis; draft action plan (May 2019);
  - implementation phase 1 by 2020,
  - phase 2 by 2021–2023
- Build standard for data sharing process and mechanism

Long term Plan Up to 2030:

- Contribute to S&T roadmap (OSS contributes to many actions under the different priorities for action).
- Foster the establishment of science and policy coordination framework at the local and national level
- Develop the integrated scientific knowledge base through the promotion of interdisciplinary research.

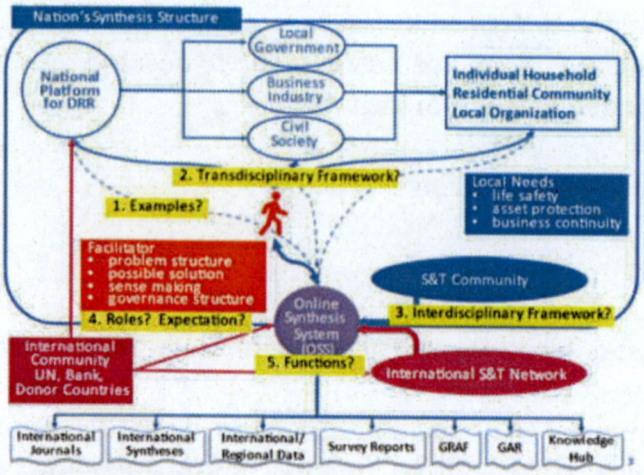

Copy from PPT

Session II-D: Research Funding - where is the funding for disaster risk reduction activities coming from and is this invested to the right causes and the areas that are most needing support?

Chair: Prof. Kaoru Takara, Dr. Takahiro Sayama, Dr. Subhajyoti Samaddar

Co-chairs: Prof. Charles Scawthorn, Prof. Mahua Mukherjee, Dr. Sameh Kantoush

Increase of research funding for science and technology promotion in disaster risk reduction (DRR) is critical for promoting safer and resilient communities. There are two aspects of research and development, considered most critical in DRR discourse—first, the distribution of funding; second, the access and information of funding. The systematic and coherent discussion of these core aspects of funding for R & D in DRR is critical to promote integrated disaster risk management.

The distribution of research funding is unequally distributed in terms of hazards, management cycle (phrases), nature of science (hard and soft sciences), and regions (country and regional variation). The funding generally varies across types of research-basic, applied, and development-oriented sciences. The funding distribution can be further grouped and segregated according to hard and soft sciences or physical and social sciences. The inclination of the funding agencies towards one science to another is historically evident, although plural and integrated approach for DRR has been advocated.

The nature of funding also decides the priorities of disaster management works ranging from response and relief, reconstruction and rehabilitation works to preparedness and resiliency building works. Ideally more funding should be allotted and used on preparedness and resiliency building, however, it remains the least invested area whereas the bulk of the funding goes for disaster relief and rescue. As a result, the countries and regions are not only extremely vulnerable to disasters, but also their science and technologically are way behind to uphold the future DRR works. The funding also varies across hazards depending on frequency, severity, and vulnerability of the disaster.

The distribution of research funding allocation, its information and accessibility are found to be challenging and uneven. In developing country context specially, a very limited funding is available for DRR research and development. Even whatever funding available, it has skewed distribution, and is less open to everyone.

However, from the perspective of donors and funders, it is equally important to make sure that research funding is given to priority and apt areas, efficiently used and benefiting the wellbeing of the common citizen. The academic communities must be responsible to inform funders and donors about the critical and emerging research areas for funding, its significance and potential future impacts on societies and communities. The collaboration and deliberative discussion between researchers across regions, countries and disciplines is therefore important to define the scope, allocation and usage of research funding in DRR.

- To identify the gaps in research funding mechanism and strategies.
- To identify the key priority areas and regions for research funding in DRR.
- To systematically examine how research funding could be better utilized, aptly distributed across sciences and used for the betterment of common citizens and societal development.
- To find out the mechanisms and strategies to increase the accessibility of research funding information across regions and disciplines.
- To find out the role of GADRI in DRR funding advocacy.

Key topics to be discussed:

- The gaps in current research funding mechanism.
- Identify key priority areas of research.
- The value of fundamental science vs. the importance of development or need based research.
- The availability of funding.
- Information provision of research funding.
- The provision of international collaboration for obtaining fund for greater purpose.
- Distribution of funding.
- Appropriate usage of funding.
- Research collaboration and advocacy for more appropriate, equal and just funding.

Key questions to be addressed:

- What is the recent status of funding on disaster risk reduction (DRR) field (status quo)?
- How GADRI member institutes obtain research funding on DRR projects?
- What types of research funding are available?
- Donners: international organizations, national government, private
- Purpose: scientific studies, application studies, implementations
- Target: research institutes, collaboration with private sectors, international collaborations
- What kind of issues we face with respect to research funding?
- Any particular study area in DRR requiring large-scale research funding?

- With such a funding opportunity, what kind of new research can be conducted?
- Should GADRI strategically try to obtain research funding as a group? If so, for what and how?

Results and recommendations:

It was observed that the primary sources of funding for DRR are from the national government agencies. At least, this is the case for certain developed and industrialized countries. International funding is least available to developed countries as well as for some developing countries. The private sector is least interested in DRR as it could prove to be a risk on the return on investment. An area to be explored is the corporate social responsibility (CSR) funding which could prove to be interested in funding for DRR efforts.

The nature of funding focus was on hazards and aids for response and relief, particularly on engineering-based solutions or engineering-based research and funding on fundamental science. For DRR social science, it was noted that the difficulties faced in obtaining funds for research. Funding, in general, is very few for DRR activities as most of the funding is allotted for disaster relief and response; afterward, only a few budgets get sanctioned for preparedness, rehabilitation, and resiliency.

In developing countries, there is a lack of funding for DRR and certain independence to conduct own research. Depending on the funding agency, it was observed that most funds are spent on post-disasters in developing countries rather than pre-disaster initiatives. For example, funding is used for catastrophic events by donors as propaganda to promote the business and attract the media. DRR as a component of the development is ignored or perhaps not considered by the policymakers or funders. Most of the fund goes to response and relief operations or this kind of short-term benefit research.

As the role of GADRI, the following highlight a few suggestions:

- GADRI to advocate for DRR funding. For example, request a certain percentage from GDP to be given to DRR research activities. For that, maybe, we can show some examples of what are the costs and benefits if the funders are funding the DRR sectors.
- Expand focus on institutional capacity building such as knowledge-based development or implementation research.
- GADRI may encourage crowd-funding and initiate building partnerships with other crowd-funding agencies to raise the budget for DRR research.
- GADRI would promote and showcase the importance of DRR research through adopting selected cost–benefit analytical research on disaster and development.
- GADRI may carry out a comprehensive survey on funding availability for various DRR activities in different geographical regions. The survey may include the process and criteria of the funding.
- GADRI Database: The participants propose to develop an international online cyber base. GADRI would promote its networks for training and academic exchange through regularly held workshops and seminars involving donors and policymakers to discuss and exchange ideas and expose the priority areas in DRR that they should fund.

- GADRI can promote and endorse selected training courses or online DRR courses as GADRI approved projects that may in the future encourage and influence the donors to render funds.

II-E: Data Working Group - what are the real strengths and weaknesses of data and risk management?

Chair: Profs. Andrew Collins, Irina Zodrow

Co-chair: Prof. Jörgen Sparf, Prof. Jean-Paul Pinelli

A UNISDR STAG Data Working Group (G-STAG DWG) has been initiated to coordinate an examination of how data is contributing to the expected outcomes of the Science and Technology Road Map of the Sendai Framework for Disaster Risk Reduction (SFDRR). The DWG aims to provide a steer to risk-informed development and science-based decision-making at all levels. It recognizes that data contributing to the implementation of SFDRR stems from varied methodologies and sources including numerical, narrative or visual, digitally, or otherwise.

Discussion around data for implementing SFDRR has so far largely centered on issues of data capture, interoperability, minimum standards, and capacity for data sharing. Challenges include varying cultures of data and information production and usage across public, private, and hitherto underrepresented groups. Some advances are being made in respect of acting on opportunities to improve data and the availability of technologies. A DWG objective, however, is to improve DRR data implementation and engagement across varying cultural and scientific interpretations.

The DWG purpose shares orientations of G-STAG and the Global Risk Assessment Framework (GRAF) toward improving processes of risk reduction, and a call from the International Science Council (ISC) for science as a global public good including defragmentation of science, so that it is more inclusive, integrated and engaged.

Overall, an initial rationale and driver of the DWG is therefore "data for action", whereby data gathering, dissemination and uptake can be actualization of a voice for disaster risk reduction at local level whilst transmissible and interpretable by wider groups of decision-makers at all levels. A listening approach is essential because both the quantification and qualification of disaster as experienced and forthcoming variously remains unknown to potential data user groups. Gaps span impact monitoring data (primarily loss related), exposure and vulnerability data (primarily risk related), emergency preparedness and response data (mitigation, response, and adaptation capacity related). Better understanding of action data points to the need for conceptualization, compilation, interpretation, and dissemination across a multisector environment with better facilitation of all of society engagement.

Key topics to be discussed:

To this end the Global STAG DWG requests members of GADRI and participants of the 4th Global Summit to reflect and react to the following initial topics:

- The heavy weight on "hard" data on risk and vulnerability in relation to data on root causes and risk drivers for disaster impact, above the standard DRR-related measures.

- The applicability of different data in different parts of the world.
- The lack of use of risk creation data, e.g., on social inequity, and on illegal and corrupt practices by elites and officials.
- The user perspective, i.e., the many different stakeholders/practitioners on various administrative levels might have very different needs for data, both in form and substance.

Key questions to be addressed:

- What data do you consider to be already contributing to the progress of the Sendai Framework Priorities for Action?
- What other existing data is there that could contribute to the progress of the Sendai Framework Priorities for Action?
- Which cooperative actions enable data to progress the Sendai Framework Priorities for Action?
- How is data a *voice* for those exposed to disaster risks and for facilitators of the Sendai Framework Priorities for Action?

Results and recommendations:
Current Situation of Big Data Characteristics

Scientific papers may not have complete data but need to provide frameworks for data.
Value is produced when the data output is put into action.
Quality, decentralized, and diversity issues need revisiting:
well-formed data.
Missing values.
cleanliness.

In each country there are networks collecting data and it varies from country to country with type of data:

France all centralized national agencies.
Switzerland only environmental data is centralized.
Standardization of data to fulfil the fill data sets-not for all data.
Office of national emergencies and of statistics can be political.
Minimum standards for data sharing are important-should journals and research funders have an impact?

Data Characteristics Issues

Vulnerability indices are often most effective at local level not national level-This is the area of work for action data theme.
Boarders between privately owned data.
Interesting case from Taiwan:
Building data comes from building tax data.
Maintained by local government from every county or city.
Combine data, interpreting, and share with water companies.

Who is responsible for local warning if nothing happens or someone dies in the evacuation?
Given so many ways to gather data from satellites and other sources, what applications are there for data-mining and processing the good from the bad data?

Issues are gaps in data on the one hand and risks associated with "big data" on the other.

- What does the policy maker to determine how it will affect people?
- There are tools for people to express themselves, so we need to think about how we communicate better,-e.g., e-health has transformed health-is this a model we could use??? Citizen science is a huge movement in itself.
- Networks that we all belong to-if someone is ordered to evacuate-if the network is local such as Hurricane Katrina then people may not evacuate- a photo and a video from the top may have no impact on those at risk?
- Is soft data being pushed out by hard data?-how should soft data be included and gender relevant-a million women is very different to men.
- An apparent difference between "developed and developing" world-implications for growing data inequality

What are the Challenges?

- Deep learning and machine learning rapidly developing-in five years things have evolved very fast;
- How to get the influence of indigenous knowledge and how to get engagement of local resilience in all different disciplines
- How to coordinate local data-be it territorial or indigenous knowledge++
- Migration may make data collection more complex
- Strategy and intervention-do not construct wadi or dam in this place and indigenous knowledge is key
- Type of soft data that links to memories for migration or staying and trying to use past data and this is difficult to find and is not collected routinely
- Local populations might not have knowledge of the local hazard risk: they might include a majority of recent immigrants; return period of events might be long; there might be a large number of tourists.

Proposal to GADRI

- Promote the flow of data between all parties including academics and to local communities as well as to government;
- GADRI could provide a platform for sharing information for the alliancing context
- Need to link/partner/coordinate with other database initiatives (DesignSafe, ISC CODATA and other networks)
- Agreement on classifications and taxonomies
- Agreement on principles and standards for data loss collection, recording, and reporting:
- Need for comparability of disaster loss data:
- Agreement on data exchange and standards databases:

References

GADRI. http://gadri.net/

IRDR. https://www.irdrinternational.org/

Nations Synthesis Framework. https://www.scj.go.jp/ja/member/iinkai/kagakubosai/pdf/kagakubosai-siryo2404-4a.pdf

Report of the 4th Global Summit of Disaster Research Institutes

Role played by science and technology in disaster risk reduction: from framework planning to implementation. https://rikaindia.com/wp-content/uploads/2019/07/Science-Tech-in-DRR.pdf

Sendai Framework Agenda 2015–2030. https://www.undrr.org/publication/sendai-framework-disaster-risk-reduction-2015-2030

STAG of UNDRR. https://www.undrr.org/implementing-sendai-framework-partners-and-stakeholders/science-and-technology-action-group

Science and Technology Roadmap. https://www.preventionweb.net/files/45270_unisdrscienceandtechnologyroadmap.pdf

The Tokyo Statement 2017. chrome-extension://efaidnbmnnnibpcajpcglclefindmkaj/viewer.html?pdfurl=http%3A%2F%2Fwci.t.u-tokyo.ac.jp%2FResilienceForum2017%2Fpdf%2F01_Tokyo%2520Statement%25202017.pdf&clen=182649&chunk=true

Part IV
Selected Presentation Papers (Arranged by Country Order)

Chapter 9
Community Resilience Against Flood: The Case of the LIFE PRIMES Project

Eleonora Gioia, Alessandra Colocci, and Fausto Marincioni

Abstract Natural forces pose threats to human communities. Ongoing changes to the environment worsen these hazards, causing the associated risk to increase. Whilst the number of disasters is observed to decline worldwide, the costs of the disastrous events are increasing, showing how higher and higher human values are exposed and vulnerable to hazards. There is a growing acknowledgement of the need to enhance the resilience of local communities. At the same time, it is important to act accordingly to a common and broadly-shared strategy. International and regional organisations can be very effective in supporting local projects, providing solid knowledge and valuable funds. The European Commission manages the LIFE funding programme, aimed to support projects addressing environmental and climate challenges. It started in 1992 and until now it has supported more than 4600 projects Europe-wide. Amongst them is the PRIMES (Preventing flooding RIsks by Making resilient communitiES) project that ran from October 2015 to December 2018. The LIFE PRIMES project intended to increase the resilience of local communities against floods by strengthening risk awareness and promoting the involvement in innovative participatory tools. Ten flood-prone municipalities within three Italian Regions were involved in the activities, along with Regional Environmental Institutions, local Authorities and Academic Institutions. The activities included assessments of the project's impacts. The analysis suggests a sensible increase of local communities' resilience, thus highlighting the success of the PRIMES project and encouraging a further implementation of its methodology in other communities, to help build societies better prepared to cope with extreme events.

Keywords Resilience · Flood · Local community · EU funds · European Commission

---

E. Gioia (✉) · A. Colocci · F. Marincioni
Università Politecnica delle Marche, Ancona, Italy
e-mail: e.gioia@staff.univpm.it

H. Tatano and A. Collins (eds.), *Proceedings of the 4th Global Summit of Research Institutes for Disaster Risk Reduction*, Disaster and Risk Research: GADRI Book Series,
https://doi.org/10.1007/978-981-19-5566-2_9

9.1 Introduction

Natural forces, like floods, earthquakes and landslides, have always represented dreadful threats for humans. When they occur amidst a fragile urbanised area, such hazards may yield a disaster. Changes affecting the global climate system are now altering hazard scenarios, and thus associated disaster risk (UNISDR 2015). The future we are facing is hardly predictable (IPCC 2014), but it is a fact that human societies need to prepare to face growing risks. Recent evidences show that in the last years disasters are becoming less in number but more severe in toll (Below and Wallemacq 2018). Efforts are thus required to address the vulnerability and exposure of local communities, in order to mitigate the effects of disasters (Wallemacq et al. 2018). The United Nation International Strategy for Disaster Risk Reduction (UNISDR) promotes a paradigm shift towards recognising that risks are inherently part of societies and it is within societies that risks must be mitigated (UNISDR 2015). As a consequence, Disaster Risk Reduction (DRR) strategies should be rooted in all the human (social, economic, political, and structural) processes that are developed within an urbanised area. Along with the resistive approaches intended to oppose natural disturbances, DRR strategies are increasingly focussing on the risk perception of local communities and on the triggering factors of their willingness to mitigate the possible disastrous effects of an extreme event (Cutter et al. 2014). In other words, the reduction of disaster risk, especially that emerging with climate changes, is more and more founded on community resilience (Alexander 2013) that is defined as the ability of a group or a community to cope with external disturbances caused by social, political or environmental changes (Adger 2000). As a consequence, an urgent demand for community resilience metrics emerges, in response to the need of tackling drawbacks and benefits of the applied risk reduction measures (Attolico 2014; Kwok et al. 2016; Toseroni et al. 2016). A challenging ground for experimentation was the LIFE PRIMES (Preventing flooding RIsks by Making resilient communitiES—LIFE14CCA/IT/001280) project: it allowed the implementation of innovative community-based resilience strategies and the following assessment through a series of qualitative and quantitative indicators of achieved effectiveness in enhancing the capacity of local communities to face extreme events.

9.2 The LIFE Funding Programme and the LIFE PRIMES Project

The LIFE PRIMES project (Fig. 9.1) is one of the outcomes of the commitment of the European Commission in furthering innovative research and applied solutions to environmental changes and risk reduction. This engagement is embodied by a series of initiatives, such as the LIFE funding programme. Since 1992, four phases of the LIFE funding programme have been completed and one is still ongoing, for more than 4600 co-financed projects so far (European Commission https://ec.europa.eu/easme/en/section/life/life-history-life). The PRIMES project was approved within the funding period 2014–2020, through the sub-programme concerned with adaptation strategies to climate changes. The PRIMES project ran from October 2015 to December 2018 and received an EU co-funding of 1,085,761 €, corresponding to 46% of the total budget (LIFE PRIMES http://www.lifeprimes.eu/index.php/il-progetto/?lang=en). The partners involved in the project were: the Civil Protection Agencies of three Italian Regions, namely, Emilia-Romagna, Marche and Abruzzo; the Environmental Agency of Emilia-Romagna; the Università Politecnica delle Marche (LIFE PRIMES http://www.lifeprimes.eu/index.php/il-progetto/?lang=en).

The aim of the PRIMES project was to mitigate the impacts of floods on local communities through the improvement of shared early-warning systems and through

Fig. 9.1 Frontpage of the informative brochure of the LIFE PRIMES project (www.lifeprimes.eu)

Fig. 9.2 Amongst the Italian Regions, those involved by the PRIMES project are highlighted in green: from North to South, Emilia-Romagna, Marche, Abruzzo. The ten pilot areas included in the project are identified by blue pins

the direct engagement of local populations. As a consequence, a total of ten municipalities were selected within the above-mentioned three Regions, on the basis of their vulnerability and exposure to flood risk (Fig. 9.2).

A multifaceted set of initiatives was dedicated to local stakeholders; activities ranged, *inter-alia*, from illustrative workshops, to civil protection drills, to a dedicated theatrical reading or to the so-called Civic Adapt Action Plan (CAAP). The latter is a web-based questionnaire that first assesses the risk awareness of the users and then allows them to build their own flood adaptation plan that is a list of actions to undertake in order to reduce disaster risk.

Amongst the ten pilot areas of the PRIMES project, this study focusses on the Municipality of Senigallia, located in the Northern area of the Marche Region (Fig. 9.3). River flooding is the main hazard for this Municipality, caused by the Misa River that crosses the City of Senigallia down to its estuary in the Adriatic Sea. However, as a consequence of laying in front of the Adriatic Sea, Senigallia may be affected also by coastal floods. Actually, the Municipality has suffered several river and coastal floods, for example on 3rd May 2014, when a flood from the Misa River inundated large portions of the Municipality, in particular the city centre and the neighbourhoods in the upstream section of the river (Centro Funzionale Regionale 2014), leaving behind three victims along with severe damages (Consiglio Nazionale delle Ricerche-Istituto di Ricerca per la Protezione Idrogeologica 2015).

Fig. 9.3 The Municipality of Senigallia, located in the Northern area of the Marche Region (as shown in the low-left corner), crossed by the Misa River down to its estuary in the Adriatic Sea (basemap copyright of OpenStreetMap)

9.3 Methodology

Amongst the other tasks, the Università Politecnica delle Marche was charged with the assessment of the variation in flood risk resilience of local communities due to the PRIMES project. Incremental or detrimental changes would suggest either positive or negative effects of the project activities in fostering community ability to cope with external disturbances.

The most common categories of community resilience identified in the literature (Cox and Hamlen 2015; Khalili et al. 2015; Kotzee and Reyers 2016; Mayunga 2007; Saja et al. 2018; Sharifi 2016) were summed in three criteria and adopted within this study (Fig. 9.4). They are:

- *social capital*, defined by the social ties within the community (*bonding*), the networking abilities amongst people from different groups (*bridging*), and the interaction between community and authorities (*linking*)
- *social values*, encompassing the level of community engagement, the community efficacy, the community shared goals and priorities, the community joint actions, and the community competences towards flood risk resilience
- *social culture*, reflecting the local cultural and religious beliefs or norms.

Fig. 9.4 The criteria adopted to describe community resilience

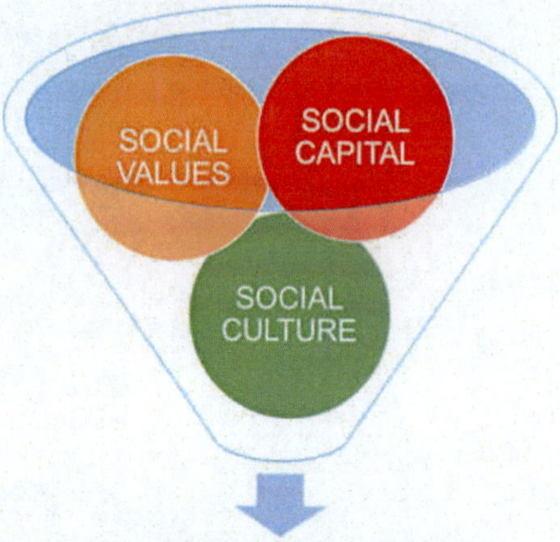

Each criterion was further described by specific indicators and sub-indicators, in a hierarchical framework. Overall, 10 community resilience indicators and 28 sub-indicators were considered, as represented in Table 9.1.

Since the proposed sub-indicators cannot be directly observed, related proxies were identified and measured through questionnaires. Such questionnaires were delivered to adult citizens and included 86 questions. The answers encompassed different forms: psychometric scale; multiple choice, including the possibility to add personalised options; open-ended.

The sampling method chosen was the probabilistic stratified sampling method. According to this technique, the studied population is divided into smaller groups (*strata*) based on the accordance to a fixed condition (e.g. in this case, residing in a flood-prone area); afterwards, the sample is randomly selected from the group of interest. The questionnaires were delivered before (*ex-ante*) and after (*ex-post*) the activities of PRIMES to reveal the variation in community resilience due to the project effectiveness.

After the collection of the questionnaires, each proxy was normalised, in order to quantify the qualitative information that was gathered, thus allowing a comparison between indicators and criteria. According to the normalisation procedure, each answer was given a weighting factor (W), depending on the importance attributed to such response in terms of resilience. The greater the number of respondents (R) who selected the best answer (highest W) the greater the value of the normalised score (NS). Therefore, the normalised score was obtained as the weighted sum of the percentage of respondents (1):

$$\mathrm{NS} = \sum (W \cdot R) \tag{9.1}$$

At the end of the procedure, each proxy was attributed a NS: an example in provided in Table 9.2.

Table 9.1 Collection of the criteria of community resilience adopted, with their indicators and sub-indicators

Criteria	Indicators	Sub-indicators
Social capital	Social bonding	Perceived support Received support Figures perceived pivotal
	Social bridging	Involvement in social networks Use of social media Volunteerism
	Social linking	Trust towards authorities Trust towards the community Acknowledgement of community leader
Social values	Community engagement	Political engagement Strategy for enhanced community engagement Engagement in public affairs
	Community efficacy	Collective efficacy Objectives and priorities of the community Perceived efficacy of the community
	Community shared values	Sense of belonging to the community Common values Adaptive capacity
	Community joint actions	Community disaster plans Willingness to collaborate Community-based problem solving and decision making
	Community competence	Knowledge and awareness of local risks Previous experience with disasters Information and communication processes about floods
Social culture	Local cultural beliefs/norms	Existing cultural and behavioural norms Cultural and historical heritage
	Local religious beliefs/norms	Existing religious beliefs and practices Engagement in religious principles

Table 9.2 Example of the procedure to attribute a normalised score (NS) to each question used as a proxy for the sub-indicators

Question/proxy	Type of answer	Answer	*W*	*R*	NS
Who is responsible for the management of flood emergencies in your city?	Open-ended	I don't know/No answer	0.1	39.71	27.35
		Other	0.2	1.47	
		Civil protection	0.3	4.41	
		Mayor/municipality	0.4	54.41	
Total			1	100.00	

The type of answer for the question is specified and the possible answers made explicit; a weighting coefficient (W) and the related percentage of respondents (R) were identified; their weighted sum, calculated as in (1), became the NS of the proxy

Then, mean values were calculated for each NS associated to a specific indicator, in order to quantify the indicator. Then, further means were derived from indicators to quantify each criterion. Eventually, the mean values of the three main criteria became the community resilience score.

9.4 Results and Discussion

Within the Municipality of Senigallia, a total of 138 questionnaires were collected, 68 *ex-ante* and 70 *ex-post* the PRIMES project. Table 9.3 reports the scores calculated for each indicator, criteria and eventually for the community resilience, and their variation possibly due to the activities proposed by the project. Indicators were not considered if the number of respondents was not enough significant for the analysis.

Overall, the community resilience of the Municipality of Senigallia appears to be stable and not much influenced by LIFE PRIMES. However, the criteria *social values* and *social culture* seem to have both benefitted from the project activities, since their normalised scores show a significant increment (+1.02 and +1.15, respectively). In particular, the "*community engagement*" and "*community competence*" achieved the highest increase rate (+5.85 and +2.47, respectively), along with the "*community efficacy*" indicator (+1.86), whilst being also the worst performing indicators in the *ex-ante* analysis (for the *social values* criterion). This result is particularly encouraging, because the PRIMES project was specifically focussed on reinforcing the tenets agreed upon by the local community. In this perspective, such positive variations

Table 9.3 Normalised scores for the adopted indicators, criteria and for the community resilience

Criteria and indicators	*Ex-ante*	*Ex-post*	Δ
Social capital	30.37	27.56	−2.81
Social bonding	31.49	32.64	+1.15
Social bridging	29.04	19.61	−9.43
Social linking	30.57	30.43	−0.14
Social values	30.88	31.90	+1.02
Community engagement	28.18	34.03	+5.85
Community efficacy	28.60	30.46	+1.86
Community shared values	36.02	31.81	−4.21
Community joint actions	31.77	30.91	+0.86
Community competence	29.82	32.29	+2.47
Social culture	44.02	45.17	+1.15
Local cultural beliefs/norms	44.02	45.17	+1.15
Local religious beliefs/norms	n.a.	n.a.	n.a.
Community Resilience	35.09	34.88	−0.21

Values are reported for *ex-ante* and *ex-post* results, and then for their difference (Δ)

suggest that the activities carried on during the project were effective in strengthening the community commitment and capacity of coping with external disturbances, confirmed by the enhanced awareness of the role of the community when facing a flood. Similarly, the normalised score for local cultural beliefs and norms supports the effectiveness of the LIFE PRIMES project, even though the analysis for the related criterion was affected by the low number of respondents to the proxies connected to the local religious beliefs and norms. Unfortunately, the criterion *social capital* showed a decrease (−2.81), for the most part caused by a lower score achieved by volunteerism-related answers in the *ex-post* questionnaires. However, the "*social bonding*" indicator still shows an increase (+1.15), mostly due to the risen awareness on the legitimate pivotal figure during a flood emergency.

9.5 Conclusion

This study was focussed on the case of the PRIMES (Preventing flooding RIsks by making resilient communitiES) project. The project unfolded through three years of activities (from 2015 to 2018) and it was partly sustained by the LIFE funding programme that is led by the European Commission. The initiatives involved ten flood-prone pilot municipalities chosen within three Italian Regions, with the aim to foster community resilience against floods through the direct engagement of local populations. Within the other partners, the Università Politecnica delle Marche was charged with the assessment of community resilience. In this paper, the analysis for the Municipality of Senigallia is presented. Questionnaires were delivered to adult citizens before (*ex-ante*) and after (*ex-post*) the activities of the project. Questions were used as quantifiable proxies for selected sub-indicators that were then grouped in ten indicators and further in three main criteria (*social capital*, *social values*, *and social culture*) that described community resilience. Even though community resilience does not show a significant variation, the scores achieved by indicators and criteria shed some light on the effects of the PRIMES project. The *social values* and *social culture* criteria were the most positively affected by the activities carried on, suggesting an effective enhancement of the shared knowledge about floods and of the awareness of the community towards their role when facing a flood emergency, thus their increased willingness to provide a personal contribution. Although the *social capital* criterion showed a negative variation, the increment of one of its indicators ("*social bonding*") is associated with a clearer identification of the pivotal figure in case of needed help during a flood. Consequently, the LIFE PRIMES project seems to have brought to a sounder capacity of the local community in coping with flooding events, seemingly fulfilling its original purpose. The results also hint some areas that may need additional support to develop, e.g. engagement in the activities of the community should be encouraged ("*social bridging*" indicator), as well as personal actions to undertake in case of emergency should be better pinpointed ("*community shared values*" indicator). Nevertheless, the analysed outcomes of the project strongly reassure on its effectiveness, thus promoting further implementations of the proposed initiatives in other communities, in order to significantly increase their resilience and thus reduce disaster risk.

Acknowledgements The PRIMES project (LIFE14 CCA/IT/001280) has received funding from the LIFE programme of the European Union.

References

Adger WN (2000) Social and ecological resilience: are they related? Prog Hum Geogr 24(3):347–364. https://doi.org/10.1191/030913200701540465

Alexander DE (2013) Resilience and disaster risk reduction: an etymological journey. Nat Hazard Earth Sys 13(11):2707–2716. https://doi.org/10.5194/nhess-13-2707-2013

Attolico A (2014) Building resilience through territorial planning: the experience of province of Potenza. Procedia Econ Finan 18:528–535. https://doi.org/10.1016/S2212-5671(14)00972-1

Below R, Wallemacq P (2018) Natural disasters. Lower mortality, higher cost. https://cred.be/sites/default/files/adsr_2017.pdf

Centro Funzionale Regionale (2014) Rapporto di evento 2–4 maggio 2014. http://www.regione.marche.it/Regione-Utile/Protezione-Civile/Progetti-e-Pubblicazioni/Progetti-Europei-Copy

Consiglio Nazionale delle Ricerche-Istituto di Ricerca per la Protezione Idrogeologica (2015) Rapporto Periodico sul Rischio posto alla Popolazione italiana da Frane e Inondazioni. Anno 2014. http://polaris.irpi.cnr.it/wp-content/uploads/Report-annuale-2014.pdf

Cutter SL, Ash KD, Emrich CT (2014) The geographies of community disaster resilience. Glob Environ Chang 29:65–77. https://doi.org/10.1016/j.gloenvcha.2014.08.005

Cox RS, Hamlen M (2015) Community disaster resilience and the rural resilience index. Am Behav Sci 59(2):220–237. https://doi.org/10.1177/0002764214550297

European Commission. https://ec.europa.eu/easme/en/section/life/life-history-life

IPCC (2014) Summary for Policymakers. In: Edenhofer O, Pichs-Madruga R, Sokona Y, Farahani E, Kadner S, Seyboth K, Adler A, Baum I, Brunner S, Eickemeier P, Kriemann B, Savolainen J, Schlömer S, von Stechow C, Zwickel T, Minx C (eds) Climate change 2014: mitigation of climate change. Contribution of working group III to the fifth assessment report of the intergovernmental panel on climate change. Cambridge University Press, Cambridge, United Kingdom and New York, NY, USA, pp 1–33. https://doi.org/10.1017/CBO9781107415324

Khalili S, Harre M, Morley P (2015) A temporal framework of social resilience indicators of communities to flood, case studies: Wagga wagga and Kempsey, NSW, Australia. Int J Disaster Risk Reduct 13:248–254. https://doi.org/10.1016/j.ijdrr.2015.06.009

Kotzee I, Reyers B (2016) Piloting a social-ecological index for measuring flood resilience: a composite index approach. Ecol Indic 60:45–53. https://doi.org/10.1016/j.ecolind.2015.06.018

Kwok AH, Doyle EEH, Becker J, Johnston D, Paton D (2016) What is 'social resilience'? Perspective of disaster researchers, emergency management practitioners, and policymakers in New Zealand. Int J Disaster Risk Reduct 19:197–211. https://doi.org/10.1016/j.ijdrr.2016.08.013

LIFE PRIMES. http://www.lifeprimes.eu/index.php/il-progetto/?lang=en

Mayunga JS (2007) Understanding and applying the concept of community disaster resilience: a capital-based approach. In: Summer academy of social vulnerability and resilience building, vol 1(1), pp 1–16

Saja AAM, Teo M, Goonetilleke A, Ziyath AM (2018) An inclusive and adaptive framework for measuring social resilience to disasters. Int J Disaster Risk Reduct 28:862–873. https://doi.org/10.1016/j.ijdrr.2018.02.004

Sharifi A (2016) A critical review of selected tools for assessing community resilience. Ecol Indic 69:629–647. https://doi.org/10.1016/j.ecolind.2016.05.023

Toseroni F, Romagnoli F, Marincioni F (2016) Adapting and reacting to measure an extreme event: a methodology to measure disaster community resilience. Energ Procedia 95:491–498. https://doi.org/10.1016/j.egypro.2016.09.074

UNISDR (2015) Making development sustainable: the future of disaster risk management, pp 1–316. https://gar.unisdr.org/reports

Wallemacq P, Below R, McLean D (2018) Economic losses, poverty and disasters (1998–2017). https://www.cred.be/unisdr-and-cred-report-economic-losses-poverty-disasters-1998-2017

Chapter 10
Yonmenkaigi System Method as an Educational Framework for Climate Change Within the United Kingdom: A Pilot Study

Mark Ashley Parry

Abstract Climate change can be viewed as probably one of the greatest environmental issues that society is currently facing. It is already being argued that society is already feeling the effects of climate change, with an increase of 1 °C being experienced worldwide. Climate change impacts within the United Kingdom are expected to be extensive, with an increased probability of droughts, heatwaves and flooding events by the end of the century. Despite these potential effects, there is at the moment only a limited education on the subject of climate change mitigation and adaption strategies within the United Kingdom's education curriculum. One potential option to overcome this issue is an engagement framework that can be used within schools and colleges, such as a Yonmenkaigi System Method [YSM]. Therefore, a Yonmenkaigi System Method was undertaken with sixteen first year Northumbria University Environmental Management students on the 4th October 2018; a questionnaire before and after was given to monitor the students perception of climate change throughout the engagement exercise. It was found that there was a significant change ($p < 0.10$) in the way a number of questions were answered within the questionnaire. This demonstrates that the Yonmenkaigi System Method has a potential to be used as an educational approach for educating students within the United Kingdom about environmental issues, such as climate change.

Keywords Climate change engagement · Risk reduction education · United Kingdom · Yonmenkaigi system method

M. A. Parry (✉)
Northumbria University, Newcastle upon Tyne, England
e-mail: ashley2.parry@northumbria.ac.uk

H. Tatano and A. Collins (eds.), *Proceedings of the 4th Global Summit of Research Institutes for Disaster Risk Reduction*, Disaster and Risk Research: GADRI Book Series,
https://doi.org/10.1007/978-981-19-5566-2_10

10.1 Introduction

10.1.1 Background

Climate change can be argued as the greatest environmental and social issues that the world is facing during the twenty-first century. Global temperatures have risen by 1 °C since the start of the industrial revolution, with the rate of temperature change increasing (IPCC 2018); with 2015 been the first year to be more than 1 °C (Hawkens et al. 2017). In 2015, world leaders agreed to try and limit global temperature increases to 1.5°C (UNFCCC 2015), in order to avoid the worst of climate change. To meet this target, unparalleled action will have to be undertaken at the international level to reduce greenhouse gas emissions (IPCC 2018). However, it is estimated by the Intergovernmental Panel on Climate Change in 2018 that global temperatures were likely to reach this level by between 2030 and 2052 (IPCC 2018) depending on future greenhouse gas emissions. This is further highlighted within recent studies that estimate that there is only 1% chance that global temperatures can be limited to 1.5 °C by 2100; and only a 5% chance to limit global warming to a 2 °C (Raftery et al. 2017).

The main cause of this temperature increase since the industrial revolution is due to the increasing levels of greenhouse gas emissions, especially carbon dioxide and methane. As a consequence, the United Kingdom agreed in 2008 to reduce its greenhouse gas emissions by 80% by 2050 based on 1990 levels (DEFRA 2008) under the 'Climate Change Act 2008', thus making them the first national government to set targets to reduce greenhouse gas emissions. However, in June 2019, the government announced an amendment to the bill, which increases the reduction of greenhouse gas emissions target to 100% by 2050 (Walker et al. 2019). It should be noted that, by 2016, the United Kingdom has already reduced its greenhouse gas emissions by 41% (BEIS 2018).

10.1.1.1 Effects of Climate Change Within the United Kingdom

It is well known that climate change will have impacts, not just within the United Kingdom, but everywhere around the world. But, it should be highlighted that some of the worst effects that will impact the United Kingdom are not going to be felt fssor many decades to come.

It is difficult to actually predict the exact effects of climate change, as these will depend on the increase in global temperatures, which in turn depends on the quantity of greenhouse gases that are emitted in the forthcoming years. However, research highlights some general trends that are expected to happen in the future. This includes the suggestion that the average temperature within the South of England will be 4 °C warmer in the summer and 2.5 °C in the North by the 2080s (Kendon et al. 2018). In addition to this, it is expected that the frequency and magnitude of heatwaves are to increase within the United Kingdom due to climate change (Green et al. 2016), with projections that the heatwave that was experienced within the United Kingdom in 2003[1] could occur every other summer by the 2040s (Christidis et al. 2015).

It should be highlighted, that the United Kingdom is already starting to experience the effects of climate change. This was highlighted by the average temperature in the decade period between 2008 and 2017 that was 0.8 °C warmer than the average temperature between 1961 and 1990 and was 0.3 °C warmer than the 1981 and 2010 period (Kendon et al. 2018), with 2017 being the fifth warmest year within the United Kingdom since records began (Lowe et al. 2018). In addition, nine of the ten warmest years have occurred since 2002 (Kendon et al. 2018), with the other being in 1998 during an exceptional El Niño.

It should be noted that droughts within the United Kingdom are not that uncommon and tend to happen capriciously (Dessai and Sims, 2010), with south east England most likely to be effected (Marsh et al. 2007). Since 1900, there has been at least 11 periods of drought within the United Kingdom, with these occurring in 1890–1910, 1921–1922, 1933–1934, 1959, 1975–1976, 1988–1992, 1995–1997, 2004–2006, 2010–2012, 2018-2019 and 2022 (Perry 1976; Doornkamp 1980; Marsh et al. 2007; Kendon et al. 2013; Turner et al. 2021. Despite this, there is concern amongst some researchers about the impacts that droughts might have on the United Kingdom in the future due to climate change. Numerous studies have projected that due to climate change, droughts are more likely to increase in frequency, duration and severity (Fowler and Kilsby 2004; Vidal and Wade 2009); it is expected the type of drought experienced in 1976 is likely to change from being a 1 in 100 year occurrence to a 1 in 10 year occurrence (Burke et al. 2010).

Another major impact that climate change will have on the British climate is flooding. Due to its geography, flooding within the United Kingdom is not a new occurrence. Historical examples are the St. Lucia's Flood in 1287, Great Storm of 1703, Bristol Channel Floods of 1607, Eyemouth Disaster in 1881 and the 1928 Thames Flood. However, rainfall patterns within the United Kingdom are expected to change, with an overall increase in precipitation, higher intensity of rainfall and a potential for rainfall to become more seasonal (Maraun et al. 2008; Osborn and Maraun 2008). These changes of rainfall patterns will increase the likelihood of the frequency and intensity of flooding occurring within the United Kingdom.

The most recent significant flooding event within the United Kingdom was in 2007. During this event, the United Kingdom suffered its worst flooding in recorded history. The Environment Agency (2007) report on why this occurred, the flooding was blamed on short but intensive rainfall that kept repeating every few days, and which accumulated. For example, parts of the North Yorkshire Moors received four times the normal rainfall for June, and parts of Worcestershire received six inches of rainfall in just two days (Environment Agency 2007). During the floods, the situation was described by the Environment Agency as "critical" (BBC 2007).

---

[1] Temperatures peaked 38.5 °C, which was still the highest temperature recorded within the United Kingdom until it was surpassed in 2019 and 2022, and resulted in the death of over 70,000 people in Western Europe (Robine et al., 2008).

What this section demonstrates is that the climate within the United Kingdom already suffers from extremes, but it is expected that climate is going to become more extreme within the future, with increases in temperatures, droughts and flooding. More detailed reports about the effects that climate change will have in future within the United Kingdom can be founded within the United Kingdom Climate Projects 18 (Lowe et al. 2018). This raises the related important point as to how students should be taught about expected climate change and how best to both mitigate against and adapt to future climate change.

10.1.2 Climate Change Education

At a global level, it has long been recognised that the education of the public about climate change is important to reduce the effects and the impacts on individuals' lives. This was demonstrated in 1992, when all members of the United Nations, including the United Kingdom, signed the UNFCC agreement. Within this agreement, Article 6 is devoted to "Education, Training and Public Awareness" of climate change (United Nations 1992). This is further backed up in Article 10e of the Kyoto Protocol, which states:

> Cooperate in and promote at the international level, and, where appropriate, using existing bodies, the development and **implementation of education and training programmes**, including the strengthening of national capacity building, in particular human and institutional capacities and the exchange or second of personnel to train experts in this field, in particular for developing countries, and facilitate at the national level public awareness of, and public access to information on, climate change
>
> United Nations (1998, p.10)

As demonstrated within the Kyoto Protocol, the United Nations highlights the need for education and training programmes towards the public in order to help develop their understanding of the climate change issue, which builds the national capacity. Therefore, this demonstrates that the UN believes that education has a critical part to play in the development of a climate change narrative. This belief is supported by social scientists (Cherry 2011; Sharma 2012). One study by Cherry (2011) highlighted that young people have a tendency from their studies to their parents, thus increasing the transfer of knowledge. Similar results have been found by Mochizuki and Bryan (2015).

As report by Grantham Research Institute on Climate Change and the Environment highlights that

> there can be no justification for omitting climate change from the National Curriculum, and the education of pupils would be deficient if they did not receive teaching about it … if core climate change teaching is not included as compulsory learning … there is a risk that some students would not acquire essential basic knowledge about climate change. As the UK Youth Climate Coalition points out, "climate change is too important to be left to individual teacher choice
>
> Hicks et al. (2013, p.3-5)

10.1.2.1 Climate Change Education Within the United Kingdom

Within the United Kingdom, some form of formal education is compulsory until the age of 18. However, from the age 14, the students have the ability to drop some subjects that they are not interested in, one of which is geography, which is where most of the climate change education occurs within the British education curriculum (George 2017). In the school year 2015–16, 41% of 15 and 16 years undertook General Certificate of Secondary Education [GCSE] Geography (Carroll and Gill 2017). This demonstrates that the majority of students are not receiving extensive knowledge about climate change. However, it should be noted that elements of climate change education are still available post-14, be it assiduous to carbon dioxide and methane increases and the uncertainties of the impact this has on climate change (DfE 2014a; George 2017). Whilst these are is the most concentrated and potent greenhouse gases in the atmosphere, respectively, there are not the only greenhouse gases that contribute to climate change. But, the basic education neglects to go into the details about these other greenhouse gases.

It should be noted, that the education of climate change is important in trying to combatting climate change, as it allows students and other young people to become informed about the sources, impacts and different strategies to both mitigate and adapt, with the hope that they start to adopt more environmental friendly behaviour (UNESCO No Date).

A number of countries and regions globally, such as Brazil, China, Denmark, Singapore, South Africa, South Korea and Vietnam, have adopted climate change in the national curriculum in recent years (Han 2015; Læssøe and Mochizuki 2015; Sung 2015; Trajber and Mochizuki 2015; Chang and Pascua 2017). However, despite this progress, a number of countries, such as New Zealand, are still teaching limited amount of climate change, if any at all (Eames 2017).

In the United Kingdom, climate change first appeared on the national curriculum in 1995 and has featured regularly since despite the rumours in the majority of major national newspapers that it was under threat of been taken off the curriculum in 2013 (DfE 2013a; Coughlan 2017).

Within the current geography curriculum, climate change is taught at different levels. Within Key Stage 3,[2] the national curriculum states that students should be taught about key processes within physical geography, such as "weather and climate, including the change in climate from the Ice Age to the present; and glaciation" and "understand how human and physical processes interact to influence, and change landscapes, environments and the climate; and how human activity relies on effective functioning of natural systems" (DfE 2013b, p.2). This demonstrates that students should be given the basic fundamentals of climate change as they start secondary school within England. Within the Key Stage 4[3] curriculum, whilst there is education of climate change, it is very limited and is very similar to the Key Stage 3 with only one brief mention, which states:

[2] Key Stage 3 within the United Kingdom is usually for children between the ages of 11 and 14.

[3] Key Stage 4 within the United Kingdom is usually for children between the ages of 14 and 16.

> Changing weather and climate – The causes, consequences of and responses to extreme weather conditions and natural weather hazards, recognising their **changing distribution in time and space** and drawing on an understanding of the global circulation of the atmosphere. The **spatial and temporal characteristics, of climatic change and evidence for different causes, including human activity**, from the beginning of the Quaternary period (2.6 million years ago) to the present day.
>
> DfE (2014b, p.6)

This demonstrates that the amount of climate change education within the national curriculum is very limited. Further, the inclusion of climate change on the national curriculum within the United Kingdom has not been without controversy. During the mid-2000s, the United Kingdom government started to increase the amount of climate change included within the national curriculum. This is demonstrated when the government announced that the academic year 2006/07 would be the "Sustainable Schools Year of Action," with the underlying aims to promote sustainable development and other environmental concerns, for example climate change (DfES 2006). As a consequence, this lead to all schools within England, Scotland and Wales to be given a free copy of Al Gore's "An Inconvenient Truth." In addition, the government announced guidance notes and resources about how climate change fits within the national curriculum alongside the Al Gore documentary (WalesOnline 2007). However, in May 2007, a court case was brought by Stewart Dimmock, who took legal action against the Secretary of State for Education and Skills, in an effort to stop Al Gore's "An Inconvenience Truth" being shown in schools. It should be noted that Stewart Dimmock had "backing" from Viscount Monckton (Leake 2007), who is a prominent climate change sceptic. The court ruled that the film "is substantially founded upon scientific research and fact, albeit that the science is used, in the hands of a talented politician and communicator, to make a political statement and to support a political programme." (Dimmock vs. Secretary of State for Education and Skills 2007, Online). Therefore, he ruled that an amendment is needed to the notes to highlight that there are alternative points of views and that teachers need to highlight the nine mistakes within the film.

Currently, the majority of this teaching is done by conventional methods, such as movie and students listening to a teacher. The problems of these methods are they are prone to student losing attention thought the lesson, which leads to the question as to how much of the information they obtain, let alone personalise and act upon?

10.1.3 Attention Rates of Students

There is a growing body of research that student's attention within classes declines throughout a lesson (Davis and Buskist 2002; Bradbury 2016). However, there are slightly conflicting results as to when and the length of time it takes for students to lose concentration. Sousa (2006) suggests that the unmotivated students will start to lose focus within the first ten to twenty minutes of a lesson. This finding is backed up by similar observations of Davies (1993), Benjamin (2002) and Wankat (2002).

However, other research has found that that there are cycles in concentration of students' lapse (Johnstone and Percival 1976; Bunce et al. 2010). For example, Johnstone and Percival (1976) state that laps within student's lapse of attention tend to last between two and four minutes, as student settle into the lecture, and then between ten and eighteen later. In addition, they found that by the end of lesson, the average lapses of attention by the students were occurring every three to four minutes.

Despite these differences in opinion amongst researchers, one thing that can be agreed upon by nearly all researchers is that the attention rate in the class declines throughout the class, with attention lapse becoming more frequent and longitudinal. Therefore, methods are needed to try and keep the student's attention rate throughout the time period.

One suggestion by Sousa (2006) is that classes should be split up into segments of between 15 and 20 min, as a method of keeping the students attention throughout the lesson. Other suggestions are that alternative features should be used within lessons to keep the attention of the students (Bunce et al. 2010).

10.1.4 Aims of the Study

The main aim of the study is to determine whether an engagement framework, such as Yonmenkaigi System Method (YSM), can work as an educational tool in relation to climate change to increase more understanding and awareness of climate change.

10.2 Methodology

10.2.1 A Brief Introduction to the Yonmenkaigi System Method

The participatory method that is been used within this study is called YSM, which was originally developed in the mountainous region of Chozi, Tottori Prefecture in the mid/late 1980s (Okada 2018) and become more extensively used after the 1995 Great Hanshin Earthquake that had a significant impact on the southern part of the island of Honshū in Japan (Na et al. 2009). The relevance of this approach to this research was that the YSM had developed as an action with local communities through interactive workshops (Na et al. 2009) and offered particular consideration of engagement processes. It is recognised that there are a variety of participatory methods available that claim engagement, though this one has been used particularly in disaster risk settings. Samaddar et al. (2015) describe how the YSM framework allows stakeholders to debate an issue through face-to-face communications during which they learn about other people's point of view on these issues. A normal YSM workshop usual has between eight and sixteen stakeholders and the facilitator (Na et al. 2009). When using an YSM, there are usually five main steps that need to be followed. However, during the pilot study, there were two additional steps, with the inclusion of a second questionnaire. This was done for a before and after comparison of the participants. Each of the seven steps is highlighted within Fig. 10.1.

The first and final steps within the pilot YSM is a questionnaire. The aim of the first questionnaire is to see how stakeholders perceived climate change before the YSM process. The aim of the final questionnaire is to see how the students perceived climate change after completing the YSM process and to determine how as a group their perception changed during the pilot study. The first questionnaire is made up of ten questions, whilst the second is made of just nine.

The next step within the YSM is the SWOT analysis. Na et al. (2009) highlight that SWOT analysis is an important element within YSM, as it helps the participants to see the present and future risks to their community and therefore helps them recognise that action is needed to reduce these risks. Within this pilot study, it is used to provide the students the opportunity to share their view on the chosen issue, in this case climate change. The SWOT analysis within the YSM consists of strengths (*S*) and weaknesses (*W*), which are classed as internal factors.

Within this pilot study, the SW allowed the students the opportunity to review the community and government's strengths and weaknesses in the approaches to both mitigating and adapting to climate change.

The SWOT analysis also consists of opportunities (*O*) and threats (*T*), and these are considered as external factors.

Within this pilot study, the OT allowed the students the opportunity to review the opportunities that climate change will have and the threats that it could cause on the population of Newcastle-upon-Tyne in the medium to long term future.

After completing the SWOT analysis, the students were asked to identify main themes that are important to reduce the impact and effect of climate change within Newcastle-upon-Tyne, the student came up with nine main themes. From this, the students were asked to narrow this down to four main themes, which were as follows:

1. Communication and Engagement [Green].
2. Mitigation [Blue].
3. Resources and Logistics [Yellow].
4. Research and Data [Red].

The students were then divided into four groups of four each group having and allocated role, and then asked to choose which theme they wanted to debate and investigate further. After this, each of the students was given colour cards which represented the theme that they were working with, as demonstrated in the list previously. Then, students were given a Yonmenkaigi Chart, as demonstrated in Fig. 10.2;

Fig. 10.1 Yonmenkaigi system method's seven steps (author)

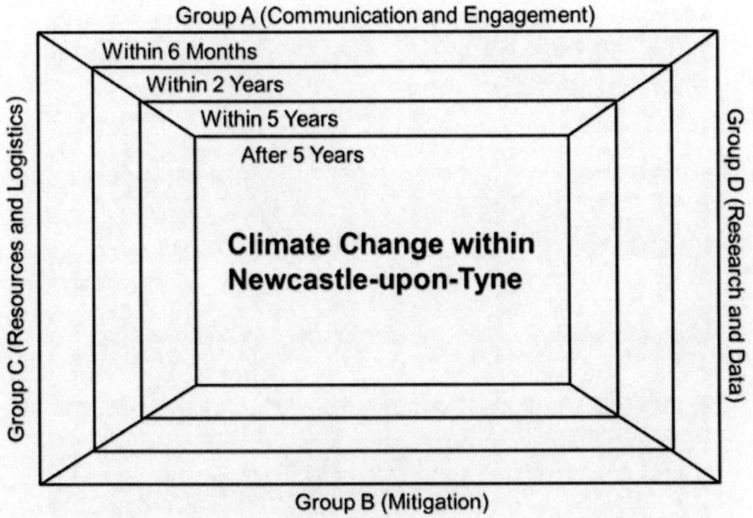

Fig. 10.2 Application of Yonmenkaigi Chart (author)

the students were asked to identify the appropriate timescales for the theme. During an YSM, differing timescales are needed, depending on the scale of the problem that participants think the issue concerns. This was demonstrated by Samaddar et al. (2015) who allocated a timescales of within 3 months, 6 months, 1 year and after 1 year; whereas, Na and Okada (2011) allocated a timescales of within 3 months, 6 months, 1 year and 2 years. Within this pilot study, the students choose timescales of within 6 months, 2 years, 5 years and after 5 years.

Once the roles and timescales are allocated, the students were then given the opportunity to suggest actions that could be undertaken in accordance with the roles that they were allocated and within the timescale on the Yonmenkaigi Chart. The purpose of the Yonmenkaigi Chart is to illustrate the student's actions in which they believe they will reduce either the effect or impact of climate change which can be undertaken at the local scale in Newcastle-upon-Tyne.

The next stage of the YSM process, as demonstrated in Fig. 10.1, is debating. The purpose of the dating is to provide a platform to process all the ideas by each group, which allows the ability of the students to further develop a potential combining of ideas together (Samaddar et al. 2015). Within the YSM process, two types of debating are available to the group; these are general debating and inverse debating (Na et al. 2009). For general debating, two groups debate with each other to defend their own ideas (Samaddar et al. 2015). By way of contract, inverse debating is when two groups swap position and start debating in relation to their new group (Na 2016). For example, if group A and B are debating, when inverse debating occurs, group A must defend the ideas of B. The purpose of the inverse debating is that it challenges their thinking about the idea, especially when it is not their own. As the students have

Fig. 10.3 Students debating during the Yonmenkaigi system method (author)

written their ideas on card, when debating, the students can move the cards around, including adding and removing cards when appropriate.

After the debates are completed, a collaborative action plan is then established by the students using the Yonmenkaigi Chart. The action plan chart is classified in accordance to the time frame and each of the four roles (Samaddar et al. 2015).

10.2.2 Sample

The study was conducted on the 11th October 2018 at Northumbria University within the North East of England. The students who participated in this study were sixteen first year undergraduate students on the environmental management course, as demonstrated in Fig. 10.3. The study was conducted on the first week of study, and therefore, the students were limited to very little to no information about climate change. As a result, the students would only have the knowledge that they were either self-taught or were taught within formal education.

10.2.3 Data Analysis

Inferential statistical tests were undertaken including use of the mean, standard deviation and chi-square test. The mean helped show what the group of students as a

Table 10.1 Levels of significance (author)

Per cent significance (%)	*p*-value	Meaning
> 99	0.000–0.015	Large significance
95	0.016–0.055	Significance
90	0.056–0.105	Slight significance
< 90	0.106–1.000	No significance

whole thought about climate change before and after the YSM. The standard deviation showed the range of responses. The chi-square test was used to test whether there was a significant difference before and after the test.

Within this paper, a significance level of 95% has been the guide (Craparo 2007), but if data had a p-value between 0.056 and 0.105 or 0.00 or 0.015, then a significance level of 90% and 99% has been acknowledged, respectively. Table 10.1 describes the meaning of the 90, 95 and 99% significance within this paper.

10.3 Results and Discussion

10.3.1 Yonmenkaigi Chart

Overall, 32 actions were brought up by the students for which a council, such as Newcastle, needed to be doing something in terms of combatting climate change. The breakdown of the number of actions for each category depending on different timeframes can observed within Table 10.2, with more detailed analysis in Table 10.3. It is demonstrated that most of actions are likely to occur within the first two years (Table 10.2).

Table 10.3 demonstrates that the majority of ideas thought up by the students were how best adapt Newcastle and its citizens to the effects of climate change. But, what it does highlight as well is that the students felt that climate change is a multi-organisation issue, with the mention of firefighters, local authorities, national government, National Health Service (NHS), police and universities. When comparing this

Table 10.2 Number of cards within the Yonmenkaigi Chart for each role

Time frame →	Within 6 months	Within 2 years	Within 5 years	After 5 years
Roles↓				
Resources and logistics	3	1	2	1
Communication and engagement	3	4	3	2
Mitigation	1	2	1	2
Research and data	5	1	1	0

Table 10.3 Yonmenkaigi Chart (author)

	Within 6 months	Within 2 years	Within 5 years	After 5 years
Resources and logistics	• More training for police in disaster response • More equipment for services (Police, firefighters, etc.) • Budgets for local councils to address green issues	• Emergency food supplies	• More pedestrianised areas in city for better air quality around densely populated areas • Prepare safe zones	• No fossil fuels on public transport
Communication and engagement	• Increase public/local awareness • Extreme weather information, e.g. snow storms and heatwaves • Inform/protect elderly community (most at risk)	• Government involvement • Online resources • Knowledge based on prediction from data • Local authorities preparedness	• Preparation, e.g. disaster plans • New layout proposals to prevent damage from floods • Make public health system aware of future changes	• Full national awareness and preparation plans for worst case • Secure/suggest funding for NHS
Mitigation	• Penalise non-green behaviours/incentives for green behaviours such as cheaper parking for electric cars	• Train people in disaster recovery in preparation for a disaster/extreme weather event • Ban vehicles from certain area, congestion charging	• General legislation against climate change, such as taxations on business in terms of emissions	• Flood/sea level barriers along the River Tyne and surround bodies of water • Pass laws on poor behaviour/environmental issues, e.g. banning plastics straws

(continued)

Table 10.3 (continued)

	Within 6 months	Within 2 years	Within 5 years	After 5 years
Research and data	• Question people on their knowledge of climate change - > NHS, police and public, etc • Checking and testing air/land and water quality • Questionnaire's for local population to see what they want and need • Research on other past successful schemes • Flora + Fauna	• Begin education people on effects of climate change	• Check monthly climate for 2–5 years to act on	

with what is being taught in schools, it demonstrates that the students are taking it step further on from the science of climate change, which is predominantly being taught and starting to think and talk about the human interactions in adapting to and mitigating the effects of climate change.

Table 10.3 also demonstrates that the students recognise that the changes that are required for climate change are not going to occur in the short term, and that organisations and society need to think more medium- and long- term scenarios to both mitigation against and adaptation to climate change.

10.3.2 *Perceived Worriedness of Different Issues*

The first question within the questionnaire was asking the participant how worried they were about a series of perceived issues. They were given a list of ten perceived issues and were asked to rate their worry out of "Not at all Worried," "Not Very Worried," "Somewhat Worried," "Very Worried" and "Extremely Worried." Each of these categories was then converted into a rating, with "Not at all Worried" being rated as 1 and "Extremely Worried" being rated as 5. The results of these changes in perceived worriedness within both questionnaire 1 and 2 are demonstrated within Table 10.4.

As demonstrated within Table 10.4, before the respondents participated within the Yonmenkaigi stated, they were most worried about climate change, with an average rating of 4.25. This compares to the second most worried issue, which was deforestation at an average worriedness rating of 3.88. This represents a worriedness difference of 0.375; this is the largest difference between any two issues that were listed. This demonstrates that the sample was already eminently worried about climate change, even before completing a Yonmenkaigi System exercise. However, it was found that there was an average increase of worry about climate change of 0.188. This increase is significant at the 90% but not at either the 95 or 99%. It should also be noted that there were a number of changes that were also observed to be significant at the 90% significance, such as economic instability and espionage. The most likely reason why espionage had the largest change, and the largest *p*-value is that most students did not know what espionage was when they filled out the first questionnaire.

It is also noted that the standard deviation for climate change is slightly reduced from 0.683 to 0.629. This demonstrates that there is a reduction of spread amongst the students; therefore, the students were becoming more homogeneous in worriedness of climate change.

10.3.3 *Perception Change on Differing Themes*

Table 10.5 demonstrates changes that the students experienced in their thinking on four differing themes. Within the third question, the students were asked to rank their

Table 10.4 Perceived worriedness of different issues before and after the YSM (author)

Perceived issue	Questionnaire 1 (before)		Questionnaire 2 (after)		Mean diff	*t* value	*p*-value	Change in response	
	Mean	Standard deviation	Mean	Standard deviation				Kurtosis[4]	Skewness
Air pollution	3.438	0.629	3.438	0.892	0.000	0.000	1.000	3.297	−1.174
Climate change	4.250	0.683	4.438	0.629	0.188	−1.861	0.083	1.285	1.772
Deforestation	3.875	0.885	4.063	0.854	0.188	−1.145	0.270	−0.373	−0.197
Diseases	2.813	0.655	2.750	0.683	−0.063	0.565	0.580	3.616	−0.392
Economic instability	3.000	0.516	3.188	0.544	0.188	−1.861	0.083	1.285	1.772
Espionage	2.125	0.806	2.500	0.894	0.375	−2.087	0.054	1.699	1.731
Genetic engineering	2.375	0.885	2.375	0.806	0.000	0.000	1.000	0.027	0.000
Migration	2.250	0.931	2.500	0.730	0.250	−1.732	0.104	−0.660	0.000
Nuclear weapons	3.563	1.031	3.313	1.014	−0.430	1.291	0.216	0.598	−0.492
Terrorism	3.375	0.885	3.063	0.854	−0.312	1.576	0.136	0.126	−0.254

[4] Standard Error of Kurtosis is 0.564.

Table 10.5 Perception change on differing themes before and after the YSM (author)

	Questionnaire 1 (before)		Questionnaire 2 (after)		Mean diff	*p*-value
	Mean	Standard deviation	Mean	Standard deviation		
Personal responsibility	7.375	1.500	7.750	1.528	0.375	0.054
Global impact	7.625	2.778	8.250	2.176	0.625	0.129
UK impact	6.813	2.287	7.125	1.962	0.313	0.208
Likely to take personal action	8.000	1.789	8.063	1.564	0.063	0.855

own personal responsibility on a Likert scale from 1, meaning "not at all," and 10, meaning "a great deal." For the four and fifth questions, the students were asked how bad or good that there think that impacts of climate change will have people globally and within the United Kingdom, respectively. This was again done on a Likert scale from 1, meaning "Very Good," to 10, meaning "Very Bad." The final question, asked the students "how likely are you to take personal action to fight climate change in the next six months?" This used a Likert scale with 1 meaning "unlikely" to 10 meaning "very likely."

It can be found that by participating in the exercise, the personal responsibility towards climate change of the students increased by 0.375. However, the standard deviation amongst the students has slightly increased by 0.028, which means there an increase spread in the responses that the student gave back. However, when referring to Table 10.1, it can be observed that a *p*-value of 0.054, when rounded up, means that there is a significant difference between the first and second questionnaire, at the 95% significance.

Table 10.5 also demonstrates the belief amongst the students that climate change is more likely to have greater impact to people outside the United Kingdom compared to those within the United Kingdom, but the UK is still likely to experience negative impacts. Research demonstrates that the students were right to believe this, with increase emigration from coastal areas and places that already are suffering from high temperatures and water shortages.

It is also observed that students increase their belief in the extent to which they think that climate change will impact people within both the United Kingdom and, globally, with the greatest difference being observed for the global impact which increase 0.625 compared to 0.313 for the United Kingdom. However, it is observed that despite this large increase in perception for both questions, it observed that neither was significant. This is likely due to a few students experiencing a large change in their perceptions, whilst the majority did not change their perception in these questions.

Lastly, a clear majority of the students stated that they were highly likely to undertake personal action in the next six months to help combat climate change. It is noted that there is very little difference (non-significant) between before and after

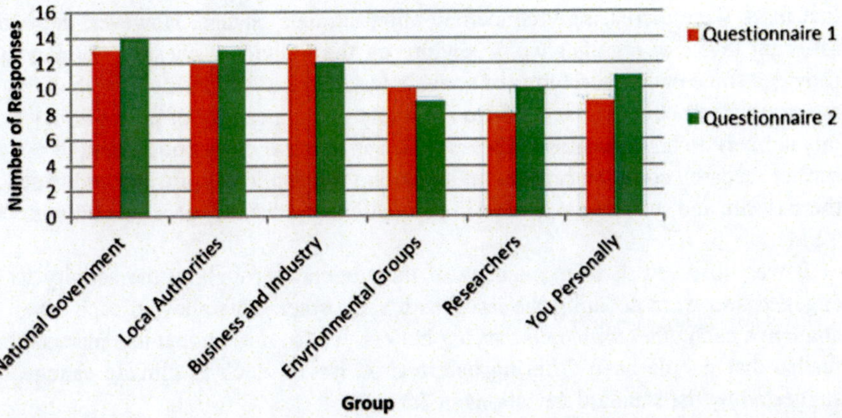

Fig. 10.4 Views of who is responsible in combatting climate change before and after the Yonmenkaigi system method. *Source* Author

the YSM. This is likely due to the already high value of 8.000 to start with, meaning little opportunity for change.

10.3.4 Views on Responsibility for Tackling Climate Change?

In question eight of the surveys, the students were asked which of six groups should be responsible for tacking climate change. The groups that were chosen within the survey were "national government," "local authorities," "business and industry," "environmental groups," "researchers" and "you personally."

As demonstrated within Fig. 10.4, it was found that number of students that felt they are have a responsibility for tacking climate change increased from 9 to 11. This represents the equal largest response increase between the two questionnaires; the other being "researchers." As a consequence, out of these six groups shown in Fig. 10.4, the "you personally" response was a greater change than responses indicating "environmental groups."

10.4 Conclusion and Limitations

10.4.1 Conclusions

As demonstrated within both Tables 10.4 and 10.5, participant's responses indicate change in their responses to questions about climate change pre and post the YSM exercise. The data demonstrate that after the questionnaire, the students show signs that there were increasingly concerned about climate change. However, there are differing levels of significance depending on the individual question. Whilst this provides some opening in terms of a viable learning and engagement methodology, the results are not being used here to definitely prove a change of perception, albeit this is likely to have occurred. The approach is a useful contribution around which further research could be carried out, to extend the application further and to see if there is true and sustainable change in perception as a result of engagement with the YSM.

It was observed that the majority of the students throughout the activity were engaged and were debating the issue with each other. This allowed each other to challenge each other with their varying beliefs. It is suggested that this makes them realise that people have differing opinions on issues, such as climate change, as suggested by the standard deviations in Table 10.4.

10.4.2 Limitations

Only, two hours were allocated to complete a full YSM during this pilot study. On reflection, this was not enough time to allow a full discussion amongst the stakeholders; going forward, a minimum of two and half to three hours is recommended.

This was one pilot study, to fully determine whether this would be successful methodology to adopt for use within pre-18 education groups within the United Kingdom, which the group using it here were post 18, albeit only just so in most cases. Further research is needed across this targeted group.

References

BBC (2007) Flood crisis grows as rivers rise. In: BBC News, 23rd July. Available at: http://news.bbc.co.uk/1/hi/uk/6911226.stm. Accessed 26 Feb 2019

BEIS (2018) Final UK greenhouse gas emissions national statistics: 1990–2016. Department for Business, Energy and Industrial Strategy, London, United Kingdom. Available at: https://www.gov.uk/government/statistics/final-uk-greenhouse-gas-emissions-national-statistics-1990-2016. Accessed 24 Oct 2018

Benjamin LT (2002) Lecturing. In: Davis SF, Buskist W (eds) The teaching of psychology: essays in honor of Wilbert J. McKeachie and Charles L. Brewer. Mahwah, New Jersey, United States: Lawrence Erlbaum Associates, Inc. pp 57–67

Bradbury NA (2016) Attention span during lectures: 8 seconds, 10 minutes, or more? Adv Physiol Educ 40(4):509–513. https://doi.org/10.1152/advan.00109.2016

Bunce DM, Flens EA, Neiles KY (2010) How long can students pay attention in class? A study of student attention decline using clickers. Chem Educ Res 87(12):1438–1443. https://doi.org/10.1021/ed100409p

Burke EJ, Perry RJH, Brown SJ (2010) An extreme value analysis of UK drought and projections of change in the future. J Hydrol 388(1–2):131–143. https://doi.org/10.1016/j.jhydrol.2010.04.035

Carroll M, Gill T (2017) Uptake of GCSE subjects 2016. Statistics Report Series No. 114. Cambridge Asssessment, Cambridge, Cambridgeshire, United Kingdom
Chang C-H, Pascua L (2017) The curriculum of climate change education: a case for Singapore. J Environ Educ 48(3):172–181. https://doi.org/10.1080/00958964.2017.1289883
Cherry L (2011) Young voices on climate change: the Paul F-Brandwein 2010 NSTA lecture. J Sci Educ Technol 20:208–213. https://doi.org/10.1007/s10956-011-9280-9
Christidis N, Jones GS, Stott PA (2015) Dramatically increasing chance of extremely hot summers since the 2003 European heatwave. Nat Clim Chang 5:46–50. https://doi.org/10.1038/nclimate2468
Coughlan S (2017) Did Michael Gove really try to stop teaching climate change?. In: BBC News, 12th June. Available at https://www.bbc.co.uk/news/education-40250214. Accessed 26 Feb 2019
Craparo RM (2007) Significance level. In: Salkind NJ (ed) Encyclopaedia of measurement and statistics. SAGE Publications, Thousand Oaks, California, United States, pp 889–891
Davies BG (1993) Tools for teaching. Jossey-Bass, San Francisco, California, United States
Davis SF, Buskist W (2002) The teaching of psychology: Essays in honor of McKeachie WJ, Brewer CL. Lawrence Erlbaum Associates Publishers, Mahwah, New Jersey, United States
DEFRA (2008) Climate change Act 2008. London, United Kingdom: Department for Environment, Food and Rural Affairs. Available at: http://www.legislation.gov.uk/ukpga/2008/27/contents. Accessed 17 Oct 2018
Dessai S and Sims C (2010) Public perception of drought and climate change in southeast England. Environ Hazards 9(4):340–357. https://doi.org/10.3763/ehaz.2010.0037
DfES (2006) Sustainable schools: for pupils, communities and the environment. Department for Education and Skills, London, United Kingdom
DfE (2013a) Climate change in the draft national curriculum. Available at: https://www.gov.uk/government/news/climate-change-in-the-draft-national-curriculum. Accessed 26 Feb 2019
DfE (2013b) Geography programmes of study: key stage 3. National Curriculum in England. Department for Education, London, United Kingdom. Available at: https://assets.publishing.service.gov.uk/government/uploads/system/uploads/attachment_data/file/239087/SECONDARY_national_curriculum_-_Geography.pdf. Accessed 26 Feb 2019
DfE (2014a) Science programmes of study: key stage 4. National curriculum in England. Department for Education, London, United Kingdom. Available at: https://assets.publishing.service.gov.uk/government/uploads/system/uploads/attachment_data/file/381380/Science_KS4_PoS_7_November_2014a.pdf. Accessed 26 Feb 2019
DfE (2014b) Geography: GCSE subject content. Department for Education, London, United Kingdom. Available at: https://assets.publishing.service.gov.uk/government/uploads/system/uploads/attachment_data/file/301253/GCSE_geography.pdf. Accessed 26 Feb 2019
Dimmock vs. Secretary of State for Education and Skills (2007) Case no. CO/3615/2007
Doornkamp JC, Gregory KJ, Burn AS (1980) Atlas of drought in Britain 1975–1976. Institute of British Geographers, London, United Kingdom
Eames C (2017) Climate change education in New Zealand. Curr Perspect 37(1):99–102. https://doi.org/10.1007/s41297-017-0017-7
Environmental Agency (2007) Review of 2007 summer floods. Environment Agency, Bristol, United Kingdom
Fowler HJ, Kilsby CG (2004) Future increases in UK water resource drought projected by a regional climate model. In: Hydrology: science and practice for the 21st Century, vol 1, pp 15–21
George M (2017) Government is failing to ensure pupils are taught real-world impact of climate change, MPs told. In: Tes, 21st December. Available at: https://www.tes.com/news/government-failing-ensure-pupils-are-taught-real-world-impact-climate-change-mps-told. Accessed 26 Feb 2019
Green HK, Andrews N, Armstrong B, Bickler G, Pebody R (2016) Mortality during the 2013 heatwave in England—How did it compare to previous heatwaves? Retrospective Observ Stud Environ Res 147:343–349. https://doi.org/10.1016/j.envres.2016.02.028

Han Q (2015) Education for sustainable development and climate change education in china: a status report. J Educ Sustain Dev 9(1):62–77. https://doi.org/10.1177/0973408215569114

Hawkens E, Ortega P, Suckling E, Schurer A, Hegerl G, Jones P, Joshi M, Osborn TJ, Masson-Delmotte V, Mignot J, Thorne P, van Oldenborgh GJ (2017) Estimating changes in global temperatures since the preindustrial period. Am Meteorol Soc 98:1841–1856. https://doi.org/10.1175%2FBAMS-D-16-0007.1

Hicks N, Ward B, Lester S (2013) Climate change in the national curriculum in England: submission to a consultation by the department for education. Grantham Research Institute on Climate Change and the Environment, London, United Kingdom

IPCC (2018) SR15 headline statements. Available at: http://report.ipcc.ch/sr15/pdf/sr15_headline_statements.pdf. Accessed 18 Oct 2018

Johnstone AH, Percival F (1976) Attention breaks in lectures. Educ Chem 13(2):49–50

Kendon M, Marsh T, Parry S (2013) The 2010–2012 drought in England and wales. Weather 68(4):88–95. https://doi.org/10.1002/wea.2101

Kendon M, McCarthy M, Jevrejeva S, Matthews A, Legg T (2018) State of the UK climate 2017. Int J Climatol 38(S2):1–35. https://doi.org/10.1002/joc.5798

Læssøe J, Mochizuki Y (2015) Recent trends in national policy on education for sustainable development and climate change education. J Educ Sustain Dev 9(1):27–43. https://doi.org/10.1177/0973408215569112

Leake J (2007) Please, sir—Gore's got warming wrong. In: The sunday times, 14th October, p 4. Available at: https://www.thetimes.co.uk/article/please-sir-gores-got-warming-wrong-swxn7kbrc9w. Accessed 7 Jan 2019

Lowe JA, Bernie D, Bett P, Bricheno L, Brown S, Calvert D, Clark R, Eagle K, Edwards T, Fosser G, Fung F, Gohar L, Good P, Gregory J, Harris G, Howard T, Kaye N, Kendon E, Krijnen J, Maisey P, McDonald R, McInnes R, McSweeney C, Mitchell JFB, Murphy J, Palmer M, Roberts C, Rostron J, Sexton D, Thornton H, Tinker J, Tucker S, Yamazaki K, Belcher S (2018) UKCP18 science overview report. Met Office, Exeter, Devon, United Kingdom

Maraun D, Osborn TJ, Gillett NP (2008) United Kingdom daily precipitation intensity: improved early data, error estimates and an update from 2000 to 2006. Int J Climatol 28:833–842. https://doi.org/10.1002/joc.1672

Marsh T, Cole G, Wilby R (2007) Major droughts in England and Wales, 1800–2016. Weather 62(4):87–93. https://doi.org/10.1002/wea.67

Mochizuki Y, Bryan A (2015) Climate change education in the context of education for sustainable development: rationale and principles. J Educ Sustain Dev 9(1):4–26

Na J-I (2016) The Yonmenkaigi system method for disaster restoration of a local community in Korea. Soc Behav Sci 218:76–84. https://doi.org/10.1016/j.sbspro.2016.04.011

Na J-I, Okada N (2011) Implementation of the Yonmenkaigi system method for capacity building on disaster risk management in local community of merapi volcano. Annu Disaster Prevent Res Instit Kyoto Univ 54(2):157–164

Na J-I, Okada N, Fang L (2009) A collaborative action development approach to improving community disaster reduction using the yonmenkaigi system. J Nat Dis Sci 30(2):57–69

Okada N (2018) Adaptive process for SMART community governance under persistent disruptive risks. Int J Disaster Risk Sci 9(4):454–463. https://doi.org/10.1007/s13753-018-0204-7

Osborn T, Maraun D (2008) Changing intensity of rainfall over Britain. Clim Res Unit Inf Sheet 15:1–2

Perry AH (1976) The long drought of 1975–76. Weather 31(10):328–336. https://doi.org/10.1002/j.1477-8696.1976.tb07448.x

Raftery AE, Zimmer A, Frierson DMW, Startz R, Liu P (2017) Less than 2°C warming by 2100 unlikely. Nat Clim Chang 7:637–641. https://doi.org/10.1038/nclimate3352

Robine J-M, Cheung SLK, Le Roy S, Van Oyen H, Griffiths C, Michel J-P, Herrmann FR (2008) Death toll exceeded 70,000 in Europe during the summer of 2003. CR Biol 331(2):171–178. https://doi.org/10.1016/j.crvi.2007.12.001

Samaddar S, Choi J, Misra BA, Tatano H (2015) Insight on social learning and collaborative action plan development for disaster risk reduction: practicing Yonmenkaigi System Method (YSM) in flood-prone Mumbai. Nat Hazards 75:1531–1554. https://doi.org/10.1007/s11069-014-1380-4

Sharma A (2012) Global climate change: What has science education got to do with it? Sci Educ 21:33–53. https://doi.org/10.1007/s11191-011-9372-1

Sousa DA (2006) How the brain learns, 3rd edn. Corwin Press, Thousands Oaks, California, United States

Sung J (2015) Climate change education and education for sustainable development in the republic of Korea: a status report. J Educ Sustain Dev 9(1):78–89. https://doi.org/10.1177/0973408215569116

Trajber R, Mochizuki Y (2015) Climate change education for sustainability in Brazil: a status report. J Educ Sustain Dev 9(1):44–61. https://doi.org/10.1177/0973408215569113

Turner S, Barker LJ, Hannaford J, Muchan K, Parry S, Sefton C (2021) Weather 76(8): 248–253. https://doi.org/10.1002/wea.4003

UNESCO (No date) Climate change education. United Nations Educational, Scientific and Cultural Organisation, Paris, France. Available at: https://en.unesco.org/themes/education-sustainable-development/cce. Accessed 30 Oct 2018

UNFCCC (2015) Adoption of the Paris agreement. United Nations Framework Convention on Climate Change, Bonn, North Rhine-Westphalia, Germany

United Nations (1992) Agenda 21. United Nations, New York City, New York, United States

United Nations (1998) Kyoto protocol to the united nations framework convention on climate change. United Nations Framework Convention on Climate Change, Bonn, North Rhine-Westphalia, Germany

Vidal J, Wade S (2009) A multimodal assessment of future climatological droughts in the United Kingdom. Int J Climatol 20(14):2056–2071. https://doi.org/10.1002/joc.1843

WalesOnline (2007) Schools to be given free copies of Gore film. Wales Online, 13th March. Available at: http://www.walesonline.co.uk/news/wales-news/schools-given-copies-gore-film-2266318. Accessed 9 July 2016

Walker P, Mason R, Carrington D (2019) Theresa may commits to net zero UK carbon emissions by 2050. Available at https://www.theguardian.com/environment/2019/jun/11/theresa-may-commits-to-net-zero-uk-carbon-emissions-by-2050. Accessed 4 June 2020

Wankat PC (2002) The effective efficient professor: scholarship and service. Allyn and Bacon, Boston, Massachusetts, United States

Chapter 11
Personalizing a Low Cost Weather Forecasting System

Muditha Dantanarayana, Srikantha Herath, and Dilum Bandara

Abstract Performance and cost of rainfall observation systems use in the study phase of Centre for Urban Water (CUrW) of Metro Colombo Urban Development Project (MCUDP) were compared and a low cost observation system that can be used to customize rainfall forecast was tested. System includes an IOT-based real time rainfall observing station which can be configured personally and access data directly along with a computational system. Weather Research and Forecasting (WRF) model with freely available Global Forecasting System (GFS) data, is used to forecast a set of possible rain systems and real time observations are used to select the most suitable forecast in Kelani river basin which effects the water level variations in the Metro Colombo area.

Keywords Weather observation · IOT · Weather forecast

11.1 Introduction

Climate change has changed the seasonal weather patterns being experienced for centuries, and adverse effects can be witnessed in the traditional and well-established weather related social and economic activities such as plantation and the occurrence of disasters (Filed et al. 2012).

The uncertainty imposed by the climate change leads to difficulty in making timely decisions on future activities which will be otherwise effectively made based on historical patterns (Lizumi and Ramankutty 2015). In terms of plantation, the

M. Dantanarayana · S. Herath (✉)
Centre for Urban Water, Metro Colombo Urban Development Project, Ministry of Megapolis and Western Development, Battaramulla, Sri Lanka
e-mail: srikantha@heraths.com

M. Dantanarayana
e-mail: muditha.danta@eng.pdn.ac.lk

D. Bandara
Department of Computer Science and Engineering, University of Moratuwa, Moratuwa, Sri Lanka
e-mail: dilumb@cse.mrt.ac.lk

H. Tatano and A. Collins (eds.), *Proceedings of the 4th Global Summit of Research Institutes for Disaster Risk Reduction*, Disaster and Risk Research: GADRI Book Series,
https://doi.org/10.1007/978-981-19-5566-2_11

well-established crop schedules are failing due to changing patterns and distribution of air quality, temperature, humidity, solar radiation, precipitation and other weather parameters (Vijayan et al. 2013). Increase in Disaster risk due to regular occurrence of severe weather conditions in unsuspected times as such currently occurring in Metro Colombo area makes the current manner of resource utilization for disaster response much ineffective.

This paper discusses system that could be individually implemented to assess the forecast weather in local area so that the near future activities could be planned effectively.

11.2 Methodology

Centre for Urban Water (CurW), the organization working on flood early warning and water management in the Kelani River Basin, Sri Lanka (Fig. 11.1) is involved in deploying different types of weather stations systems covering the entire catchment and use these data to select a reliable weather forecast (Herath 2017).

The study was conducted to assess the performance of a variety of rain monitoring devices deployed under MCUDP based on accuracy. Five tipping bucket rain gauges namely, TR-525M and TR-525USW from Texas Electronics Inc., Vantage Pro2 from Davis Instruments, WS-601SS2 from Beijing Guoxinhuayuan and Technology and WS-2600-1 from Misol International were tested in field. The configuration of the rain gauges are as in Table 11.1.

First the resolution of rain gauge was checked by measuring the water volume consumed at each tip and deviding by funnel area to calculate the rain depth as in Eq. 11.1.

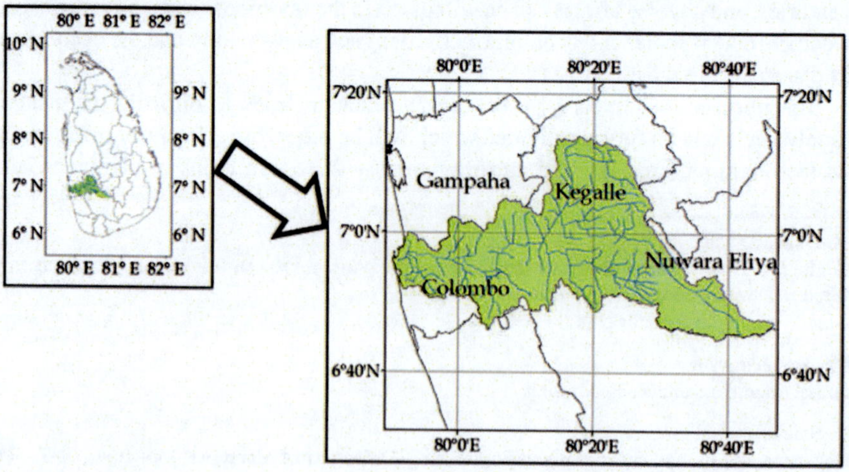

Fig. 11.1 Kelani river basin

Table 11.1 Rain gauge configuration

Rain gauge model	Resolution (mm)	Funnel diameter (mm)	Wall height (mm)
TR-525M	0.1	245	50
TR-525USW	0.2	203	50
Vantage Pro2	0.2	200	25
WS-601SS2	0.2	200	110
WS-2600-1	0.3	120	5

Table 11.2 Transmission module configuration

Trans. sys.	Com. method	Rate (s)	Power
A&T	GSM	360	Solar
LeeCom	GSM	360	Solar
Davis	Wi-Fi	5	AC
MISOL	Ethernet	16	AC

$$\text{Rain\_Depth(mm)} = \frac{\text{Consumed\_Volume(ml)} \times 10^3}{\frac{1}{4}\pi \times \text{Funnel\_Diameter(mm)}^2} \tag{11.1}$$

Then all the stations were deployed in field at same place with a standard rain gauge to assess the accuracy.

Four data transmission systems stated in Table 11.2, provided by A&T Labs Pvt. Ltd., Leecom Scada Systems Pvt. Ltd., Davis Instruments and Misol International were used with the rain gauges to detect the tips and transmit to the local server. Transmission system from A&T labs was coupled with rain gauges from Texas Electronics and the other rain gauge data were transmitted from the transmisson sytem from the same provider. The reliability and and cost of the complete systems were tested for the performance assessment.

Data transmitted in different rates were processed and recorded in a common database at interval of 5 min, Reliability of data were assessed from the percentage of successfully transmitted records from all records of each system.

Currently the CUrW uses Weather Research and Forecasting (WRF) model instances calibrated for different types of weather patterns around Sri Lanka to produce rainfall in 3 km spatial resolution and 1 h time resolution, downscaled from 0.5° resolution 3 h time step Global Forecast System (GFS) produced by National Centres for Environmental Prediction (NCEP). These WRF model outputs differ from each other with the different schemes used for downscaling in terms of microphysics, planetary boundary layer and land surface, cumulus convection, atmospheric radiation and the effect of clouds and aerosols. Suitable parameters for four weather types namely, Anti-Cyclonic, Cyclonic, East and South East prominent wind direction (Vuillaume and Herath 2017) were used for the comparison. A 3 day forecast

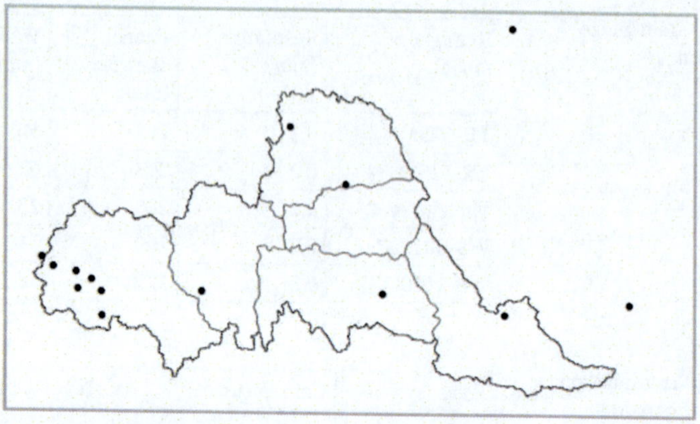

Fig. 11.2 Rain gauge locations

completes processing within 5 h in a workstation with Intel® Xeon® E5-2650 v4 @ 2.20 GHz *24 processor and 16 GB RAM run on Ubuntu 16.04 LTS.

Suitable configuration for the Kelani basin is to be selected based on the temporal distribution and magnitude of rain events in the forecasts and observed data obtained from the field observation stations shown in Fig. 11.2 where majority is located in MCUDP project area. As the first step a scoring criteria will be developed to rank the WRF models with respect to the timing of rain events. Observed and forecast time series of hourly resolution were used and a mark was given if rain or no-rain condition is matching in each hour. Ranking was the percentage of matching hours.

11.3 Results and Discussion

All weather stations perform with over 90% accuracy which means only less than 10% of rain volume will be missed from the rain gauge. Such errors might be due to high winds and changes in funnel design. LeeCom Transmission system shows the highest reliability as it is independently powered with solar and not effected by the power shortages, as in the case of Davis and MISOL Transmission systems which are powered by AC supply.

Rain gauge model	Trans. sys.	Accuracy (%)	Reliability (%)	Cost (USD)
TR-525M	A&T	94.1	91.5	846
TR-525USW	A&T	96.6	91.5	835
Vantage Pro2	David	95.6	94.0	1655
WS-601SS2	LeeCom	92.7	99.2	1100
WS-2600-1	MISOL	91.7	92.2	200

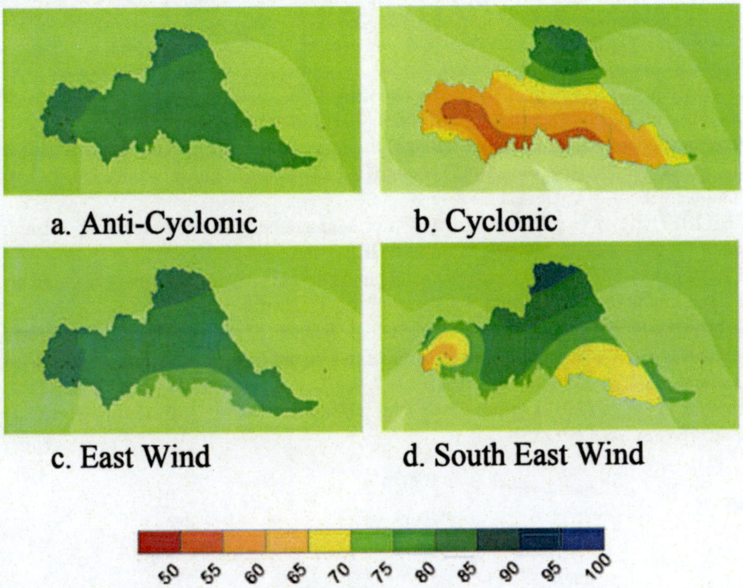

Fig. 11.3 Temporal performance of WRF models

WRF models configured for weather types of Anti-Cyclonic and East Wind direction shows the best performance in the entire catchment being over 70% matching in terms of the temporal distribution of rain events. These models can be used for the corrections in magnitude with bias correction to deliver an accurate forecast (Fig. 11.3).

11.4 Conclusion

MISOL weather stations proved to deliver reliable weather data at low cost and convenient installation facility at personal scale. Where as TR-525USW rain gauge coupled with LeeCom transmission provides more robust and accurate system for MCUDP implementations. Data gathered from such weather stations can be used to identify suitable forecasting models for specific regions and produce an effective forecast to be incorporated in decision support systems for regional disaster response in Metro Colombo area.

Acknowledgements The authors are grateful to the CUrW of MCUDP under the Ministry of Megapolis and Western Development for providing instrumentation and computational resources.

References

Field CB, Barros V, Stocker TF, Qin D, Dokken DJ, Ebi KL, Mastrandrea MD, Mach KJ, Plattner G-K, Allen SK, Tignor M, Midgley PM (eds) (2012) Managing the risks of extreme events and disasters to advance climate change adaptation. A special report of working groups I and II of the intergovernmental panel on climate change (IPCC). Cambridge University Press, Cambridge, UK, and New York, NY, USA, p 582

Herath S (2017) Keynote address: Urban Water Management in Metro Colombo, Engineering Research Conference (MERCon), Moratuwa. IEEE, pp xxi–xxii

Lizumi T, Ramankutty N (2015) How do weather and climate influence cropping area and intensity? Glob Food Sec 4:46–50

Vijayan L, Abdullah AT, Noor EFED (2013) Significance of meteorological parameters in the implementation of agriculture engineering practices in and around Tabuk region, KSA. Int J Appl Sci Technol 3(5):53–65

Vuillaume JF, Herath S (2017) Improving global rainfall forecasting with a weather type approach in Japan. Hydrol Sci J 62(2):167–181

Chapter 12
Transforming Flood Risk Management: A Case of Kakching, Manipur, India

Vickyson Naorem, Mahua Mukherjee, and Atul Kumar

Abstract North-eastern regions of India are progressing slower in terms of development than in India's mainland since the Independence of India in 1947. Moreover, the most triggering and frequently occurring factor, natural disasters, especially floods, adds more burden to the Local Government. Despite many natural disasters in recent years like floods and landslides, no or few disaster-related planning projects have been conducted so far to participate in any local or regional development projects to avoid the vulnerability of the people. On the other hand, Master Plan, Kakching (draft) n.a. has not been revised for the last three decades by the concerned authorities. To help in the development, we have demonstrated the assessment of green vegetation cover and impervious surface for the near future, benefit the labour cost and take less time in strengthening the upcoming plans in the case of Kakching, Manipur, India. This study analysed the spatio-temporal of vegetation cover and built-up area for the last two decades in the study area. Furthermore, we studied the trend analysis of vegetation and built-up area by comparing it with the flood extent. It is found that there is an enormous loss of vegetation cover and proliferation of urban areas in last two decades are detected. Therefore, we have recommended that the concerned authorities should integrate to strengthen and shape the future for better development plans to build urban resilience in changing climate.

Keywords Co-designing strategies · Kakching · Urban floods · Urban resilience

V. Naorem (✉) · M. Mukherjee
Centre of Excellence in Disaster Mitigation and Management (CoEDMM), Indian Institute of Technology, Roorkee, India
e-mail: vnaorem@dm.iitr.ac.in

M. Mukherjee
e-mail: mahuafap@iitr.ac.in

A. Kumar
Architecture and Planning Department, Indian Institute of Technology, Roorkee, India
e-mail: akumar5@ar.iitr.ac.in

H. Tatano and A. Collins (eds.), *Proceedings of the 4th Global Summit of Research Institutes for Disaster Risk Reduction*, Disaster and Risk Research: GADRI Book Series,
https://doi.org/10.1007/978-981-19-5566-2_12

12.1 Introduction

The eight North-Eastern states of India have the heaviest rainfall globally because of the topographical impacts plus the original Monsoon: Pre-Monsoon in March and April every year visiting this region and the fall fledged summer Monsoon visiting the same in the first half of June every year. Up to 1950s and early 1960s, Cherrapunjee was receiving the highest rainfall above 600 inches, according to the Meteorological Department of Assam state. But from the middle of 1960s, Mawsynram, on the boarders of modern Bangladesh got the heaviest rainfall, even heavier than Cherrapunjee. But the Deccan and Coromandel coasts enjoyed two monsoons every year in summer and winter (Mahanta et al. 2013). There is a less scientific study in the region (Singh et al. 2018). Kakching, the headquarter of the Kakching district, has been taken as the study area. Main settlers of Kakching are agricultural farmers and 30% of rice production in the valley of Manipur state is contributed by Kakching alone. Kakching is known as the Granary of Manipur. There are reasons behind being so. The first reason is that double or triple cropping is practised here very widely. Secondly, the township has vast agricultural and arable land bigger than anywhere in the valley of Manipur. Thirdly the arable land is almost fed by a perennial irrigation system for three hundred years and more of its settlement by the intelligent ancestors under pre-planning. They were also experts in black smith and in farming the land.

Moreover, Kakching was one of the highest population densities settled in a compact, having the number of 4590 per square kilometre in the state of Manipur, India.

With the increasing availability of remote sensing data, the researchers and authors have extracted the information in large areas with less time and lesser human capital cost. For the last two decades, this advantage has helped us to assess the spatio-temporal analysis (trend analysis) of vegetation cover and built-up area. Thus, the trend analysis of vegetation and built-up area may help finding the casual factors affecting the recent flood events.

In this study, we investigate the causal factors of flood vulnerability because the vulnerability has spatial and temporal phenomena and is dynamic to understand the nature of flood vulnerability. Eventually, knowing the nature of vulnerability would help understand how the weather and climate events cause the happenings of disaster and define risk management strategies, and foster effective adaptation. Remote sensing techniques like indices of normalised difference vegetation index (NDVI) and non-vegetation area would help understand the interaction of ecosystem services in building urban resilience.

12.2 Materials and Methods

Kakching is a small town with a population of 32,138 as per the Indian Census 2011. And the town is considered a Class-III town as per Indian Census 2001, which is surrounded by agricultural fields and situated at the state of Manipur, India, with the exact boundary of Municipal Council. The majority of the Kakching population engaged in agricultural occupation. Ancestors had planned and designed the residential buildings horizontally perpendicular to the main road passing through the primary market of Kakching town. The whole town looks like an iron grid shape town, and the main road is passing through the centre of the town. However, the pattern of what the town was, is not visible any more due to increase in population tremendously in the last three decades as shown in Table 12.1. The town has become so congested due to the absence of planning laws, regulations and policies. Ultimately, people started converting agricultural fields close to residential buildings causing a reduction of yields in crop production. Additionally, the concerned authorities have not revised the Master Plan, Kakching (draft) n.a. for the last three decades. To assist revising the Master plan for the benefit of people and authorities, we conducted this study taking Kakching, Manipur, India, as a study area as shown in Fig. 12.1.

In this study, the authors had developed the conceptual framework so that the concepts could be achieved through methods elaborated in the flowchart of the methodology shown in figure. The framework defines the factors of vulnerability of flood events to build urban resilience because unless we understand the factors of vulnerability, we could not proceed further to build resilience. The factors are socio-economic, physical, cultural, and ecological. To achieve the factors of vulnerability, we adopted the methods in Fig. 12.2.

Out of the total population, 14,867 were engaged in work or business activity. Of this, 8458 were males whilst 6409 were females. In a census survey, a worker is defined as a person who does business, job, service, cultivator, and labour activity. Of the total 14,867 working population, 84.61% were engaged in Main Work whilst 15.39% of total workers are engaged in Marginal Workers, which reflects the poor

Table 12.1 Population trends of Kakching

Year	1991	2001	2011
Population	24,437	28,724	32,138

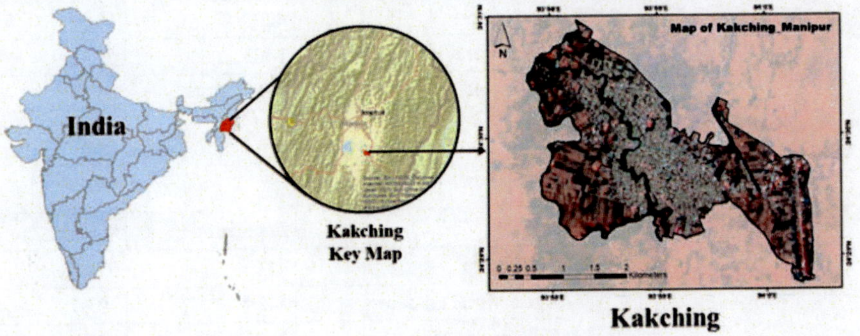

Fig. 12.1 Maps showing study area from India map (left to right direction)

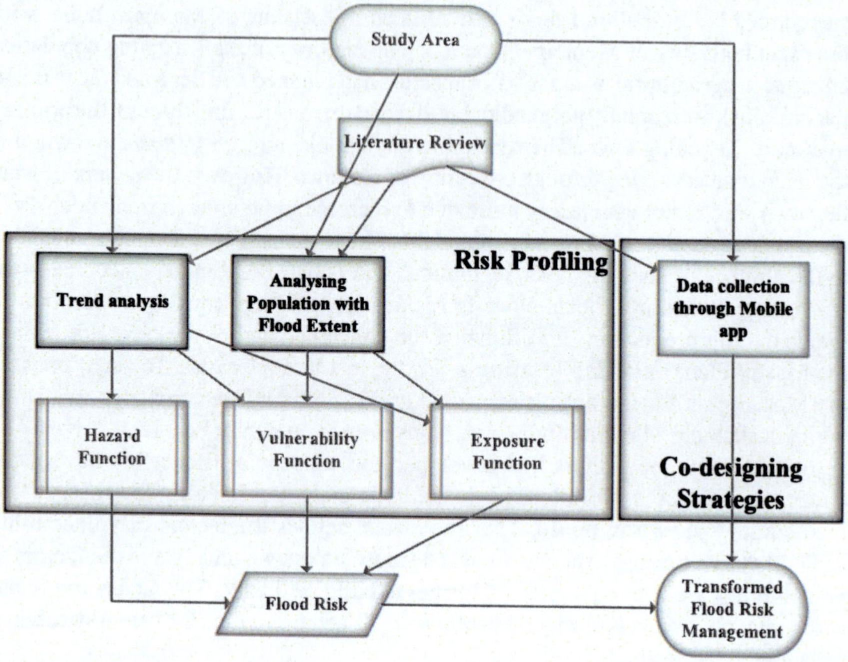

Fig. 12.2 Flowchart of the methodology

economy of the town (8.7% Marginal workers in India, as per Indian Census 2011). Occupation patterns of the town are shown in Tables 12.2, 12.3 and 12.4, where agricultural-related occupations are dominant.

This paper discussed a specific flood risk management case of Kakching Town in Manipur, India. Methodology (Fig. 12.2) in the paper had two distinct stages: (1)

Table 12.2 Distribution of marginal workers in Kakching

	Population	% total
Cultivators	420	18.36
Agricultural labourers	1403	61.32
Household	126	5.51
Other	339	14.82

Table 12.3 Distribution of main workers in Kakching

	Population	% total
Cultivators	4740	37.68
Agricultural labourers	1179	9.37
Household	382	3.04
Other workers	6278	49.91

Table 12.4 Distribution of type of worker in Kakching

	Population	% total
Non-workers	17,271	53.74
Main workers	12,579	39.14
Marginal workers	2288	7.12
Total workers	14,867	100.00

Risk profiling of town at present and (2) Co-designing strategies highlighting the role of nature-based resilience and community participation. Spatio-temporal changes in vegetation cover and non-vegetation for the last two decades were studied for trend analysis using Landsat 5 and 8. Primary data were collected through a survey questionnaire using a mobile app, Collector for ArcGIS. Data used for preparing questionnaire of key interviewees' interviews are as follows:

1. Interviewee's Occupation (Ordinal data)
2. List of consequences due to "Increasing population in town" (Ordinal data)
3. Does Kakching face floods frequently? (Interval data)
4. Source of floods (Nominal data)
5. List of recent past flood events? (Nominal data)
6. List Traditional Wisdom (Nominal data)
7. Locations and patterns of bamboos grow and ponds inside settlements? (Interval data)

Socio-economic factors (Tables 12.2 and 12.3) were predominant in occupation related to agricultural activities. During the trend analysis, firstly, we filtered the remote sensing data acquired from the spaceborne source by the pre-processing stage. The stage had included the flow of the radiometric correction and pan-sharpening processes for spatio-temporal analysis of vegetation and built-up area for the last two decades since 1995 with a time interval of five years. Because the activity of urban areas and vegetation cover in the urban area took time to change. Secondly, we processed the data by using NDVI and NDBI for mass information extraction of vegetation cover and built-up area, respectively, from the processed data. These outcomes were used for the temporal analysis of ecological and physical factors associated with the spatial phenomenon. We collected the socio-economic factors and cultural factors through stakeholders (local people) interactions. We also collected the socio-economic factors through secondary data of the Indian Census 2011. We collected qualitatively to cross-check and find the vulnerable areas by overlay analysis with the flood extent of years 1997, 1999, and 2000 from the secondary document of Manipur Science and Technology Council (2008). The results of overlay analysis contributed in identifying the causal factors of vulnerability and proceeded further for the discussion and conclusion of building urban resilience and shaping the future shape of the town, Kakching, Manipur, India, through integrating with the existing development planning processes.

This study assessed the trend analysis of vegetation cover and impervious surface for one decade in our study area and by comparing with the flood extent. One limitation in the study is that the trend analysis of water footprints was done using coarse resolution of satellite imagery which cannot detect smaller water footprints.

12.3 Results

Figure 12.3 showed how trend analysis of vegetation highlighted *diminishing vegetation* by comparing since 1995. Horizontal bars represent the year of satellite data used to analyse NDVI maps selected for each of the categories listed on the y-axis. Each of these bars is then split into what actual land cover of vegetation and non-vegetation had become. The spatial maps were also shown for 1995, 2005, 2015, and 2018 in Figs. 12.4, 12.5, 12.6 and 12.7, respectively. It was found that, in the town of Kakching, the area of vegetation loss were aligned with the location of the bamboo plantation, as shown in Fig. 12.11. So, this plantation seemed to be part of livelihood preventing floods, bringing eco-live livelihood and improving the liveability of the place. And *key interviewees* particularly mentioned that if the amount of bamboo in the last few decades were preserved, then recent flood events might not have occurred.

Overall, for the town of Kakching, the study found that vegetation conformity in the core urban areas remained moderate to low, ranging from around 58 to 27% conformity with the original urban plan. Meanwhile, due to the enormous loss of vegetation in the core urban areas as shown graphically in Fig. 12.3 and physically in the maps of Figs. 12.4, 12.5, 12.6 and 12.7 with respect to the selected years of the study, the frequency of recent past floods, as shown in Figs. 12.8 and 12.10, had increased storm water because of less penetration into groundwater. In Fig. 12.11, the pattern of bamboo plantations was seen *behind the houses* where the local people plant the bamboo aligned the houses. This pattern of bamboo plantation allows water to absorb water below the soil and release it slowly into the stormwater. An important finding from the study is that flood extent of the town collected through means of the mobile app showed better conformity than flood data obtained from earlier decades where community opinion is not taken into consideration highlighting the *significance of co-designing strategies* leading to transforming flood risk management.

12.4 Discussion

In the last few decades, disasters had made massive changes in the quality of life in local communities and the level of countries as a whole (United Nations 2015b). Disasters had shuffled environments and lives of over 700 thousand people, injured over 1.4 million and brought approximately 23 million homeless. Lastly, the disasters

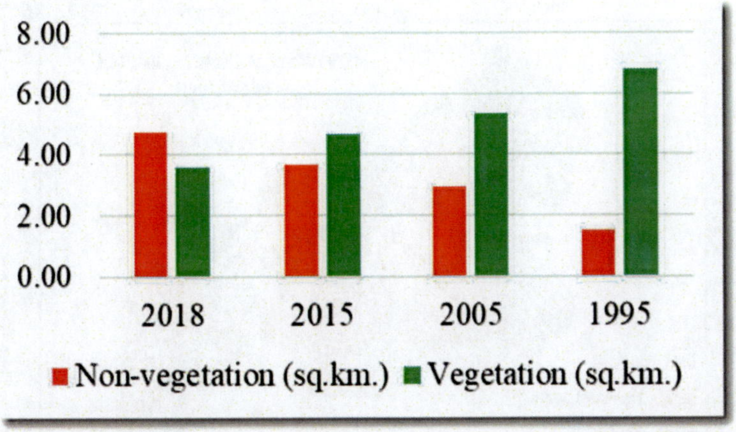

Fig. 12.3 Vegetation versus built-up growth

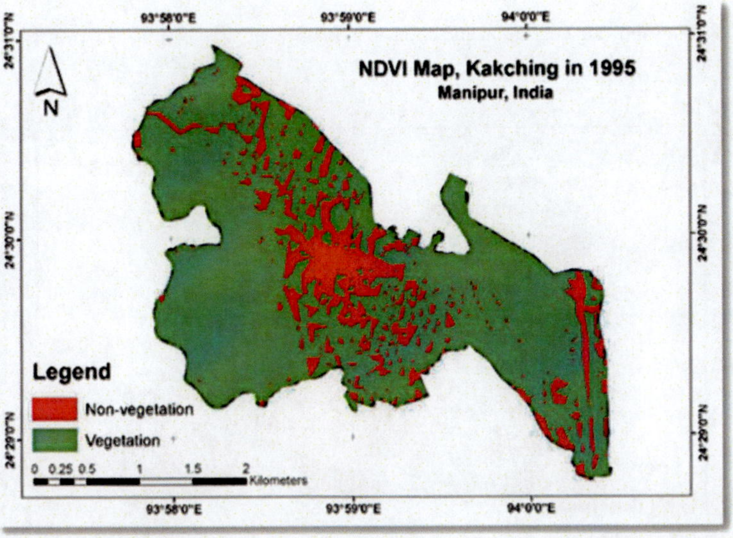

Fig. 12.4 Map showing vegetation and built-up in 1995

had affected more than 1.5 billion people to the most vulnerable of the vulnerable, especially with old-aged persons, women, and children (The Global Bank Group 2014). Alarming, various disasters had affected the total economic loss of more than 1.3 trillion and displaced 144 million people between 2008 and 2012 (United Nations 2015b). Moreover, disasters are exacerbated by climate change, thereby impeding progress towards sustainable development. In short, the exposure to the disaster of persons and their assets in almost all countries had proliferated faster than expected; thus, the concerned departments should give priority to age groups, and

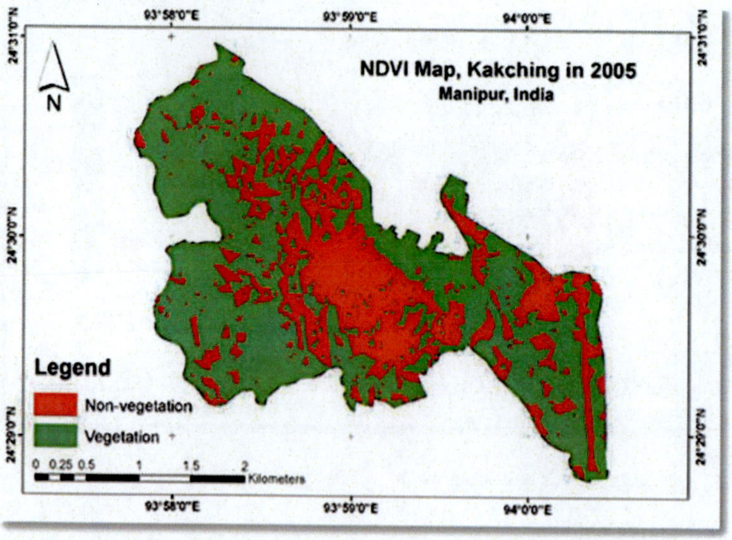

Fig. 12.5 Map showing vegetation and built-up in 2005

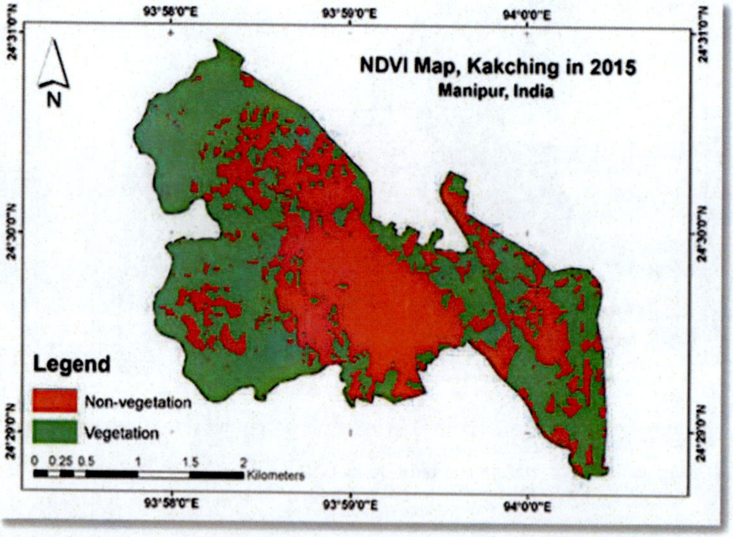

Fig. 12.6 Map showing vegetation and built-up in 2015

economic aspects should be considered whilst planning for reducing new risks of a significant economic, social, health, cultural, and environmental impact (United Nations 2015b). Whilst planning the risk reduction strategy, one should start from the most vulnerable sections like old-aged persons, women and children, resource

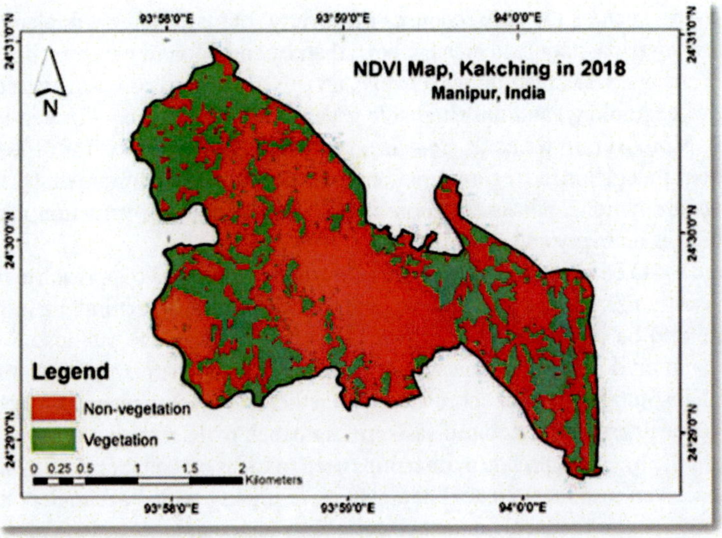

Fig. 12.7 Map showing vegetation and built-up in 2018

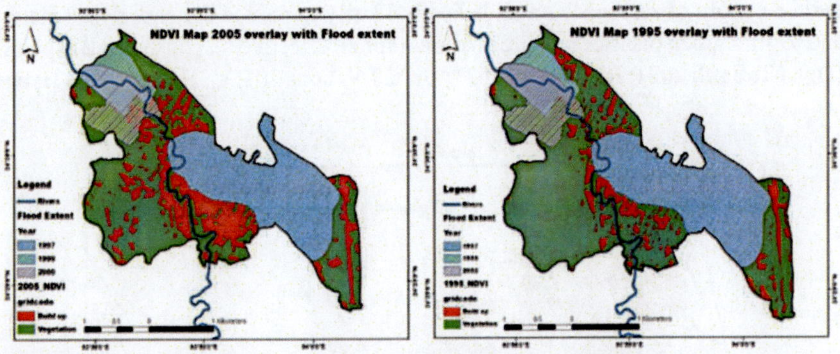

Fig. 12.8 Map showing flood extent for the years of 1997, 1999 and 2000

allocation of infrastructure and services in the proper places, and handle carefully with proper management and risk communication, which were pointed out in this study to be considered. It is an urgent and critical to anticipate a plan to reduce disaster risk to more effectively protect communities and countries, their livelihoods and health, socio-economic assets and ecosystems and thus strengthen their resilience (Economic and Social Commission for Asia and the Pacific 2015; United Nations 2015b).

As shown in Fig. 12.9, planning for disasters needs enhanced efforts on all levels to reduce exposure and vulnerability and prevent the creation of new disaster origins. As more dedicated action needs to be focussed on tackling underlying disaster risk drivers, such as the consequences of poverty; climate change, unplanned and rapid urbanisation, poor land management, thereby compounding factors: like demographic change, weak institutional change, non-risk informed policies, limited availability of technology, and unsustainable uses of natural resources (United Nations 2015b). Moreover, in terms of area size, good governance is vital for disaster risk reduction strategies at the regional, national and global levels for improving preparedness in coordination for disaster response, rehabilitation and reconstruction, and using post-disaster recovery and reconstruction.

There has to be a broader and more people-oriented preventive approach to disaster risk. Disaster risk practises need to be multi-hazard and multisectoral, inclusive and accessible to be efficient and effective. There is a must for the public and private sectors and civil society organisations who help with academia and scientific and research institutions to work more closely together and create prospects for collaboration for businesses to integrate disaster risk into their management practises (United Nations 2015b). Thus, private partnerships need to focus on collaborating in building resilience even in a local context and also co-designing with stakeholders brought more insights into the local context of any events happening around the region.

Overall, the Hyogo framework for action place the priorities in 2005–2015 is "to ensure disaster risk reduction in national and local priority to implementation and to identify, assess and monitor disaster risks using innovation and education to build a culture of resilience on all levels and lastly to reduce the underlying factors and to strengthen disaster risk preparedness for an effective response on all aspects" (DownToEarth 2021; Rahul et al. 2021; United Nations 2015a). However, the Hyogo

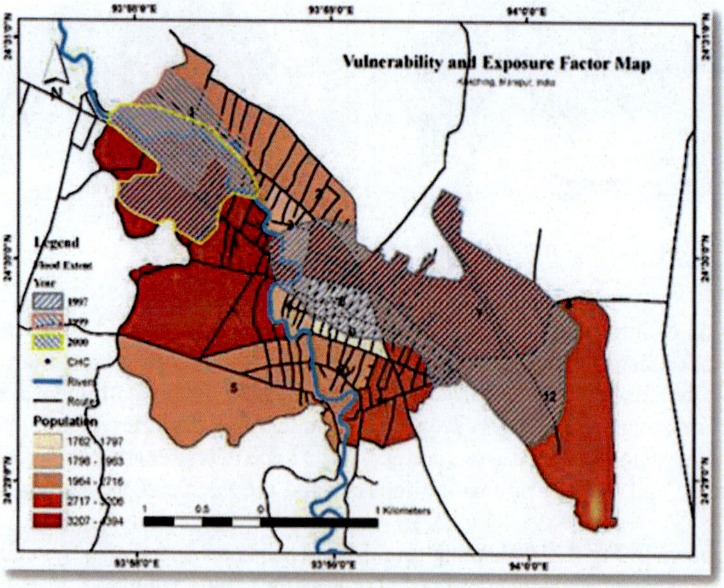

Fig. 12.9 Vulnerability and exposure factor map

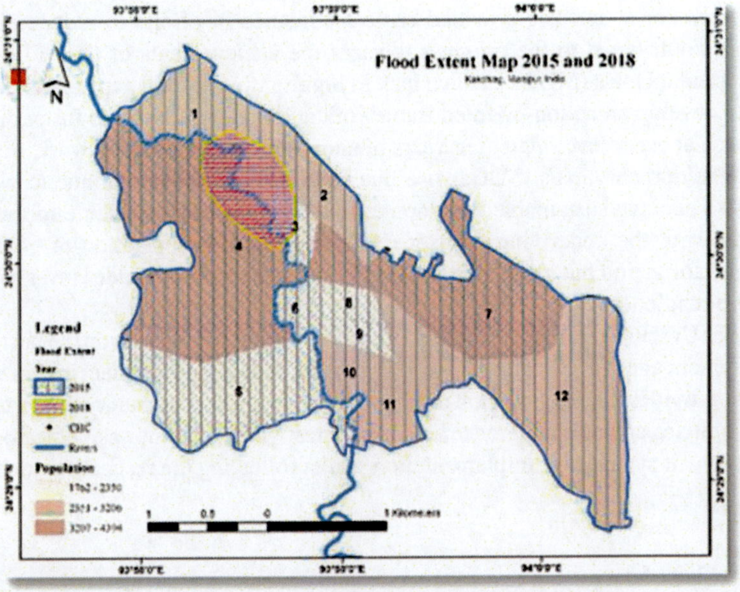

Fig. 12.10 Flood extent map for 2015 and 2018 through mobile app

Fig. 12.11 Photo showing the pattern of bamboo plantation behind the houses

framework for action has provided critical guidance in efforts to reduce disaster risk and contributed to the progress towards the achievements of the Millennium Development Goals (MDGs), which lack in highlighting several gaps that indicate a need to develop an action-oriented framework and mainly in need to foster disaster resilience at every level. After ten years of adopting the Hyogo framework, Sustainable Development Goals (SDGs) rise due to disasters, which continue to weaken efforts to achieve sustainable development. SDGs not only contribute to the gaps in addressing the underlying disaster risk factors in the preparation of goals and priorities for action but also helps to identify disaster risks and guides investment to improve resilience.

As per United Nations (2015a), the inter-government negotiation on the post-2015 development agenda, climate change, disaster risk reduction and financing for development provides the international community into a small community with a unique opportunity to enhance coherence across policies: goals, institutions, indicators and measurement systems for implementation whilst following the respective directives.

12.5 Conclusions

The preliminary study investigates in two stages—in the first stage, risk profiling had shown that recent flood events in Fig. 12.10, as compared to Fig. 12.9, were able *to gather more recent data* through community participation by *mobile apps* and in the second stage, *co-designing strategies* had provided *locations of bamboo plantation* which grew along each drainage right behind the houses. In contrast, the study compares and shows the significance of community participation in building urban flood resilience that *mobile app-based data* shows better information and details than literature-based data.

The enormous loss of vegetation cover and proliferation of built-up areas (impervious surfaces) had taken place in the last two decades. The affected area in Kakching could have been reduced risk by incorporating the ecological factors. Ancestors planted deep-rooted plants like sugarcane, bamboo and also paddy in fields because local people knew that such plants absorbed more water as compared to any other type of crop. We can replant more deep-rooted plants like bamboo, sugarcane, and paddy, which can also help in eco-livelihoods to bear the soil and contribute to the speed of excess runoff. Advantages of planting bamboos, paddy fields, sugarcane are for flood protection by holding the soils as well as reducing the speed of runoffs as suggested by Mukherjee 2018 and Saikia 2018.

Finally, this study can help in understanding the value of ecological and cultural factors, interaction with the physical and socio-economic factors. The results of this study suggested that the ideas of cultural and ecological factors can be used to explain variation in vulnerability and economic values for adapting and building urban resilience. Therefore, we recommend that the concerned authorities should take up to integrate for shaping the future of better development plans.

Acknowledgements We would like to especially thank family members for their help in taking photographs and collecting valuable data of bamboo plants in the photograph. A special thank goes to VN's father, Naorem Ibopishak Singh and mother, Naorem Tomchoubi Devi, for helping with the explanation of the plantation timing of bamboo for the local context of Kakching. We also acknowledge the support of critics from Climatology Lab. (Architecture and Planning Department) for providing the opinion for this research.

References

DownToEarth (2021) Sendai framework on disaster risk reduction disappoints. https://www.downtoearth.org.in/blog/sendai-framework-on-disaster-risk-reduction-disappoints-49067

Economic and Social Commission for Asia and the Pacific (2015) Post-2015 framework for disaster risk reduction. https://www.unescap.org/sites/default/files/12_INF(8)_Post-2015_Framework_DRR.pdf

Indian Census (2011) Primary census abstract—data highlights, NCT of Delhi. In: Census of India 2011. http://webcache.googleusercontent.com/search?q=cache:mIoKDiDf9EIJ:www.censusindia.gov.in/2011census/PCA/PCA_Highlights/pca_highlights_file/Delhi/DATA_SHEET_PCA_DISTRICTS_NCT_OF_DELHI.pdf+&cd=1&hl=en&ct=clnk&gl=nl

Mahanta R, Sarma D, Choudhury A (2013) Heavy rainfall occurrences in northeast India. Int J Climatol 33(6):1456–1469. https://doi.org/10.1002/joc.3526

Manipur Science and Technology Council (2008) Flood hazard zonation in Manipur valley

Mukherjee M (2018) Blue-green network and urban infrastructure development. AMCDRR 2018. https://www.preventionweb.net/files/56219_amcdrr2018eventsandpptsnewfinal.pdf

Rahul A, Edupuganti SR, Naorem V, Mukherjee M, Brooks T (2021) Urban risk assessment tools and techniques for ecosystem-based solutions. pp 253–278. https://doi.org/10.1007/978-981-16-4815-1_11

Saikia A (2018) Ecosystem protecting infrastructure and communities (EPIC): eco-safe roads for enhancing resilience of communities in Nepal. AMCDRR 2018. https://www.preventionweb.net/files/56219_1th03iucnsolutionsamcdrr2018290618.pdf

Singh A, Punia M, Haran NP, Singh TB (2018) Development and disaster management. In: Singh A, Punia M, Haran NP, Singh TB (eds) Development and disaster management: a study of the Northeastern States of India. Springer, Singapore. https://doi.org/10.1007/978-981-10-8485-0

The Global Bank Group (2014) The global bank disaster risk reduction and reconstruction fund. https://gbdrrrf.org/About-Global-Bank-faqs

United Nations (2015a) Resolution adopted by the General Assembly on 3 June 2015. A/RES/69/283: Sendai framework for disaster risk reduction. https://www.un.org/en/development/desa/population/migration/generalassembly/docs/globalcompact/A_RES_69_283.pdf

United Nations (2015b) Sendai framework for disaster risk reduction 2015–2030. https://sustainabledevelopment.un.org/frameworks/sendaiframework

Chapter 13
Adaptive Architecture—A Pathway to Resilient Communities

Rakhi Kashyap

Abstract Adaptive architecture can be explained wherein contemporary construction and vernacular architecture techniques can be brought together to makes its inhabitant feel safe and secured and can be adapted to the local environment. Through the evidence based on the example of the Sangkhola School built for Human Resource Development Department (HRDD), Sikkim State Government, India, in earthquake prone region of East India Sikkim, the study is intended to understand how an architectural intervention can adapt itself within the community instead of just being an architectural artefact, and how this construction can help the community feel prepared and resilient towards next earthquake. The study indicated that, with the hybrid RCC and Ikra, contemporary construction and vernacular architecture techniques brought together to create Sangkhola Primary School, it has attempted in imparting safety and comfort to its users and is adaptive to local weather and environment which is proven from the testimonies of its users.

Keywords Adaptive architecture · Resilience · Disaster resilient architecture · Resilient communities

13.1 Introduction

From 2006 to 2016, several thousand natural disasters have impacted many lives and different communities across different parts of the world. The increase in disasters have accelerated their impact on social and infrastructure systems and which has been affecting more than 75% of the world's population since 1980, also impacting the economic and human development setbacks in developing countries. Owing to this, resilient infrastructure is often very much needed in developing communities. Due to the accelerated number of disasters and number of people who are financially or

R. Kashyap (✉)
Nalanda University, Rajgir, Bihar 803116, India
e-mail: rkashyap.sees17@nalandauniv.edu.in

H. Tatano and A. Collins (eds.), *Proceedings of the 4th Global Summit of Research Institutes for Disaster Risk Reduction*, Disaster and Risk Research: GADRI Book Series,
https://doi.org/10.1007/978-981-19-5566-2_13

socially impacted, there is need for resilient and sustainable infrastructure systems—systems that not only have the capacity to adapt when future disruptions and shocks occur but are also used and maintained over time.

Infrastructure rehabilitation and reconstruction works after disasters are majorly undertaken by governments, civil society, international, and/or non-governmental organizations (NGOs) which have necessary expertise in the area. Various agencies come forward to demonstrate innovative approaches in disaster resilient construction to disseminate and demonstrate various risk reduction practices. These organizations and institutions strive to build resilience in the affected communities in the regions exposed to various disasters like earthquakes, flash floods, and cyclones where there is need to achieve resilience in terms of response to next event of disaster. Resilience for social-ecological systems is related to: (i) the magnitude of shock that the system can absorb and remain within a given state, (ii) the degree to which the system is capable of self-organization and (iii) the degree to which the system can build capacity for learning and adaptation.

Through the case study example of disaster resilient architectural intervention in earthquake prone region of East India Sikkim, this paper studies how adaptive architectural interventions can help the community gain a resilient pathway towards the impacts of disasters. The study extends to understand how an architectural intervention can adapt itself within the community instead of just being an architectural artefact and how this construction can help the community feel prepared and resilient towards next earthquake. The paper through a live case example strives to put across an idea that safe construction practices can help the community cope up the next disaster in a better way and the architectural intervention can act as a medium for community resilience in terms of disaster preparedness.

13.2 Method

As stated by Sturman, "A case study is a general term for the exploration of an individual, group or phenomenon". Creswell (2009) defines case study as "Researcher explores in-depth a program, an event, an activity, a process, or one or more individuals." Case studies can be especially helpful for understanding how different components fit together and how various elements (execution, implementation, context, and procedure) have delivered the observed impacts. To examine various benefits of the post disaster reconstruction and understand resilience through the lenses of community, it is imperative to study an example wherein disaster resilient construction has been implemented.

This paper aims at studying the case study of architectural interventions at Sangkhola Primary School build for Human Resource Development Department (HRDD), Gangtok, Sikkim State Government in India. The paper claims that the newly intervened architectural interventions in Sangkhola, Gangtok School does not alienate itself as a standalone artefact, rather has become a part of the community helping it to adapt with the region and particular geography. This small-scale intervention promises transformative changes that have made communities feel safe and secure in terms of occurrence of next event. This paper is an empirical study of the architectural intervention of Sangkhola School in earthquake prone mountain ranges of Eastern Indian Himalayan community. This study also focuses at understanding how adaptive architectural solutions can help the community that has lived under the constant threat of earthquakes, to mitigate disasters by providing an on-field intervention towards resilience.

13.3 Geographical Location of Sikkim

Sikkim one of hilly and significant state of India is located 27.04° North Latitudes and between 88.01° and 88.55° East Latitudes. The state is landlocked and surrounded by Nepal in the west, Tibetan plateau in north and surrounded by Bhutan and valleys of Tibet in the east. It is flanked by Darjeeling which is the town in West Bengal and forms southern boundary of Sikkim.

The State of Sikkim spreads across 7096 km^2 and spans over 112 km from North to South and 64 km from East to West. Sikkim is divided into four districts-East district, West district, North district and South district. The state is vulnerable various natural calamities like sever earthquakes being in Seismic Zone IV and prone to various landslides due to heavy rainfall.

13.4 Study Area

Gangtok, which is geographically located at 27.3325° N latitude to 88.6140° E, longitude. The town is situated at an altitude of about 1800 m above sea level making itself the third highest mountain (Mount Kanchenjunga) in the World (Kaur et al. 2019). Gangtok city is surrounded by two perennial rivers which converge at Ranipool gradually flowing towards the southern region of the state eventually joining the Teesta river which rises in the mountains of Eastern Himalayas. The city towards its southern end faces a slope of Eastern Himalayas and has erosional topography which is bounded and dissected with various natural springs which flow seasonally and perennially (Kaur et al. 2019). The study area experiences a high variation of rainfall throughout the year with maximum daily rainfall reaches up to several hundred mm (Bhasin et al. 2002). The East district is the most populated and urbanized amongst the four districts (North, South, East, and West) of Sikkim, and it comprises of 79.55% of the Sikkim state's urban population. Due to myriad reason urbanized population in East district is mainly concentrated at the capital city of Gangtok and its surrounding areas and puts tremendous pressure in the study area, thereby accelerating the chances of high impacts of earthquake associated which greater loss of lives and properties (Kaur et al. 2019).

On September18, 2011, the 6.9 (IIT Kanpur, 2011) Richter magnitudes earthquake jolted Sikkim, which caused damages to properties, especially pivotal public institutions such as schools. Out of 779 schools in the state, almost 682 schools were damaged in the earthquake. The need assessment by the technical team revealed that there is dearth of technical capacities in the state across levels, which had further deteriorated the overall construction scenario of the state. It was therefore imperative to intervene and demonstrate construction practices that improved construction practices in the region.

Since the extent of damages in schools was significant. However, the intervention in the schools was also done with an aim to demonstrate the key technical aspects to construct buildings, which are resilient to disasters. The technical team of the implementing agency also researched on the traditional technologies and emphasized the use of Ikra construction in the most sustainable manner such that the buildings are resilient. The non-governmental organisation, Sustainable Environment and Ecological Development Society (SEEDS) as an implementing agency promoted and advocated for "Safe Schools" in all aspects. Through the active engagement with the schools and education department, the need for "Safe Schools" was instilled through the process of construction and a general awareness on the risk reduction was generated amongst the school authorities, teachers and students.

13.5 Field Study

A qualitative case study with the help of filed study helps in examining a phenomenon within its real life stetting and context. In this study, through the field survey at Sangkhola School and vicinity, data was collected on and about single individual, groups and events.

The field study was aimed at deriving answers from the community's and users, i.e. teachers, students, school management and parents' perspective and their understanding of resilience. The process included discussions with various identified stakeholder groups and individuals. Discussions and interviews were focused on the experiences of the school users before and after school construction and people who have observed the changes since its inception and construction and are now users of the building. The data was collected using stratified sampling method selecting samples from each user group of the school. The finding section is based on information derived from the designing and implementing process in year 2011 and experiences of its users till June 2019. The findings will be analysed using vignettes privileges, which is a narrative-based mode of presenting method findings.

13.6 Discussions

13.6.1 What is the Meaning of Adaptive Architecture?

Adaptive architecture is a multi-disciplinary field concerned with buildings that are designed to adapt to their environments, their inhabitants and objects as well as those buildings that are entirely driven by internal data like its users (Schnadelbach 2010). Overall, adaptive architecture is not a well-defined field of architectural investigation but it can range from designing eco buildings, from artificial intelligence to ubiquitous computing (Tscherteu 2009; Roaf et al. 2007; Bullivant 2005; Eng et al. 2003; Rogers 2006).

All architecture is adaptable on some level, as buildings can always be adapted "manually" in some way. Brand's "How Buildings learn" provides an insight into the different levels of adaptation to be expected and how these apply over different time scales. The use of the term "Adaptive Architecture" must be seen in the overall context with buildings designed to adapt to their environment, to their inhabitants and to objects within them (Brand 1994). The concept of Adaptive Architecture is not very prevalent in disaster sector. However the concept of adaptive architecture that concerns with its environment and to their inhabitants stands very crucial for the architectural interventions in post disaster recovery phase. This paper attempts to focus on aspects of architecture that can be adapted in a given context and response to need of habitants in post disaster recovery phase. With the help of the examples of adaptive architectural solutions, the objective is to study what elements in adaptive architecture are adapted, the method for adaptation and what effect adaptations have. The evidence-based study in this paper focuses on the hybrid design amalgamating modern Reinforced Concrete Construction and traditional practice of Ikra, intervention in Sikkim, Gangtok India. It is aimed at an understanding how amalgamated techniques can help to achieve adaptive architectural solutions to help the communities mitigate disasters by developing a pathway towards resilience.

Motivations and drivers for designing for adaptiveness are explained in following ways.

- **Technical details**—Respecting the traditional practices, the age-old construction practices with time have tested and proven long-term sustenance. These age-old practices if amalgamated with the modern practices can reinforce the construction practices to adapt to new and better ways (Schnädelbach et al. 2014).
- **Spatial features**—In the adaptive architecture, spatial features can be transformed depending upon various location, topology of a place, orientation to sun path diagram, to adaptability of various vernacular form and the link between inside and out spaces (Schnädelbach et al. 2014).
- **Environmental**—Adaptive Architecture can be designed to react to its exterior environment. Adaptive elements are designed so that they cope up with their surrounding environment, for example ensuring the interior temperatures

controls and creating a thermally comfortable space and interior environment for the inhabitants (Schnädelbach et al. 2014).

- **Cultural**—Adaptive spaces for cultural production are clearly extended from the history. One of the most prominent societal reasons for the design for adaptiveness is lifestyle. E.g., a courtyard spaces in Indian architecture are integral part of a house since it accommodates various cultural activities (Schnädelbach et al. 2014).
- **Inhabitants**—The natural design adaptation of buildings spaces depends largely upon the inhabitants, their occupations, behaviour and building and expanding spaces based on necessary requirements. Architects focus their design efforts designing spaces for the convenience and comfort of the inhabitants of an adaptive building. However, with the time, individuals gradually start to bring in changes to the designed architectural layout manually as per the growing and needs and requirements. Certainly, in the context of this framework, if not in the entire field of adaptive architecture, the effects on inhabitants are not very well explored in the disaster response area; hence this deserves some further investigation (Schnädelbach et al. 2014).

As per UNESCO, adaptive architecture can be explained wherein contemporary construction and vernacular architecture techniques can be brought together to create buildings that are disaster resilient and adapted to the local environment. UNESCO recommends that such construction practices are more economically and culturally sound, reflecting local customs and traditions and using local materials. Such buildings take local climate and environment into consideration leading to more energy efficiency, thereby generating a smaller ecological footprint, whereby encouraging communities to adapt to such construction practices.

Architecture can be called or described adaptive when it interacts with the community as a functional entity instead of just being an architectural artefact (UNHABITAT 2010). Buildings in this context are described as flexible, interactive or dynamic instead of standalone artefact (Schnädelbach 2010). Adaptive architecture looks to the potential of its immediate environmental adaptivity considering its inhabitants as a key concern (Slovák et al.). As per UNESCO, adaptive architecture can be explained wherein contemporary construction and vernacular architecture techniques can be brought together to create buildings that are disaster resilient and adapted to the local environment (UNESCO 2018).

13.7 Performance of Ikra Construction and Conventional Buildings in the 2011 Earthquake

Most of the traditional houses in Sikkim are constructed privately and are typical single-story bamboo houses which are known commonly as the Ikra houses (Fig. 13.1). Ikra type modular timber or wooden houses are commonly constructed in Sikkim due to its various advantages in regard to its light weight, easily availability of materials, etc. The overall behaviour and performance of traditional Ikra houses was noticed to be quite good earthquakes of 2006 and 2011 experienced in the region (Kaushik and Dasgupta 2012).

Reinforced concrete buildings are quite commonly found engineered building across all districts of Sikkim. The challenge in the region is that Sikkim which is in highly seismic zone, in present days, buildings are commonly constructed without proper design and without any significant involvement of structural engineers. Poor quality of materials and workmanship, and thumb rules are used for design and construction, which is mostly done by masons. This could be few of the reasons

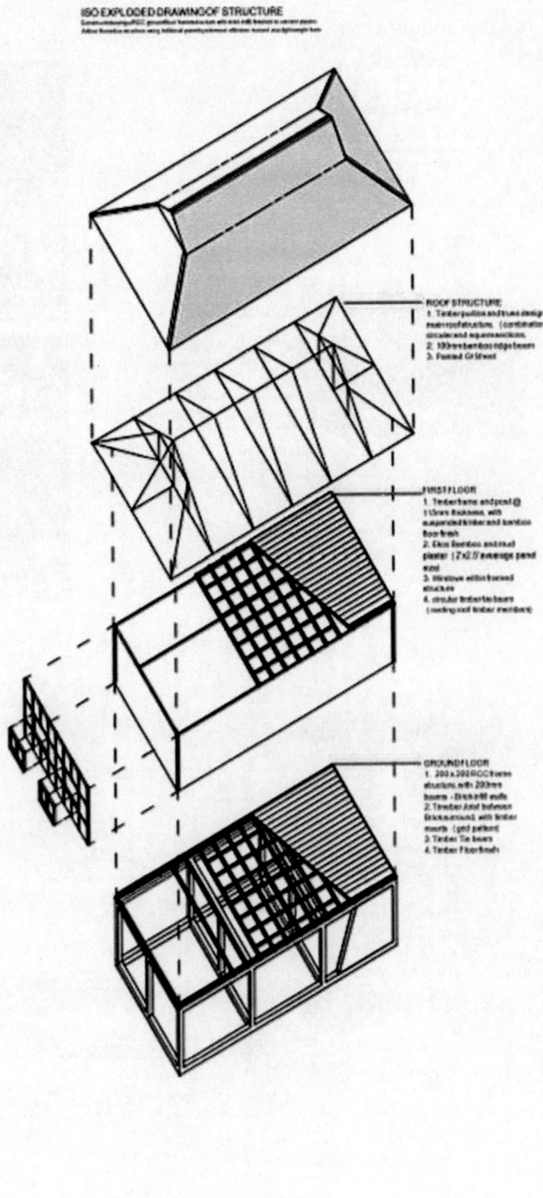

Fig. 13.1 Single Storey Ikra house in Sikkim. *Image source* SEEDS

that large number of school buildings reported damages during the 2011 earthquake (Kaushik and Dasgupta 2012).

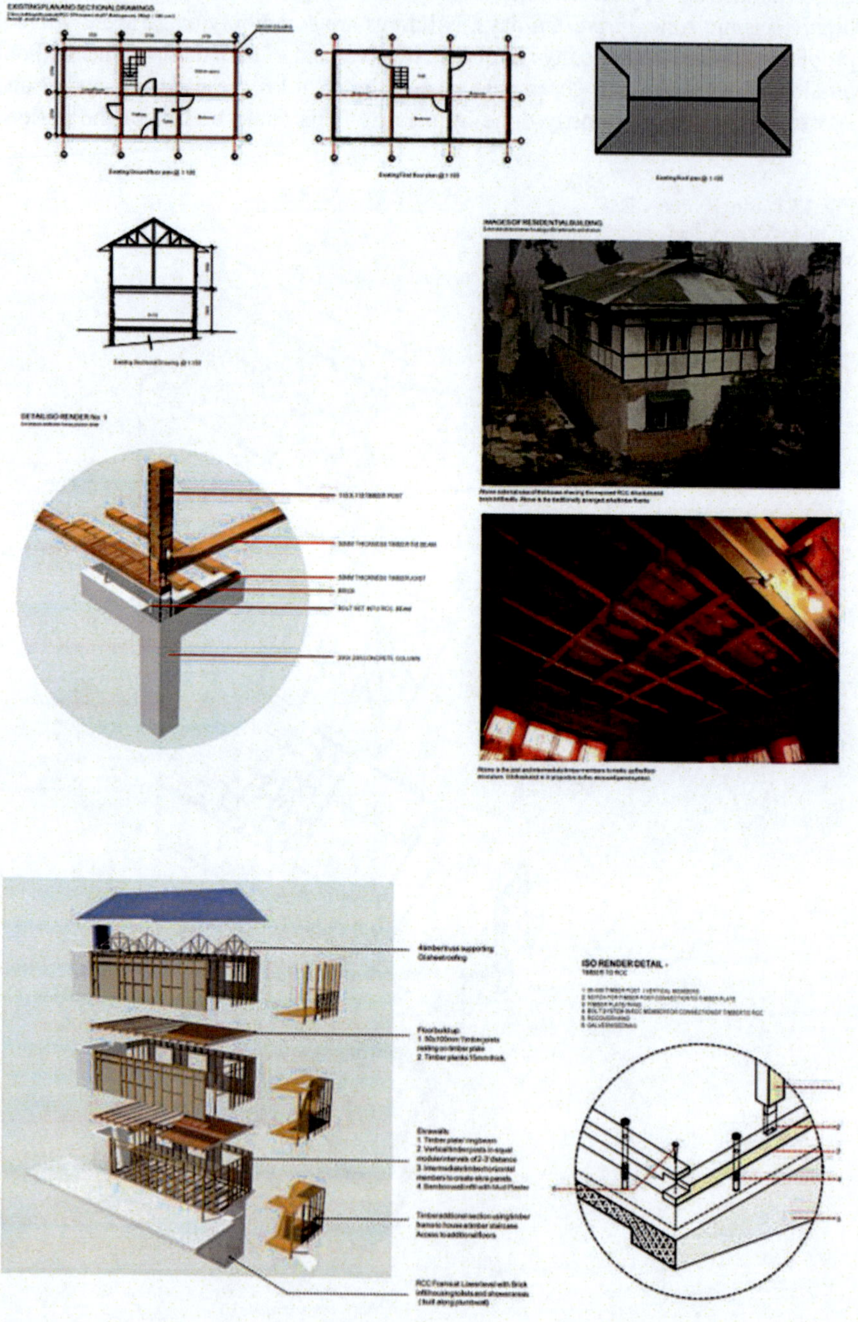

13.8 Translating the Knowledge of Ikra into Adaptive Architecture

Reconstruction was done in one of the schools-Sanghkhola Primary School. SEEDS initiated the process by understanding the local Ikra buildings-traditionally used by the communities. The technical team evolved a strategy to use Ikra panel such that it is more sustainable, thermally comfortable and appropriate to the local culture. The Ikra panel was used in the RCC main frame. Ikra infills are prevalent in the state. However, over a period of time, the use of Ikra has been replaced by RCC, more so because of the sustainability issues of Ikra. However, Ikra buildings are known for having thermally comforted the communities over the years. Moreover, the local masons have knowledge of building Ikra buildings. The biggest challenge is, however, to the use of Ikra in a manner which would make the building sustainable. The school therefore was constructed to make use of Ikra panel into RCC framework. Besides the core technical nuances of the construction, the design played an important role, which created emphasis on creating vibrant learning environment. The design demonstrated the initial concepts of creating a slender ground storey school building. The use of Ikra panel replaced from brick infill ensured that the structure is heavy and stable. This would enable the structure to withstand the jolts of earthquake.

The conventional RC main frame was completed as per all the structural design and specifications in 1:1.5:3 concrete with all ductile detailing in steel works. Constructing 10″ thick brick wall in 1:4 mortar works, till Cill Height is necessary to avoid dampening and provide longevity to wooden-bamboo Ikra panels. In order to make Ikra panels, RC frame and brick works structurally intact, horizontal wooden members of Ikra panel essentially fixed with 6 mm steel rod into brick work constructed.1:2:4 PCC mix is laid above and in between the brickworks to ensure fixing of rod in bricks. "Katus-Castanopsisindica" the locally available species of Sikkim and suggested by locals and department authorities was used for fixing Ikra wooden frames. The infill of the RC main frame is constructed with Ikra panels, in order to build up a light weight structure which minimizes the usage of brick and cement 0.5″ thick Ikra panels will help to enhance the resilience of the building in contrast to the rigid brick wall infill. Such technology ensures cost effectiveness in case of collapse and if required to be replaced.

As adaptation of local technology using local bamboo 'Melli Baan's from Sangkhola itself, are fixed in the wooden frame, later to be plastered with cement-sand mix because bamboo needs protection from rainwater and dampness for long lasting. Light weight Ikra infill demonstrates a great resilience to earthquake resistance, due to its strength to weight ratio of the timber members. Ikra panels are smaller components of interlocking panels, ability to be rigid and flex under earthquake forces and portions and modular design (length, width and height).

13.9 Sangkhola Primary School as Adaptive Architecture

In the case of Sangkhola Sikkim school construction, the design integrates various contemporary and traditional architectural solutions like adapting vernacular practices of Ikra construction with RC frame. Figure 13.2 is an example of this.

The heavy bulky brick walls are replaced with bamboo and cement plastered walls which is the design exercise practiced by local communities to build their houses. The structure was designed as an adaptation of lightweight Ikra panels systems using reinforced cement concrete (RCC) as main frames and local bamboo and timber to reduce the life risk of the school children and teachers. This design idea also helped in reducing overall concrete usage and building weight by distributing the load via smaller slender intermediate columns. The idea was to build a school building adapting Ikra panel systems using RCC as the main frame. RCC was for the reason that it has been enforced at the government level in the structures for the schools to be constructed from RCC for strength, longevity and safety. RCC was a desired form of construction amongst government construction practices as it reflects something which is modern and forward thinking as per the community.

Traditionally and historically Ikra has demonstrated a great resilience to earthquake resistance due to:

1. Its strength-to-weight ratio of the timber members.
2. Smaller components of interlocking panels.
3. Ability to be rigid and flexible under earthquake forces.

Fig. 13.2 Sanghkola primary school RCC and Ikra construction. *Image source* SEEDS

4. Potions and modular designs (length, weight and height).

The Sangkhola school was an idea to attempt to introduce principles of Ikra systems principles in RCC frame with the intention in reducing overall concrete usage and building weight by distributing the load via smaller slender intermediate columns. The implementation of this structure was intended to use medium rise structure to architectural theory—medium rise and medium design with rural areas prone to earthquake damages in Sikkim. In the existing practices of Sikkim, people have been practising the use of building brick infill in the RC frames which makes the structure heavy and bulky and unsafe for the residents. The implemented design in Sangkhola school demonstrates design concept of creating slender frame for G + 1 school form. The building has been designed with an Ikra panel systems rather than brick infill, which would make the structure heavier. The Ikra panels constructed with local timber species saving on cost and energy on brick infill construction. Traditional construction in Sikkim consists mostly of typical bamboo houses, known locally as 'Ikra' and also known as Assam-type housing. Figure 13.3 is an indicative of local Ikra architecture of Sikkim.

Fig. 13.3 Study of vernacular G+3 traditional Sikkim house. *Image source* SEEDS

Ikra houses are single-storey structures consisting of brick or stone masonry walls up to about 1 m above the plinth. This masonry supports the walls consisting of bamboo woven together with a wooden frame, and plastered with cement or mud plaster. The roofs are usually made up of corrugated galvanised iron (CGI) sheets supported on wood/bamboo trusses, which laterally connect the parallel walls. Bamboo superstructure is connected to the masonry foundation walls using steel angles, and flats with bolts and nails. There were no reports of any significant damages to Ikra structures during this earthquake (Kaushik et al. 2006).

The work of strengthening of schools followed the ideology of SEEDS of disaster resilient safe schools and use of local technology and materials for construction. In the course of construction, the agency also strived to build the capacities of local masons in safe construction techniques to ensure sustainability and wide dissemination of the processes. The capacity building programme also followed a long standing practice of building such programmes into its construction projects. In order to conform to its ideology and meet the needs of safe construction using locally available material, the architects and engineers of the implementing agency decided to combine traditional and modern construction techniques to evolve the building form. With these considerations in mind, an adaptation of Ikra panel system into RCC main frame was conceptualized. This type of construction provides strength, safety, and longevity and was used extensively in Sangkhola Primary school. The Ikra which was modified for the school is a traditional form of construction which consists of a timber framework with infill panels made using locally available Ikra reeds. The panels were plastered over with cement and the entire construction is known to be light weight and resistant to earthquakes. Considering the advantages of Ikra construction, the implementing agency decided to use it for restoration of school buildings to demonstrate the possibilities in this type of construction. The design demonstrated the concepts of creating a slender double storied school building form developed using modern and traditional construction practices. The implementing agency team worked with a team of local masons for adaptation of the local techniques.

The gradual development of design which integrated the local technology-ikra into the modern RCC and evolved a composite type of construction, "hybrid-ikra as an earthquake resilient model", provided some answers to my question of how to address building safety in the region and provided me with some options of addressing the existing challenges. Looking at the Ikra frame installation, locals also came forward for hands-on support and contributed in the construction with an enthusiasm and curiosity to learn. Raajan, a local and who sometimes helped on site voluntarily said,

> I studied in this school, my children are presently in the same school, and after seeing the kind of construction happening here I have a very strong hope that even my grandchildren will study in the same school.

The intervention in the schools was also done with an aim to demonstrate the key technical aspects to construct buildings, which are resilient to disasters. The technical team also researched on the traditional technologies and emphasized the use of Ikra construction in the most sustainable manner such that the buildings are resilient. Through the active engagement with the schools and education department and through the process of construction, the need for safe schools was instilled and a general awareness on the risk reduction was generated amongst the school authorities, teachers and students.

The implementing agency deployed its in-house technical team to define the designs of restoration of each school. The structural engineers and architects worked hand in hand to ensure that the designs are locally appropriate, technically feasible and cost effective. The agency's technical team was present on the ground to supervise and take appropriate technical decisions to support the speedy recovery of the construction work. A structural consultant was deployed to assess the extent of appropriate structural construction process right from foundation to slab work for construction. Throughout the process, the government's key technical officials and engineers were kept informed on the progress and the changes that evolved during the construction. The architects and structural engineers initiated the process by understanding the local Ikra buildings-traditionally used by the communities. The technical team evolved a strategy to use Ikra panel such that it is more sustainable, thermally comfortable and appropriate to the local culture. The Ikra panel was used in the RCC main frame.

Ikra infills are prevalent in the state. However, over a period of time, the use of Ikra has been replaced by RCC, more so because of the sustainability issues of Ikra. However, Ikra buildings are known for having thermally comforted the communities over the years. Moreover, the local masons have knowledge of building Ikra buildings. The biggest challenge is, however, to the use of Ikra in a manner which would make the building sustainable. The school therefore was constructed to make use of Ikra panel into RCC framework. Figure 13.4 indicates the RC frame with slender RC columns and Ikra panels in the RC frame.

Besides the core technical nuances of the construction, the design played an important role, which created emphasis on creating vibrant learning environment. The school designs highlighted the following:

Fig. 13.4 Indicates the RC frame with slender RC columns and Ikra panels in the RC frame. *Image source* SEEDS

1. Adopting holistic approach.
2. Enhancing natural environment physically.
3. Creating thermally comfort education spaces in the schools.

The design demonstrated the initial concepts of creating a slender ground storey school building. The use of Ikra panel replaced from brick infill ensured that the structure is heavy and stable. This would enable the structure to withstand the jolts of earthquake.

13.10 Community Resilience Through the Sangkhola School Implementation

After the event of a disaster or a conflict, following the damages that occurs in the existing buildings, design and architecture plays very crucial role to reconstruct the damaged and lost infrastructure to respond to the need for comfort and safety for the affected people. The intervention in architectural designs after a disaster can be considered effective only if it can meet short-term and long-term needs. Short-term needs can be catered as the immediate shelter design and long-term needs can be referred to reconstruction and rehabilitation (Valencia 2021). If the focus is about preparing communities and inhabitants of the buildings for disaster threats, then the importance must be re-enhanced on building resilience of human societies. Resilience the term, which is derived from the stream of physics, describes the ability of a system which is capable to come back to its original form after it is exposed to an outward stress (SGI Peace Proposal 2014). If quoting another example and meaning of resilience, the terminology has been used in a across various fields that expresses ability and capacity of societies that can recover from extreme shocks of environmental destruction or economic crisis (Daisaku 2014). When considered the case of disasters that occur naturally, improvising resilience takes the explanation and meaning towards enhancing the recovery processes to prevent and mitigate damages that often usually takes long and laborious process of recovery (SGI Peace Proposal 2014).

The concept of resilient designs inculcates technical and social strategies which is aimed at increasing resilience of a community (Oliver et al. 2019). When resilient designs are developed in the post-disaster environments it must essentially address the safety standards issues related to the quality of infrastructure. It must inclusively also address social issues such as community participation that enhances the design processes as it becomes contributively essential factor for long-term survival of people in the community (Oliver et al. 2019). In the case of Sangkhola, community participation was essentially involved right from the designing until the implementation.

13.11 Social Process

"Safe Schools are Safe Havens"—the schools are often used as relief centres and considered to be the safe evacuation point for the communities. It was therefore thought that school structures should make the users of the school building strong and should be capable of holding communities during emergencies. While constructing resilient building is one part of it, it is required that the communities are involved in the overall risk reduction measures of the schools. Through RTE [Right to Education], school management committees (SMCs) were empowered to take actions which are necessary to protect children and provide Safe Environment.

In order to make the school building adaptive instead of just a standalone artefact, the agency and school management committee together strived to inculcate culture of "School Safety" through school restoration and training and capacity building in the earthquake affected schools of Sikkim. The school management committee, community members and the implementing agency together strived to advocate safe school concept to the government departments through demonstration of the Sangkhola School. The SMCs were therefore involved in the construction process right from the designing stage. Figure 13.4 indicates the community and SMC during design discussions. The meetings were conducted in the schools to discuss the designs, the method of construction and the modus operandi of the construction work. The process also helped the communities to be proactive and take the ownership of the project. SMC vetted the final completion of the targeted schools. During the construction process, the SMCs were involved in day-to-day monitoring, resolving issues related to the material transport, providing storage of the materials and also the labour management.

Local masons whose children were enrolled in the school being constructed were involved in construction practices and were trained on the safe construction practices through hands-on training. The project aimed at building a culture of safety amongst the school authorities and the education department. The project involved active engagement with the education department, HRDD department, school principal, school management committee and the neighbouring communities. The local masons and contractors were involved in the process of restoration of the depleted buildings and were trained "hands on" to carry out the construction work. The department facilitated the process of intervention in the schools and their technical teams were involved in periodic monitoring and evaluation of the construction work. Over all the project left behind key learning's for the locals on the safe construction practices and nuances of the safe construction.

13.12 Analysis

In the previous section, it is learnt that use of local technologies and materials like bamboo and timber, helped to impart a sense of ownership to the local people, which steadily made it easier to implement hybrid construction which otherwise would have been relatively difficult. Since the technology is easy to replicate and is based on local knowledge, it found immediate acceptance amongst the local population. Stephane Hallegatte, Jun Rentschler, Brian Walsh in 2018 report to The World Bank as a part of building back better (International Bank for Reconstruction and Development 2018) states that following dimensions of building back better after reconstruction pertaining to post disaster situation can lead to shorter recovery duration and can make the recovery more efficient and resilient:

Building back stronger can reduce well-being losses as it can ensure safety of reconstructed resilient infrastructure that can withstand events with much more intensity in the future.

Building back faster helps in reducing impacts of disasters as the reconstruction accelerates.

Building back more inclusively aspect an inclusive approach so that post-disaster support can reaches the affected groups.

As per Kaushik Deshmukh (2010), he states that the newly developed reconstructed structures need to be strong to become less vulnerable if there is an occurrence of future shocks. This will allow people to quickly get back to their normal lives, and the recovery process will ensure that no one is left behind. This leads community to better adapt and recover from the disasters impacts which becomes essential part of achieving resilience in the community (Deshmukh 2010). UNISDR defines resilience, "as the ability of a system, community or society exposed to hazards to resist, absorb, accommodate and recover effects of hazards in a timely and efficient manner."

This claim needs are analysed and verified through the users' perspective during the field survey to understand if it has helped the community feel prepared and resilient towards next earthquake. The findings are analysed to study resilience in terms of infrastructural development, i.e. the resilience in school building in the current times, and social resilience to understand how the school building users relate with the school building and if they feel any safe in the existing building as compared to earlier school building that got damaged during 2011 earthquake.

13.13 Accruing Resilience Through Sangkhola the Case of Sangkhola, Gangtok School

Survey at Sangkhola Primary School was conducted to study perspectives of various users like teachers, students, school management committee, parents who are also community residents with the help of semi structured survey questions. The questionnaire mainly comprised of questions that inquired about the way these users relate the school building as factor of safety that exists within the community. The resilience is measured in existing situation by understanding if community believes that this particular school building:

a. Can resist intensive events if occurred in future and help the community build back stronger.
b. Can reduce disaster impacts.
c. Can ensure safety to the community in terms of next event.

Taking cue from the definition of resilience by Carle, 2002 and explanation by Stephane Hallegatte (2018) the aspect of resilience is analysed through users' perspective.

The schoolteachers including the school principal mentioned that the old school building which was made up of stone got fully damaged and was not suitable to the run classes. Dismantling of the damaged stone building was essential to build "earthquake proof" building in the same location. All the school teachers including the community mentioned that the school enrolment has increased after building the new school. For the school students, teachers and communities believes that the Sangkhola school building is stronger and will be able to resist future earthquake event as it is built with RCC and local bamboo. HRDD department also conducts regular mock drill trainings for preparing the residents if an earthquake would strike the region again. There is an increase in school enrolment after the construction of new school building, which was mentioned by almost all the respondents. Nearly, all the school management committee members, community members, school students said that they feel "safe and comfortable" in the new construction of the school building as compared to the old stone and mud building.

> "During 2011 Sikkim Earthquake, the old school building which was made up of mud and stone were fully damaged and was not suitable to run classes. The old school building was dismantled to construct two room earthquake proof building in the same location. This new building makes us feel safe and good. Because of the new building construction there has been increase in school enrolment," said Mr. Tashi Sherpa, Sanghkhola, the Primary School Headmaster.

The idea with which the school construction was conceptualised to integrate RCC construction with redundant slender RC columns and Ikra with local bamboo and timbers makes the user believe that the new school building is safer as compared to old stone building construction. They relate it with the factor of "feeling good and safety" factors in terms of school building. They also measured this safety factor by considering this new building "as more reliable" in terms of earthquake safety.

> "This new school building constructed by the Ngo makes is safer than the earlier school building. This new school building is more reliable and can ensure safety for my child who is studying in the Sangkhola School", quoted Mr. Topgyan Tamang, Father of Sankghkola Primary School Student.

Carle defined resilience in the aspect of (i) the intensity and magnitude of the shock which the system can absorb and remain stable within a given state, (ii) the capability of the system to self-organise and (iii) the ability of the system to learn and adapt. The Sangkhola School can be validated in terms of resilience in first two aspects explained by Carle. The redundant slender RCC design of school building with lightweight Ikra walls will allow the school building to remain within a given state when it would face earthquake forces in the future. As the earthquake forces can travel through the redundant slender columns to the ground. The lateral forces that may affect the walls of the school building may not cause any live losses within the school building. Even the users validate the school building's strength and reliability as it's a new construction in RCC instead of old construction in stone and mud.

> "This new construction as compared to old school building which was in stone and mud makes us feel safe and comfortable. I teach here and feeling of safety from any undesirable earthquake is very important for us as we spend more than half of our daily routine here in the school building."—Doma Tamang, School Teacher and School Management Committee, Sanghkola Primary School.

Since the school conducts regular mock drill, it is preparing the school occupant to safely evacuate the school building and avoid any live losses in future event of earthquake. Secondly, the school building will be in a position to suffer not severe damages in the next event of earthquake forces, it will help the school users to restore the activities within short duration of time. This aspect also validates, "Build Back Stronger" as it will help the community and school occupants resist more intense events in future and help the community and its users.

The third aspect explained by Carle the ability of the system that can learn and adapt seems to be pertinent in case of Sangkhola School. During the field study, it was also enquired if the school construction system has influenced any local construction practices, the respondents said that they did not notice any such construction that is similar to Sankhkhola School. They also indicated that there has been any impact on local construction practices influenced by the school demonstration. This aspect may need more institutional support and advocacy in terms of influencing local construction practices. This is the transition period between the last earthquake and next event to occur. What can be gathered from user's perspective that they relate factor of safety and reliability with new construction. This school building in terms of technology performance will be tested in future event of any disaster. However, at this present situation, the reliability within school users due to new construction indicates that a safe construction practice can ensure feeling of safety factor amongst its users. The users appreciate the new constructions and do not indicate any aversion. This school building makes them feel safer. These factors cannot be ignored in terms of disaster preparedness and having a factor that the community can bounce back faster if they have strong construction within the vicinity.

As stated by UNESCO, adaptive architecture can be explained wherein present-day and vernacular architecture techniques can be married together as hybrid architecture that can create buildings which are resilient to disasters and adapt to the neighbouring environment. The region faced another moderate 5.1 magnitude earthquake in year 2013 post Sankhgkhola School reconstruction. However, the school construction did not incur any damages, as reported by school authorities. The Sangkhola School demonstrates this as a model example of construction practices which can stand test of the time and intensity of next event that can occur. With the community and school users responding with an optimistic note on how the school building makes them safer than the earlier construction, the Sangkhola School, as a school building

Fig. 13.5 School management commitee discussion during designing process. *Image source* SEEDS

has succeeded in being a part of the community. It has ensured a feeling of safety and inculcating a culture of safety and preparedness instead of just being a standalone artefact and hence it stands as a validity of adaptive architecture (Fig. 13.5), wherein contemporary construction techniques of RCC and vernacular architectural practices of Ikra construction are amalgamated to create disaster resilient buildings.

13.14 Conclusion

Through the evidence-based example of disaster resilient architectural intervention in earthquake prone region of East India Sikkim, I aimed at studying how adaptive architectural interventions can help the community gain a resilient pathway towards the impacts of disasters. The study extended to understand how an architectural intervention can adapt itself within the community instead of just being an architectural artefact and how this construction can help the community feel prepared and resilient towards next earthquake. I argued in the paper that safe and adaptive architectural practices can help the community cope up the next disaster in a better way and the architectural intervention can act as a medium for community resilience in terms of disaster preparedness. The Sangkhola School case stands as evidence to this claim.

With the hybrid RCC and Ikra, present-day construction and vernacular architecture technology which are married together to create Sangkhola Primary School has stood earthquake of 5.1 magnitude in year 2013 after its construction in 2011 and has become part of the community. It has succeeded in imparting safety and comfort to its users and is been to adapt to local weather and environment as proved by the testaments from its users.

Opdyke claims that infrastructure and constructions which are build and frequently used, can be maintained by communities, and which can prerogative and resilient to hazards. Resilient infrastructures which can be maintained by communities are very much needed in the developing nations in regions which are frequently prone to disproportionate damages due to natural disasters, cause mortality and hamper development efforts (Opdyke 2017). Through the evidence-based example of Sangkhola School it can be stated that adaptive architecture can help in creating resilient and sustainable infrastructure systems. With the help of the example of Sanghkhola School it can be understood that this system of adaptive architecture can help to build the capacity to adapt to future events of disasters. Such adaptive architecture can become an integral part of the community to help them to be prepared and incur lesser disruption in future event of any disaster.

References

Deshmukh A, Hastak M (2009) A framework for enhancing resilience of community by expediting post disaster recovery, pp 362–370

Daisaku I (2014) East Asia should build resilience through disaster-relief cooperation. https://www.japantimes.co.jp/opinion/2014/03/06/commentary/japan-commentary/east-asia-should-build-resilience-through-disaster-relief-cooperation/

Folke C, Carpenter S, Elmqvist T, Gunderson L, Holling CS (n.d.) Resilience and sustainable development: building adaptive capacity in a world of. 31(5):437–440

Gautam D, Prajapati J, Paterno KV, Bhetwal KK, Neupane P (2016) Disaster resilient vernacular housing technology in Nepal. Geoenviron Disast 1–14https://doi.org/10.1186/s40677-016-0036-y

Ghosh A, Paul S, Bhattacharya S, Sasikumar P, Biswas K, Ghosh UC (2018) Calcium ion incorporated hydrous iron (III) oxide: synthesis, characterization, and property exploitation towards water remediation from arsenite and fluoride (Iii)

International SG (2014) 2014 Peace proposal value creation for global change: building resilient and sustainable societies by Daisaku Ikeda

Kaushik HB, Dasgupta K, Sahoo DR, Kharel G (2006) Performance of structures during the Sikkim earthquake of 14 February 2006, pp 449–455

Laboy M, Fannon D (2016) Resilience theory and praxis: a critical framework for. Architecture 13(1):39–53

Nnnn O (2017) Disaster sensitive architecture for the tropics: building resilient communities in developing nations. A case study in Northern Costa Rica, pp 1–6

Opdyke AA (2017) Resilient and sustainable infrastructure systems: a comparative analysis of post-disaster shelter coordination, stakeholder participation, and training

Oliver A, Thomas I, Duchemin E, Serre D (2019) Resilient and regenerative design in New Orleans: the case of the Make It Right project. 6

Paper C, Schn H (2018) Adaptive architecture—a conceptual framework, August

Post Disaster Engineering and Construction Program and Project Management Post Disaster Engineering and Construction Program and. (2015), (September 2011)

Rai DC, Mondal G, Singhal V, Parool N, Pradhan T (2011a) 2011a Sikkim earthquake: effects on building stocks and perspective on growing seismic risk

Rai DC, Mondal G, Singhal V (2011b) 2011b Sikkim earthquake effects on built environment and a perspective on growing seismic risk

Schnädelbach H, Slovák P, Fitzpatrick G, Jäger N (2014) The immersive effect of adaptive architecture. Pervasive Mob Comput. https://doi.org/10.1016/j.pmcj.2014.11.006

Soka Gakkai International Peace Proposal, 2014

Series SE (2017) Building information modeling as a tool for enhancing disaster resilience of the construction. XII(1):9–18. https://doi.org/10.1515/tvsbses-2017-0002

Thorén H, Thore H (2014) Resilience as a unifying concept. Int Stud Philos Sci 28(3):303–324. https://doi.org/10.1080/02698595.2014.953343

Tucker S, Gamage A (2014) Some design aspects of sustainable post-disaster housing. Int J Disaster Resilience Built Environ Article information (April 2017). https://doi.org/10.1108/IJDRBE-06-2012-0019

Valencia N (2021) https://www.archdaily.com/966990/haiti-earthquake-worsens-the-urban-collapse-what-can-architects-do

Wu J (2015) Ecological resilience as a foundation for urban design and sustainability, May. https://doi.org/10.1007/978-94-007-5341-9

Zhou H, Wang X, Wang J (2016). A way to sustainability: perspective of resilience and adaptation to disaster, pp 1–14. https://doi.org/10.3390/su8080737

Chapter 14
Return Period Analysis of Major Flood Events Considering Homogeneous Regions

Nor Eliza Alias, M. Mahayuddin M. Tarmizi, and M. Syazmi Chebby

Abstract Extreme flood events cause huge amount of money and human resources in handling the catastrophic during and post to the disaster. In order to cope for these events, mitigation strategies to reduce the impact or occurrence are crucial. One of the ways is to better design the flood defense structures using possible highest rainfall values that might occur at a specific location represented by rainfall return periods. The main idea of this research is to consider using extreme rainfall homogeneous regions to help predict the best return period for designing flood defense structure using information of historical flood events occurring at the location of similar extreme rainfall homogeneous regions. The objectives of this research are to identify major flood events in Peninsular Malaysia and to map the rainfall return period associated to the flood events within similar homogeneous regions. The method of pooled frequency analysis was used to combine annual maximum series (AMS) data in each homogeneous region. The best distribution model is selected based on highest rank according to Kolmogrov Smirnov, Anderson Darling, and chi-squared. The highest return period of rainfall event occured at zone 1, zone 2, zone 3, and zone 4 is 50 years, 333 years, 500 years, and 20 years, respectively. The use of the highest return period values may be applied for designing flood defense structures located in the specific homogeneous region as those return periods have already occured within respective homogeneous regions.

Keywords Flood · Homogeneous region · Return period · Extreme rainfall · Disaster risk reduction · Peninsular Malaysia

---

N. E. Alias (✉)
Centre for Environmental Sustainability and Water Security (IPASA), Universiti Teknologi Malaysia, Johor Bahru, 81310 Johor, Malaysia
e-mail: noreliza@utm.my

M. M. M. Tarmizi · M. Syazmi Chebby
School of Civil Engineering, Faculty of Engineering, Universiti Teknologi Malaysia, Johor Bahru, 81310 Johor, Malaysia

H. Tatano and A. Collins (eds.), *Proceedings of the 4th Global Summit of Research Institutes for Disaster Risk Reduction*, Disaster and Risk Research: GADRI Book Series,
https://doi.org/10.1007/978-981-19-5566-2_14

14.1 Introduction

Extreme floods are usually associated to extreme rainfall events. In order to cope for these events, mitigation strategies to reduce the impact or occurrence are crucial. Being a yearly event in Malaysia, major flood in Malaysia showed increasing trends which usually occur during the monsoon season and the most devastating natural disaster experienced (Diya et al. 2014). Malaysia experiences two categories of flood as been classified by the Department of Drainage and Irrigation; flash flood and monsoon floods. Based on the hydrological perspectives, the clear difference between these two disasters is the period taken by the river flow to recede to the normal level. Flash floods take only some hours to return to the normal water level, while monsoon flood can last for a month (Mohamad et al. 2012). In order to help decision makers in better designing flood defense structure, use of homogeneous region may be used to improve extreme rainfall estimates. In this research, the rainfall return periods analysis of major flood events within homogeneous region in peninsular Malaysia was investigated. The objectives of this study are as folows: (1) Identifying major flood events in Peninsular Malaysia; (2) Estimating the return period of rainfall during the major flood events, and (3) Mapping the rainfall return period associated to major flood events in extreme rainfall homogeneous region. As a case study, the research focuses on the flood events in peninsular Malaysia. The analysis of study only limited to the major flood events of this region. Thus, the data of the rain gauge station associated to the flooding area will be taken into consideration for the analysis purposes.

14.2 Literature Review

In Malaysia, the weather mainly influenced by two monsoon regimes, namely the Southwest Monsoon from late May to September and the Northeast Monsoon from November to March. The Northeast Monsoon brings heavy rainfall, particularly to the east coast states of Peninsular Malaysia and western Sarawak, whereas the Southwest Monsoon normally signifies relatively drier weather. In the past, widespread floods have occurred, especially during the northeast monsoon season. Malaysia has always been complacent and perceived to be relatively free from major hydrometeorological and geological hazards (Low 2006). With rapid development in high-rise buildings and other infrastructures, the occurrences of hydrometeorological and geological hazards have been increased (Low 2006). These natural hazards include monsoon flood, flash flood in the cities, severe storms, storm surge, landslides, earthquakes, and tsunami. The government has taken the initiatives to implement various projects and activities toward disaster prevention and mitigation in the country.

Extreme rainfall events are the main cause of flooding (Syafrina et al. 2015). The changes in extreme rainfall events, such as heavy precipitation and major flood, have attracted a lot of attention because of their devastating consequences on society and economics. Increasing frequency and intensity of extreme rainfall events has raised concern that human activity might have resulted in an alteration of the climate system. It is believed that rise in both frequency and intensity of extreme rainfall events are the major impacts of global warming (Sen Roy 2009).

In a frequency analysis, estimates of the probability of occurrence of future events are based on the analysis of historical rainfall records. Extreme rainfall events are considered as random variables in frequency analysis, which come from identical distribution, and are assumed to be independent. The magnitude of the random variables can be related to their frequency of occurrence using probability distribution. There are two forms of probability distribution which is probability density function (PDF) and cumulative distribution function (CDF) (Alias and Takara 2014). PDF and CDF are described in the following equation.

$$f(x) = P[X \leq x] \tag{14.1}$$

$$F(x) = \int_{-\infty}^{x} f(x)\mathrm{d}x \tag{14.2}$$

From Eqs. (14.1) and (14.2), $F(x)$ is the cumulative distribution function (CDF), and $f(x)$ is the probability density function (PDF). The CDF is the probability distribution of a random variable x with a given probability, P found at a value less than or equal to x (Eq. (14.1)). It is also regarded as the non-exceedence probability. The CDF is also related to PDF as in Eq. (14.2). A cumulative of the PDF produces the CDF. A proper fit of PDF is the best way to describe the random variables. This is to identify which frequency analysis model the random variables belongs to; afterward, the model is then used to predict the extreme values and its frequency of occurrence or return period, T (e.g., 10 year, 50 year, 100 year rainfall) (Yevjevich 1972).

There are also several studies investigated on the distribution of rainfall, either hourly, daily, or annually. Quantitative goodness of fit tests was done to determine the probability distribution most appropriate for describing annual maximum rainfall series in Peninsular Malaysia. Eight probability distributions were investigated as potential candidates to describe the annual extreme rainfall data of Peninsular Malaysia. They were the two-parameter Gumbel and gamma, the three-parameter generalized extreme value (GEV), generalized normal (GNO), generalized Pareto (GPA), Pearson type 3 (PE3) and Log-Pearson type 3 (LP3), and the five-parameter Wakeby. For annually maximum rainfall data in Peninsular Malaysia, the generalized extreme value (GEV) distribution is the most suitable to be used (Zalina et al. 2002).

In order to overcome the lack of at-site observations, a regional approach to a frequency analysis that traded space for time was developed in the 1960s for the estimation of design floods. This approach, based on the index-flood method, has gained wider popularity (Hosking and Wallis 1993). The core idea of the regional approach is that one can obtain more reliable quantile estimates for a given site based on a multi-site analysis compared to a single-site approach. The proposed groups of sites should meet the requirement of homogeneity that is sites pooled together have to exhibit, except for a scaling factor, similar probability distribution curves of extremes. Provided the groups of sites regions are homogeneous, the regional methods not only enhance the reliability of the at-site estimates at observing stations but also allow for the estimation of design values at ungauged locations.

The homogeneous region is described as one in which a group of stations with similar at-site characteristics, such as mean annual precipitation (MAP), hydrologic, physiographic, climatic variables, and trend direction is gathered (Adamowski and Bougadis 2003). This means that any areas within the homogeneous region are considered to have similar climatic exposure, conditions, and source of extreme rainfalls. The regional homogeneity test is the first step to perform in obtaining regional probability distributions for annual maximum storm intensities and regionalized IDF. The regional homogeneity test is performed to determine whether a region or an area consisting of a few rainfall stations may be considered as a homogeneous region.

The most common means used in hydrology, to show the probability of an event, are to assign a return period or recurrence interval to the event. Return period is an annual maximum event that has a return period (or recurrence interval) of T years, if this value is equaled or exceeded once, on the average, every T year. For example, a 100 year flood has a probability, $P = 1/T = 1/100 = 0.01$ or 1.0% of being equaled or exceeded in any single year. Here, however, it is important to realize that the return period implies nothing about the actual time sequence of an event. The concept of a return period is usually found by analyzing a series of maximum annual floods and rainfalls (Huber 1948).

14.3 Methodology

The main objective of the research is to study the return period of rainfall events during major floods and associate it to homogeneous regions. For this study, historical floods and rainfall events in peninsular Malaysia are used for analysis. The methodology consisted of four key activities; (1) Identify highest ranked flood events; (2) Compute annual maximum series of rainfall data at sites associated to flood events; (3) Obtain return periods of rainfall events; and (4) Map the return periods to homogeneous regions.

14.3.1 Ranking of Flood Events

The historical rainfall data and annual flood reports of Peninsular Malaysia were obtained from the Division of Flood Management, Department of Irrigation and Drainage Malaysia (DID). There are a total of 249 flood events recorded in Kelantan (1965–2015), 130 flood events in Terengganu (1979–2014), 81 flood events in Pahang

(1979–2014), 75 flood events in Johor (1979–2014), 35 flood events in Melaka (1984–2014), 128 flood events in Kedah (1984–2014), 58 flood events in Perlis (1983–2014), 37 flood events Pulau Pinang (1983–2014), 88 flood events in Perak (1985–2014), 506 flood events in Selangor (1984–2014), 33 flood events in Kuala Lumpur (1984–2014), and 83 flood events in Negeri Sembilan (1984–2014). Despite the amount of flood events mentioned, flood events were shortlisted and ranked based on number of victims evacuated. Flood events are considered severe with more numbers of victims evacuated.

14.3.2 Rainfall Annual Maximum Data

For the analysis of an extreme event, the return period of the rainfall is determined through frequency analysis. However, the daily data for the analysis need to be converted into annual maximum series since the return period is determined in yearly basis and to represent extreme values. The number of record years available at the rainfall stations selected is more than 25 years.

14.3.3 Frequency Analysis

Four probability distributions were investigated as potential candidates to describe the annual extremal rainfall data of Peninsular Malaysia, namely generalized extreme value (GEV), generalized Pareto (GPA), Gumbel and gamma max distribution. The combined results obtained suggested the generalized extreme value (GEV) as the most appropriate distribution. GEV has the highest rank in the goodness of fit based on Kalmogorov Smirnow (KS), Anderson Darling (AD), and chi-squared (CS).

14.3.4 Homogeneous Regions

Homogeneous region in Peninsular Malaysia was divided into four zone, namely East as zone 1, South as zone 2, North as zone 3, and West as zone 4 (Refer Fig. 14.1) (Bakar et al. 2016). The North covers the state of Kedah, Perlis, and Penang, while the South covers the state of Johor and Malacca. The East covers the state of Kelantan, Terengganu, and Pahang, while the West covers the states of Perak, Selangor, Wilayah, and Negeri Sembilan. The four regions were identified using 45 rainfall stations based on the geographical locations and storm events characteristics (see Fig. 14.1) for Peninsular Malaysia (Bakar et al. 2016).

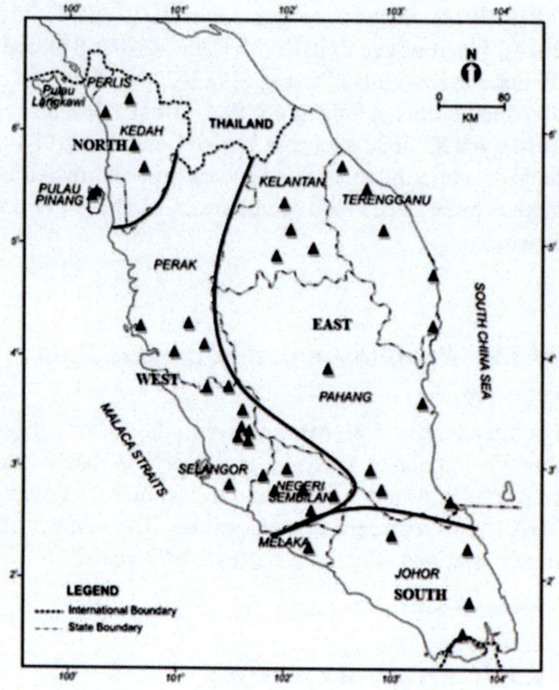

Fig. 14.1 Homogeneous regions for Peninsular Malaysia (Adapted from Baskar et al. 2016)

14.3.5 Mapping of Return Period on Homogeneous Region

Return periods of rainfall events associated to the highest ranked floods are mapped and plotted. The mapping is to determine the location of the rainfall gauge in relative to homogeneous zone.

14.4 Results and Discussion

14.4.1 Major Flood Events

The reports of flood events obtained in each zone were analyzed to identify the worst events. The events were analyzed according to number of victims evacuated, maximum rainfall, rainfall duration, flood duration, maximum flood depth, and damages as those were the variables which were recorded in the annual flood reports by DID. The data for number of victims evacuated are more complete compared to others. Table 14.1 shows the major flood events ranked highest according to number of victims evacuated in each homogeneous zone. The top six of worst flood events in zones 1 and 2, while top five worst flood events in zones 3 and 4 were chosen to be analyzed. Flood events at Tanah Merah as zone 1, Muar as zone 2, Kubang Pasu as zone 3, and Petaling as zone 4 were the highest ranked based on number of victims evacuated.

Table 14.1 Flood events according to homogeneous regions

Zone 1: East region

Year	State, District	Damages (RM)	Flood duration (days)	Rainfall duration (days)	Maximum rainfall (mm/day)	Flood depth (m)	Victim evacuated
Dec-14	Kelantan, Tanah Merah	5,002,000	4	9	1066	2	42,171
Dec-14	Terengganu. Kemaman	1,100,000	7	7	835	3	32,281
Dec-14	Pahang, Temerloh	180,000	7	6	541	3	22,765
Nov-88	Pahang, Pekan	606,392	7	8	573	7	21,360
Dec-14	Kelantan, Kota Bharu	14,585,000	5	9	713	3.5	20,379
Dec-88	Kelantan, Kuala Krai	4,289,105	5	6	669	4	16,994

Zone 2: South region

Year	State, District	Damages (RM)	Flood duration (days)	Rainfall duration (days)	Maximum rainfall (mm/day)	Flood depth (m)	Victim evacuated
Jan-11	Johor, Muar	16,480,000	2	6	352	3	21,955
Jan-11	Johor, Segamat	27,986,000	2	6	525	3	21,588
Jan-11	Johor, Kluang	3,864,000	2	6	341	1.8	11,790
Mar-11	Johor, Batu Pahat	6,430,000	2	6	472	1.5	9824
Jan-11	Johor, Johor Bahru	22,200,000	2	6	378	2	9055
Jan-11	Melaka, Melaka Tengah	0	2	4	201	1.3	4155

Zone 3: North region

Year	State, District	Damages (RM)	Flood duration (days)	Rainfall duration (days)	Maximum rainfall (mm/day)	Flood depth (m)	Victim evacuated
Oct-10	Kedah, Kubang Pasu	6,185,000	5	5	478	1.5	26,009
Oct-10	Kedah, Kota Setar	130,000	5	5	228	1.5	25,524
Dec-05	Kedah, Sungai Bata	–	9	4	219	3	20,269
Dec-05	Kedah, Jitra	–	9	4	200	2	20,269
Dec-05	Kedah, Padang Terap	–	9	4	192	3	20,269

(continued)

Table 14.1 (continued)

Zone 4: West region							
Year	State, District	Damages (RM)	Flood duration (days)	Rainfall duration (days)	Maximum rainfall (mm/day)	Flood depth (m)	Victim evacuated
Feb-06	Selangor, Petaling	–	0.125	3	209	1.3	4600
Oct-99	Perak, Kerian	–	1	4	213	0.5	4539
Jan-00	Selangor, Sungai Buloh	–	0.5	3	198	1.5	3560
Dec-96	Negeri Sembilan, Jempol	–	1	7	353	1	3165
Dec-14	Perak, Kuala Kangsar	–	2	7	264	1.8	2423

14.4.2 Rainfall Return Periods in Homogeneous Regions

The top six of worst flood events in zones 1 and 2, while top five worst flood events in zones 3 and 4 was chosen to be analyzed. Therefore, there are 22 major flood events in all zone in identifying rainfall return period. The highest value of the return period of worst flood events in a zone means that the worst flood events can occur in any area within similar zone. Zone or homogenous region means that any areas within the homogeneous region are considered to have similar climatic exposure, conditions, and source of extreme rainfalls. For example, 50 years return period, this does not mean that a 50-year flood will happen regularly every 50 years or only once in 50 years. In any given 50-year period, a 50-year event may occur once, twice, more, or not at all, and each outcome has its own probability. Note also that the estimated return period is a statistic; it is computed from a set of data observations, as distinct from the theoretical value in an idealized distribution.

Using GEV distribution models, the return periods of extreme rainfall associated to major flood events at zone 1 are calculated to be between 9–50 years, zone 2 between 4–333 years, zone 3 between 4–500 years, and zone 4 between 7–20 years. The highest value of the rainfall return period in a zone means that the worst flood events can occur in any area within similar zone. Zone or homogenous region means that any areas within the homogeneous region are considered to have similar climatic exposure, conditions, and source of extreme rainfalls. Identfying the rainfall return periods from major flood events within specific zones may provide additional information for designing flood defense structures or flood management. The highest return period experienced within the zone may be used as a guideline for designing or decision-making.

Figure 14.2 shows the return periods and its rainfall duration calculated from data at specific rain gauge located within its homogeneous zones. Analysis were conducted based on; 6-, 7-, 8-, and 9-days rainfall duration for homogeneous region zone 1-, 4-, and 6-days rainfall duration for zone 2-, 4-, 5-, and 8-days rainfall duration for zone 3; and 3-, 4-, and 7-days rainfall duration for zone 4.

In zone 1 for instant, the highest rainfall return period is observed to be 50 years for a 7-day rainfall event and 20 years for a 5-day rainfall event. Therefore, in designing hydraulic structures needing the extreme rainfall value estimates, engineers may consider using a 50 year (7-day rainfall duration) design of the hydraulic structures to be build in areas within the homogeneous zone 1. This is due to a 50 year rainfall has been experienced within the same homogeneous regions, and there would be a high possibility that similar magnitude of rainfall event may occur in other areas within zone 1. While for zone 2, for a rainfall duration of 2 days, we may design a 333 years return period for critical flood defense structure or for flood management decision-making since a 333 years return period event has occured.

14.5 Conclusion

The objective of this study is to analyze the rainfall return period of major flood events within homogeneous region. Flood reports and rainfall data at peninsular Malaysia were used for the analyses. Data from 22 rain gauge stations and 1503 flood events were analyzed. The stations were chosen according to the flood event that was ranked as the most severe based on number of victims evacuated and maximum rainfall values. GEV distribution model is the most suitable for analyzing the annual maximum series for Malaysia. The rainfall data of the selected flood events were then grouped accordingly to their respective homogeneous zone to carry out pooled frequency analysis. Identifying the return periods of historical rainfall events associated to major flood within a homogeneous region may provide additional information for decision makers. The result from this study can be used for the design of future flood mitigation structure. The estimation of the return periods of each zone may be improved using regional frequency analysis which consider a scaling factor and including more analysis of return periods from other rainfall stations within the zone.

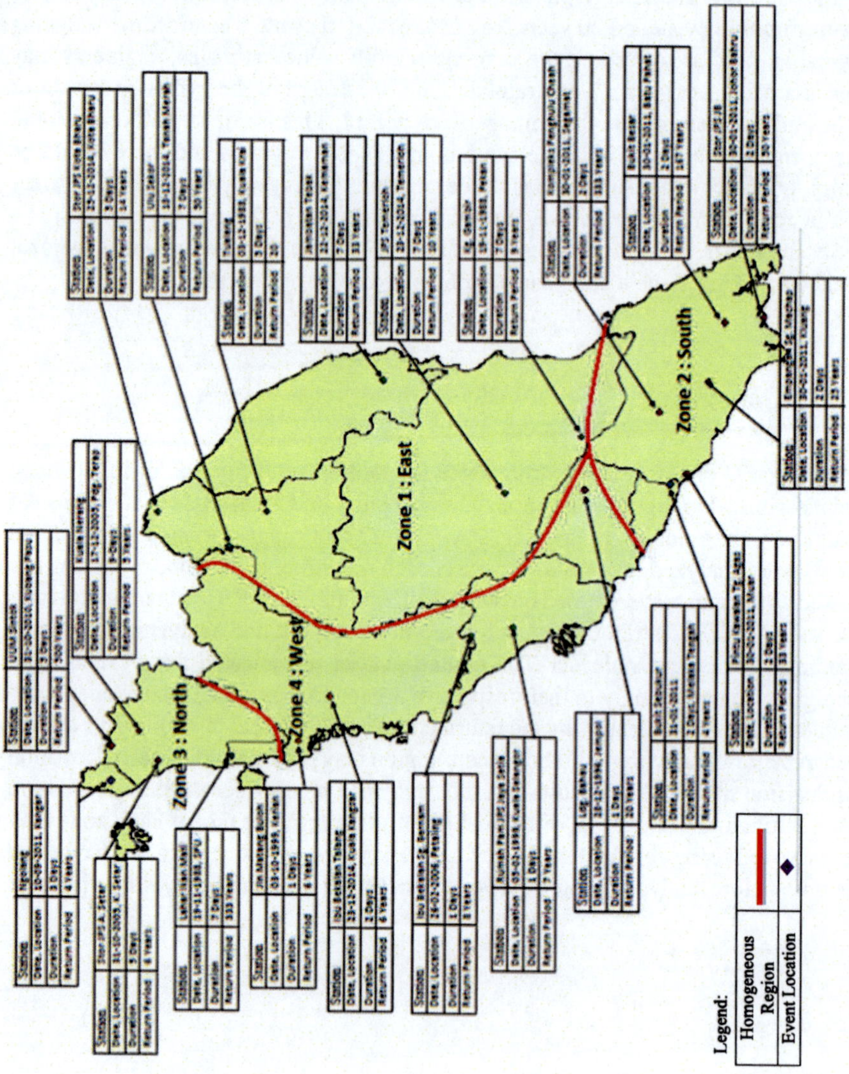

Fig. 14.2 Return period in homogeneous region in Peninsular Malaysia

Acknowledgements The authors would like to state their appreciation to the funding from the Malaysian Ministry of Higher Education (MOHE) under the GUP2018 grant with VOT No. QJ130000.2522.19H68, the special MJIIT grant with VOT No. R.J130000.7722.4J282 and JASTIP supported by Japan Science and Technology Agency (JST), Collaboration Hubs for International Research Program (CHIRP) within the framework of the Strategic International Collaborative Research Program (SICORP). The authors also thank the Department of Irrigation and Drainage Malaysia (DID) for providing rainfall data and flood reports.

References

Adamowski K, Bougadis J (2003) Detection of trends in annual extreme rainfall. Hydrol Process 17(18):3547–3560

Alias NE, Takara K (2014) Incorporating extreme rainfall behaviors into flood risk management—an assessment on probable maximum precipitation, trends and future predictions. In: DPRI annual meeting, Kyoto University, 27–28 Feb 2014

Bakar MAA, Ariff NM, Jemain AA (2016) Daily probability model of storm events in Peninsular Malaysia. World Acad Sci Eng Technol Int J Math Comput Sci 10(5)

Bedient, Huber (1948) Hydrology and flooding analysis. Addison Wesley Publishing Co

Diya SD, Mohd Ekhwan T, Muhammad Barzani G, Abdullahi MG (2014) Floods in Malaysia historical reviews, causes, effects and mitigations approach. Int J Interdisc Res Innov 2(4):59–65

Hosking JRM, Wallis JR (1993) Some statistics useful in regional frequency analysis. Water Resour Res 29:271–281

Low KC (2006) Application of nowcasting techniques towards strengthening national warning capabilities on hydrometeorological and landslides hazards. In: Public weather services workshop on warning of real-time hazards by using nowcasting technology

Mohamad S, Hashim NM, Aiyub K, Toriman ME (2012) Flash flood and community's response at Sg Lembing, Pahang. Adv Nat Appl Sci 6(1):19–25

Sen Roy S (2009) A spatial analysis of extreme hourly precipitation patterns in India. Int J Climatol 29(3):345–355

Syafrina AH, Zalina MD, Juneng L (2015) Historical trend of hourly extreme rainfall in Peninsular Malaysia. Theor Appl Climatol 120(1):259–285

Yevjevich V (1972) Probability and statistics in hydrology. Water Resources Publications, Colorado

Zalina MD, Amir HMK, Desa MNM, Nguyen V-T-V (2002) Statistical analysis of at-site extreme rainfall processes in Penisular Malaysia. In: Regional hydrology: bridging the gap between research and practice (Proceeding of the fourth international FRIEND Conference, Mar 2002. FRIEND, Cape Town, South Africa

Chapter 15
Conceptual Design for Flood Warning Study at Recreational Area—Case Study Gunung Pulai Mountain, Johor, Malaysia

Kamarul Azlan M. Nassir, Muhamad Syamil M. Zaini, and Nor Eliza Alias

Abstract It is important to conduct detail hydrological study at recreational area located at mountain sides. This is due to the fact that the geographical location and topographical conditions may post a risk of landslide dam outburst flood (LDOF). As a case for Gunung Pulai Recreation Area located at the foot hills of Gunung Pulai Mountain to be opened to public, a detail hydrological study needs to be carried out. The purpose of this study is to establish the preliminary baseline information on hydrological characteristics of the study area and to assess and analyse the existing hydrological parameter that can potentially cause flash flood within the recreational area. Then, proposing a long-term engineering solution towards prevention loss. A study area potential to be exposed to the LDOF within the recreational area needs to be determined. Hydrologic modelling is conducted to get a better understanding of the parameters in the study area. Rain gauge and water level were inserted at certain elevation level of the study area were conducted to provide the hydrological parameters needed to estimate the relationship between the rainfall and water level. Data obtained from the rain gauge and water level were plotted. The relationship between rainfall and water level is imperative to show the time needed for an alarm to be triggered. In conclusion, the parameters that can potentially indicate the occurrence of flash flood has been analysed and can be altered to make them useful against future flash flood events. A long-term solution has also been devised and ready to be deployed.

Keywords Flood warning · Flood management · Mountain · Flood modelling

K. A. M. Nassir (✉) · M. S. M. Zaini
School of Civil Engineering, Faculty of Engineering, Universiti Teknologi Malaysia, 81310 Johor Bahru, Johor, Malaysia
e-mail: kamarul.azlan@yahoo.com; kamarulazlan@utm.my

N. E. Alias
Centre for Environmental Sustainability and Water Security (IPASA), Universiti Teknologi Malaysia, 81310 Johor Bahru, Johor, Malaysia
e-mail: noreliza@utm.my

H. Tatano and A. Collins (eds.), *Proceedings of the 4th Global Summit of Research Institutes for Disaster Risk Reduction*, Disaster and Risk Research: GADRI Book Series,
https://doi.org/10.1007/978-981-19-5566-2_15

15.1 Introduction

On 27th December, 2001, tropical storm had caused massive flash flood to occur at Gunung Pulai Mountain. The tropical rainstorm occurred non-stop approximately from 2 pm, December 26th to 2 am, December 27th the next day. The two-day cumulative rainfall prior to the date of debris flow was 15 mm, and the total rain on December 27th was 90 mm. The increased in the intensity of the rain has caused the water level of the river to increase steadily. Therefore, as a result of increased in water level of the river, flash flood has occurred. The unprecedented flash flood has caused some serious casualties. Nearby residence was unable to react as they did not know that a massive flood has occurred. Not to mention, the rangers, firemen and rescuers could not take further action at such predicament. They have to wait until the flood has subsided. Therefore, until a detail hydrological study is carried out at the Gunung Pulai Recreational Area, the area is not safe for public. The Malaysia Forest Department has decided that a solution or mitigation measure must be created to overcome the potential danger due to flash flood and artificial dam burst problems within the study area.

The study area (Gunung Pulai Recreational Area) is located within the Gunung Pulai Mountain, Pontian, Johor. The recreational area is located within Level 1 and Level 2 of Gunung Pulai (Fig. 15.1) and is located within the lower level area of the Pulai Mountain. The area could be potentially exposed with frequent cases of flash flooding. Figure 15.2 shows the flooded area within the catchment during the 2001 flood. The 2001 flood causes not only losses of lives but also damages to properties and infrastructures and causes the park to be closed to the public.

Figure 15.3 shows the location of the catchment area in which the Gunung Pulai Recreational Park lies. The Pulai mountain area is situated east of the Malacca strait.

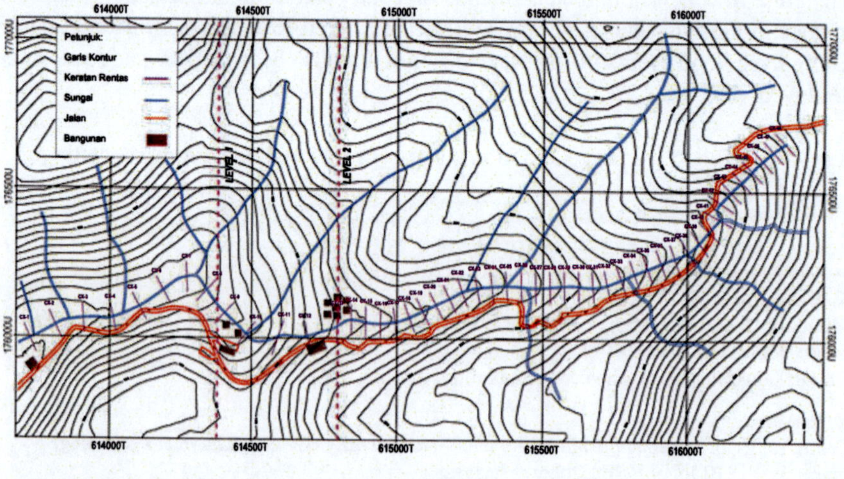

Fig. 15.1 Location of the study area—Gunung Pulai recreational area

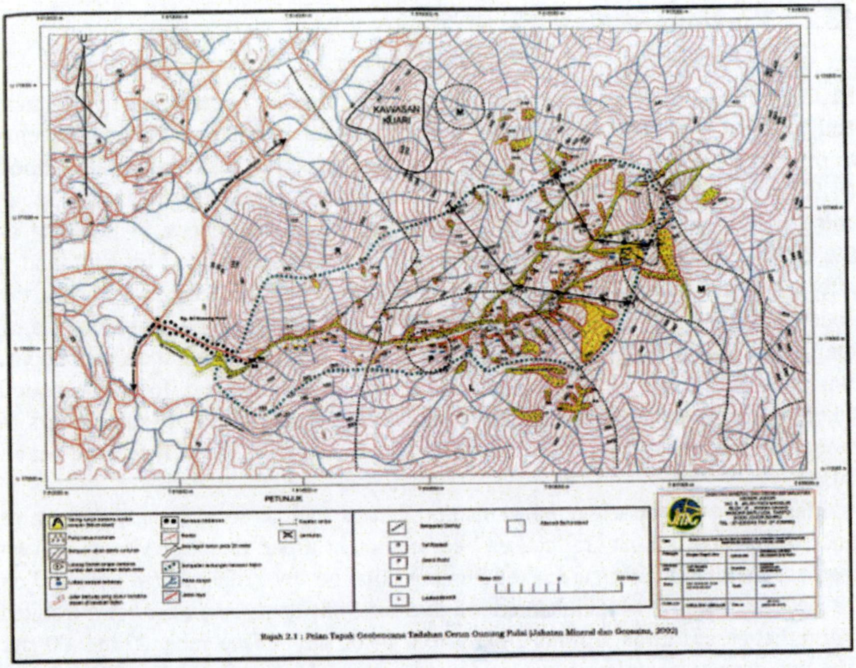

Fig. 15.2 Affected area during the 2001 flood

The catchment under study is the Sg Air Hitam River catchment which lies in Pontian, Johor, Malaysia. The Sg Air Hitam consists of three sub-catchments numbered as 1, 2 and 3.

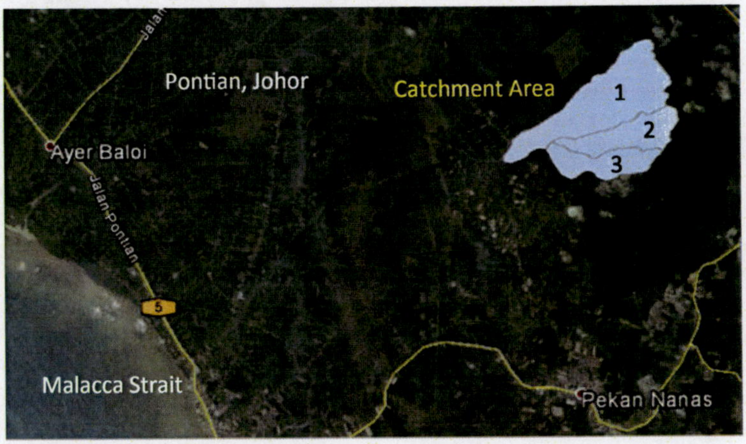

Fig. 15.3 Location of the study area (google map)

15.2 Previous Study

Malaysia is free from most of natural disasters. Accurance of earthquakes, volcanos and typhoon is some which usually does not occur in Malaysia however seems to influence us. The most severe natural disaster experienced in Malaysia is flood (Khan et al. 2014). There is a total of 189 river basins in Malaysia including Sabah and Sarawak in which 89 of them are prone to flood occurrences. 29,800 km^2 or 9% of the total Malaysia area is estimated as vulnerable to flood disaster, and it affects almost 4.82 million people which is around 22% of the total population in the country (Diya et al. 2014). There are two types of floods categorised in Malaysia by the Department of Irrigation and Drainage (DID), flash flood and monsoon floods. Based on the hydrological perspectives, these two flood types differ in the period taken by the river flow to recede to normal. Flash floods take only some hours to recede back to normal, whilst monsoon flood can last for a month (Mohamad et al. 2012).

Flash flood can be triggered by natural causes such as local weather known as line squalls and non-natural causes like inefficient urban drainage system and an increase in urban built-up areas. At Gunung Pulai however, debris flow occurred on 28 December 2001, which claimed the lives of 5 residents as a result of a tropical storm that passed through Johor Bharu from 14:00 pm on December 27 to 2:00 am. The two days of cumulative rain prior to the date of debris flow were 15 mm, and the total rain on December 28 was 90 mm. It was suggested that strong wind caused trees to fall into the river and formed a natural dam, which breached, causing debris flow (Jamaludin et al. 2014).

Other causes that can be hypothesised is the occurrence of intense rainfall such as cloudbursts. Cloudbursts can occur in remote areas as a result of topographic variations. They are generally unreported because of inaccessibility and isolation. This type of intense rainfall event may occur in the high reaches of a mountain stream and in unpopulated areas. However, it effects can produce flash flood impacting downstream communities. Prolonged monsoon rainfall and monsoon depression can also cause flash floods. Young geology formation and the instability of mountain slopes are prone to recurrent and often devastating landslides and debris flows. Such landslides and debris flows, released by torrential rain or even seismic activity, may cause temporary dams across river courses impounding immense volumes of water. Subsequent overtopping or breaking through of the earth dam can result in a landslide dam outburst flood (LDOF) in which it is difficult to predict and may cause serious loss of life and damage to property (Shrestha and Bajracharya 2013).

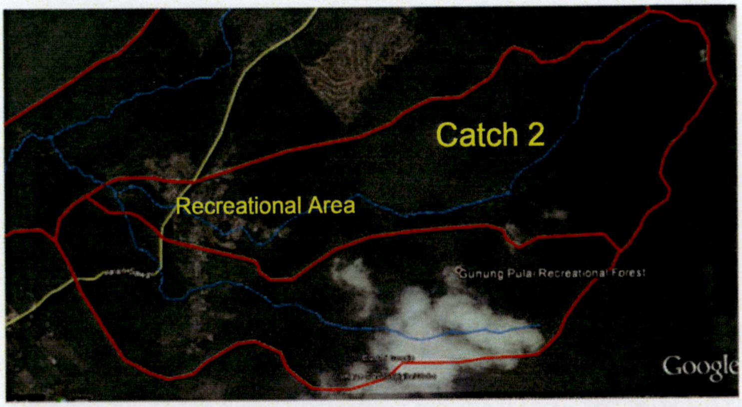

Fig. 15.4 Location of the recreational area (google map)

15.3 Methodolgy

15.3.1 Study Area

The study area (recreational area) is located within the catchment of Air Hitam River. Air Hitam River catchment lies within Pontian River Basin. Air Hitam River eventually discharges towards Straits of Melaka. Air Hitam catchment can be divided into three sub-catchment areas: Catch 1, 2 and 3. The recreational area under study is in Catch 2 as shown in Fig. 15.4.

Only, sub-catchment 2 will be considered in this study because the recreational area lies within this catchment. Surface runoff from the catchment area drained towards streams and channels flows through the recreational area before discharging towards Air Hitam River. The surface runoff discharges towards the upstream area of Air Hitam River which is about 2.5 km from the project site.

Figure 15.5 shows the existing drainage system within the study area. High water level in Air Hitam River would result in flash flood within the recreational area at Level 1 and Level 2. The labelled sub-basin in Fig. 15.5 will be used in hydrologic modelling.

15.3.2 Hydrologic Modelling

The purpose of hydrologic modelling is to estimate flow hydrograph from tributary catchment for various ARIs. The estimated flow hydrograph serves as an input to the hydraulic modelling of the study area. The hydrologic model in HEC-HMS is available and can be downloaded from the US Army Corps website. There are plenty

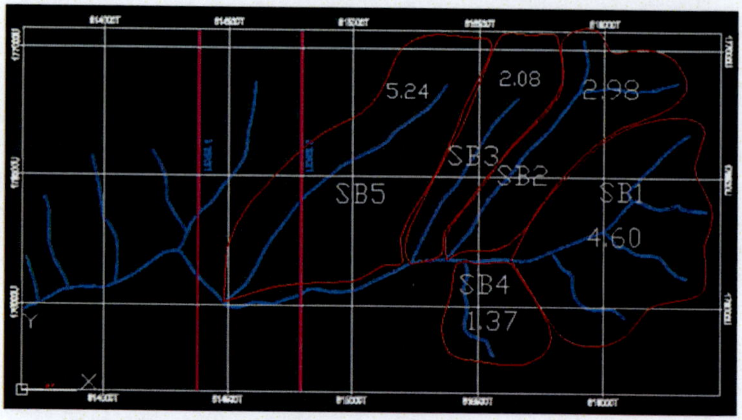

Fig. 15.5 Drainage system within catchment 2

of options available in this module for calculating catchment losses, transformation of excess rainfall and base flow estimation. Loss methods used for HEC-HMS modelling are Initial and Constant Loss method. The parameters require to complete this modelling are Initial Loss, Constant Rate and Impervious Percentage Value. For the purpose of modelling, the catchment was divided into several sub-catchments. These sub-catchments are represented as basins in HEC-HMS. The selections of the basins are based on the consideration of certain aspects of the catchment characteristics and locations where determination of flow is required. The schematic representation of the study area for the hydrologic modelling in HEC-HMS showing all sub-catchment tributaries, junctions and reaches is shown in Fig. 15.6.

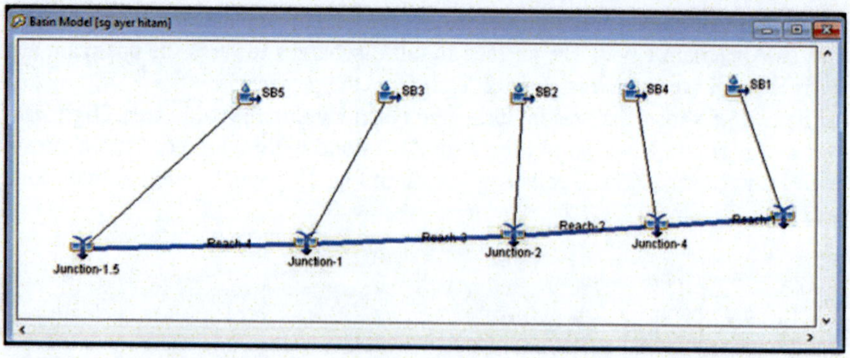

Fig. 15.6 Study area in HEC-HMS modelling. (All the basins, junctions and reaches)

15.3.3 Hydrologic Simulation

Hydrologic modelling of flows for the catchment area was carried out using HEC-HMS model. The design storm selected for this study depends on the time of concentration of the study area (t_c). The t_c was estimated by using the Barnsby William formulae (Eq. 15.1). The estimated t_c at the outlet and sub-catchment is listed in Table 15.1.

$$t_c = \frac{F_c.L}{A^{1/10}\,S^{1/5}}. \tag{15.1}$$

where

- t_c the time of concentration (minute).
- F_c a conversion factor (58.5 for area A in km^2, or 92.5 for A in ha.).
- L length of flow path from catchment divide to outlet (km).
- A catchment area (km^2 or ha).
- S slope of stream flow path (m/km).

Therefore, the design storm used in this study for the Recreational Area is 1 h. The nearest intensity–duration–frequency (IDF) curve available published by DID Malaysia for this study area is the Pekan Nenas IDF curve. Therefore, the study uses the IDF derived for Pekan Nenas for the simulation. The considered design storm is storm with the return period of 50 years ARI. The hydrologic losses for the area were based on initial and continuing loss method for forested area. The initial loss is assumed to be 15 mm, and the continuing loss is assumed to be 25 mm/hr. It is also assumed that the existing land cover consists of about 5% impervious area for the catchment area. The transformation of effective design storm to the outlet area is based on Clark time-area method. The selection of design storm was described earlier. The two parameters used for the development of this synthetic unit hydrograph are T_c and R. These two parameters can be obtained from observed hydrograph. In the absence of the observed hydrograph, the parameters can be estimated from regression equations derived areas with gauged data. The regression equation used in this study is derived from a study in small rural watersheds in Illinois, USA. The regression equations are as listed below.

$$T_c = 1.54\,L^{0.875}\,S^{-0.181}$$

$$R = 1.64\,L^{0.342}\,S^{-0.790}$$

Table 15.1 Estimated t_c values for catchment Level 1 and Level 2

Catchment	C. area (km^2)	L (km)	S (m/km)	T_c (min)	T_c (hr)
Level 1	7.5	2.5	130	45	0.8
Level 2	7	2	140	36	0.6

Table 15.2 Estimated t_c values for catchment 1 to catchment 5

Catchment	*C.* area (km^2)	*L* (km)	*S* (m/km)	T_c (min)	T_c (hr)
CATCH 1	4.6	0.916	152.8384	17	0.3
CATCH 2	2.98	1.049	152.5262	20	0.3
CATCH 3	2.08	0.765	183.0065	15	0.2
CATCH 4	1.37	0.435	218.3908	8	0.1
CATCH 5	5.24	1.247	144.3464	23	0.4

The T_c and R values within the sub-catchment are listed below in Table 15.2. L is the stream length measure along the main channel from the outlet to the watershed. S is the slope of stream flow path.

15.3.4 Field Data Collection

The approach of the current field data collection for the recreational area is to obtain lead time for flood warning purposes. Existing flood warning infrastructures within the recreational area are non-existence or almost negligible. There are two equipment being installed within the recreational area for the purposed of providing some early indication of incoming headwater. A rainfall station was installed at the highest elevation of the catchment boundary, whilst a water level was installed at Level 1. Figure 15.7 shows the location of the equipment installed within the study area.

15.4 Results and Discussion

Data obtained from field work and modelling were used to assess the hydrological characteristic of the study area.

15.4.1 Modelling

The modelling work in HEC-HMS contains 5 sub-basins, 4 reaches and 5 junctions as shown in Fig. 15.6. The results of peak discharge for 50 years design storm are shown in Figs. 15.8 and 15.9.

From the figures, we can clearly see that the highest peak discharge is 320.6 m^3/s which is at level 1 of the study area. The time of peak is 01:15. Whilst at level 2, the peak discharge is 238.9 m^3/s, and the time of peak is 01:00. This indicates that the time difference is 15 min for the water to flow from level 2 to level 1. Figure 15.9 shows the graph of flows against time for level 2 and level 1.

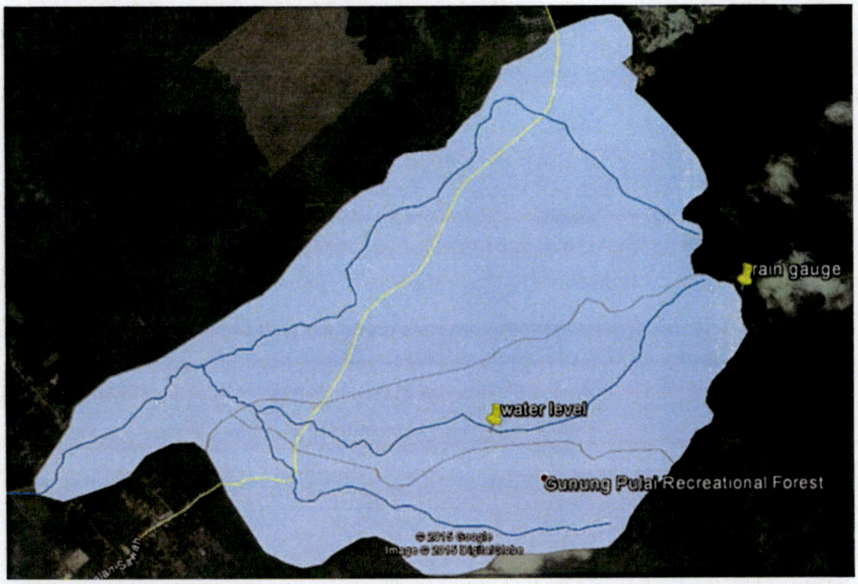

Fig. 15.7 Location of equipment installed at project site

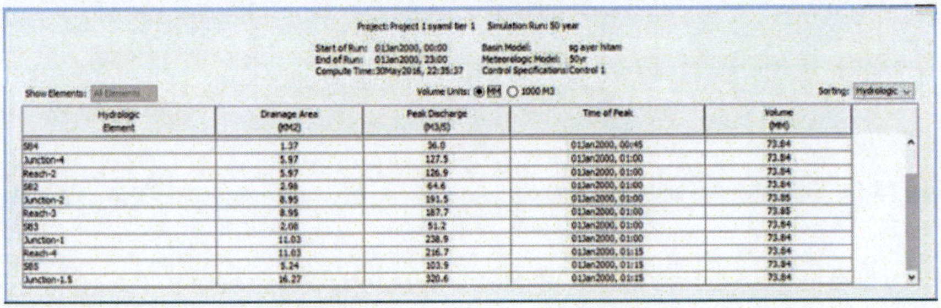

Project: Project 1 syami tier 1 Simulation Run: 50 year

Start of Run: 01Jan2000, 00:00 Basin Model: sg ayer hitam
End of Run: 01Jan2000, 23:00 Meteorologic Model: 50yr
Compute Time: 30May2016, 22:35:37 Control Specifications: Control 1

Show Elements: All Elements Volume Units: MM 1000 M3 Sorting: Hydrologic

Hydrologic Element	Drainage Area (KM2)	Peak Discharge (M3/S)	Time of Peak	Volume (MM)
SB4	1.37	36.0	01Jan2000, 00:45	73.84
Junction-4	5.97	127.5	01Jan2000, 01:00	73.84
Reach-2	5.97	126.9	01Jan2000, 01:00	73.84
SB2	2.98	64.6	01Jan2000, 01:00	73.84
Junction-2	8.95	191.5	01Jan2000, 01:00	73.85
Reach-3	8.95	187.7	01Jan2000, 01:00	73.85
SB3	2.08	51.2	01Jan2000, 01:00	73.84
Junction-1	11.03	238.9	01Jan2000, 01:00	73.84
Reach-4	11.03	216.7	01Jan2000, 01:15	73.84
SB5	5.24	103.9	01Jan2000, 01:15	73.84
Junction-1.5	16.27	320.6	01Jan2000, 01:15	73.84

Fig. 15.8 Peak discharge of 50 years ARI

15.4.2 Rainfall and Water Level

The rainfall and water level data collected from October to November are shown in Fig. 15.10. There are a few rainfall events during that period that causes water level in the river to increase during that storm events. The highest water level recorded was during the 27th October storm event.

The highest water level recorded was during the 27th October storm event. The two–hour storm event (6.00–8.00 pm) with total rainfall of 61 mm (Table 15.3) has resulted in water level to rise to about 0.45 m (Fig. 15.11). This amount of rainfall for this event is almost equivalent to 2 year ARI (70 mm). The peak water level was recorded at 7.40 pm, which is about 40 min lag. This shows the response time is quite

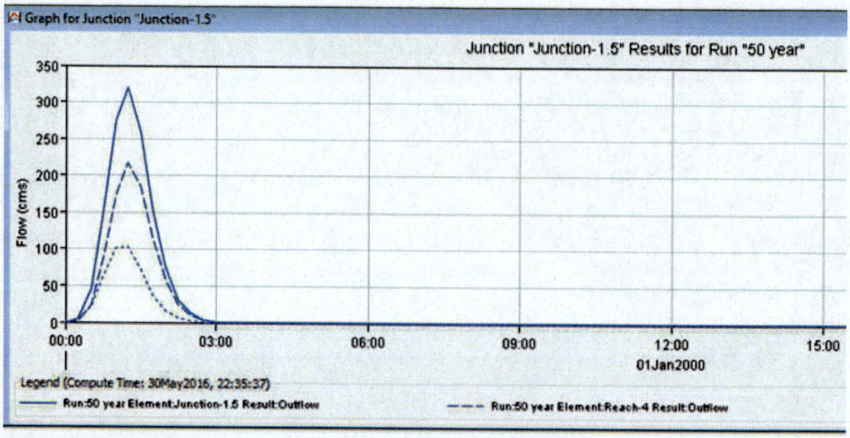

Fig. 15.9 Graph showing the peak flow against time at Junction-1.5

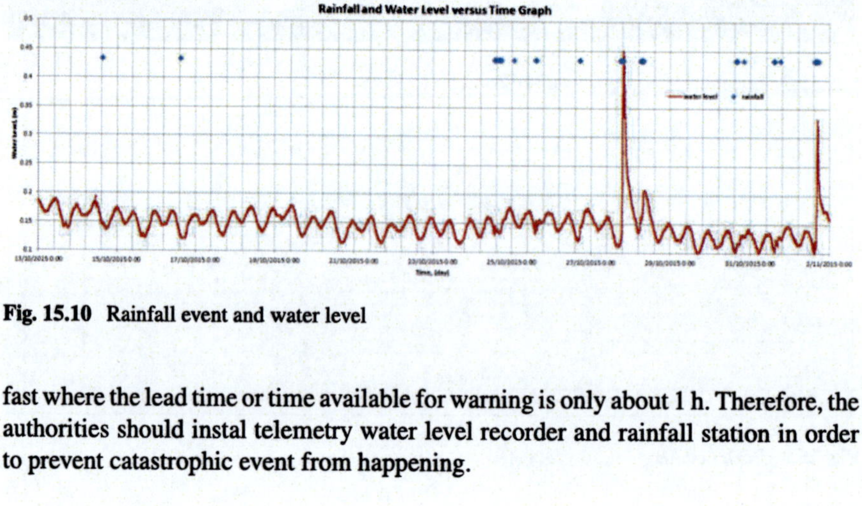

Fig. 15.10 Rainfall event and water level

fast where the lead time or time available for warning is only about 1 h. Therefore, the authorities should instal telemetry water level recorder and rainfall station in order to prevent catastrophic event from happening.

15.5 Conclusion

Based on the hydrologic modelling and field data study, the results can be used to assess and analyse the existing hydrological parameter. From the data, we can say that the modelling process helps in replicating the hydrological data and turns them into a simple flow against time graph which can help determining the response time needed for the alarm system. Not to mention, the data obtain from rain gauge and water level data set up at the study areas provide rainfall and water level data so that we know when the rain reaches certain level, the water level will rise accordingly.

Table 15.3 Rainfall on the 27th October 2015

Time	Ticking	Rainfall (mm)
11:00:00 AM–12:00:00 PM	0	0
12:00:00 PM–1:00:00 PM	0	0
1:00:00 PM–2:00:00 PM	0	0
2:00:00 PM–3:00:00 PM	0	0
3:00:00 PM–4:00:00 PM	0	0
4:00:00 PM–5:00:00 PM	0	0
5:00:00 PM–6:00:00 PM	0	0
6:00:00 PM–7:00:00 PM	167	33.4
7:00:00 PM–8:00:00 PM	138	27.6
8:00:00 PM–9:00:00 PM	0	0

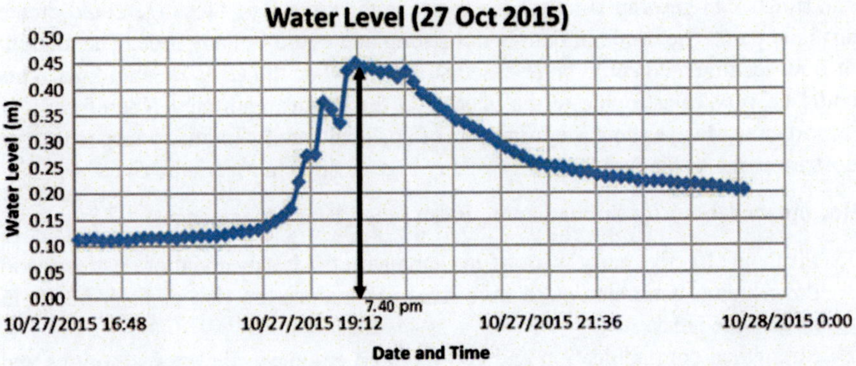

Fig. 15.11 Water level recorded on the 27th October 2015

The problem of opening the area to the public for recreational purposes especially for picnic and swimming could expose the visitors to possible danger of drowning or being washed away due to fast flowing flood as a result from heavy storm in the upstream area or artificial dam burst. The solution towards solving the impending danger to the tourist area (Level 1 and 2) due to these fast flowing flash flood phenomenon (steep slope or dam burst) could be based on various approaches which include structural and non-structural approach. The non-structural approach includes early warning system and flood zoning. The structural approach may include storage approach in the upstream area. Whilst storage approach seems to be a more appropriate solution, but site conditions (steep slope and fast flowing flood) may not permit this approach. The hilly terrain with very steep slope limits the ability of using this approach. This approach could also disturb the natural beauty of the area.

Therefore, non-structural approach such as early warning system and limited time visit approach is more appropriate for this recreational area. This early warning system approach would provide lead time (1 h) to the authorities to warn pubic and visitors of the impending danger. The limited time visit approach is to limit the access to the recreational area to during the monsoon season (October–January). This approach is more economical and allows the authorities some time to do maintenance work when the park is closed. The non-structural measures include early warning system and flood zoning. The early warning system consists of telemetry rainfall and water level recorder at various locations in the upstream area of the recreational park. The authority should provide personnel to monitor the situation at the recreational area during visiting hours by monitoring the amount of rainfall and monitoring the water level that could trigger the alarm. The visitors to the park should be warned of the impending danger whenever the amount of rainfall or water level exceeds certain level. The telemetry rainfall station should be located at the existing location whilst the water level recorder should be located at the middle stretch between the catchment boundary and the park. The upstream area of the recreational park should also be maintained regularly. Regular maintenance of the river in the upstream area by removing tree branches and tree trunks from blocking the water way. Tree branches and trunks can get trap amongst the rocks in the river. The blockage could create artificial dam. The artificial dam is not strong and could fail any time. The amount of flow accumulated could be released suddenly when the artificial dam fails. This could be very catastrophic to the visitors at the downstream area. Therefore, it is important to always avoid the formation of artificial dam by regularly inspecting the upstream area of the park and remove any logs or branches from the river.

Recommendations for Strengthening Flash Flood Risk Management:

1. Promote effective early warning systems with the involvement of upstream and downstream communities to save lives and reduce the risk of flash floods in vulnerable areas.
2. Strengthen communication and coordination amongst relevant institutions and as part of national disaster risk management strategies.
3. Conduct flash flood modelling and hazard mapping to identify hazard prone areas and develop land use guidelines and building codes.
4. Develop a standard methodology and format for documenting flash flood events and for subsequent reflection on the causes, effects, and lessons that can be derived from such events.

References

Diya SG, Gasim MB, Toriman ME, Abdullahi MG (2014) Floods in Malaysia: historical reviews, causes, effects and mitigations approach. Int J Interdisc Res Innov 2(4):59–65

Jamaludin S, Abdullah CH, Kasim N (2014) Rainfall intensity and duration for debris flow triggering in Peninsular Malaysia. In: Proceedings of world landslide forum 3, 2–6 June 2014, Beijing

Khan MMA, Shaari NAB, Bahar AMA, Baten MA, Nazarudin DAB (2014) Flood impact assessment in Kota Bharu, Malaysia: a statistical analysis. World Appl Sci J 32(4):626–634. https://doi.org/10.5829/idosi.wasj.2014.32.04.422

Mohamad S, Hashim NM, Aiyub K, Toriman ME (2012) Flash flood and community's response at Sg. Lembing, Pahang. Adv Nat Appl Sci 6(1):19–25 ISSN 1995-0772

Shrestha AB, Bajracharya SR (2013) Flash flood risk management in the Hindu Kush Himalayan region. Case studies on flash flood risk management in the Himalayas: In support of specific flash flood policies. Shrestha AB, Bajracharya SR (eds) Publisher: International Centre for Integrated Mountain Development, Kathmandu, ISBN 978 92 9115 269 8

Chapter 16
Impact of Vibration and Moisture Content to Light Non-aqueous Phase Liquid Migration in Double-Porosity Media

Mohd Fadhli Abd Rashid, Noraliani Alias, Kamarudin Ahmad, Radzuan Sa'Ari, and Mohd Zamri Ramli

Abstract Disaster such as earthquake has an impact to water resources particularly groundwater. The migration of pollutant in groundwater is crucial particularly due to earthquake impact. In addition, the soil moisture content also might influence the pollutant migration. Therefore, this paper presents the investigation of light non-aqueous phase liquid (LNAPL) migration in laterite soil with two dissimilar moisture contents with and without vibration. The study of LNAPL migration was investigated using digital image analysis. The lab-scale experiments were conducted by using soil column, mirror, LNAPL and Nikon D90 digital camera. Unsaturated laterite soil was poured in the acrylic column then compressed until 100 mm height. Then, LNAPL was poured onto the soil column surface instantaneously. The pattern and behavior of LNAPL migration in laterite soil was monitored and recorded using digital image processing technique (DIPT) at certain time intervals. The images were processed with Surfer software and Matlab routine inorder to plot the LNAPL migration pattern using hue-saturation-intensity (HSI) value. The study found that higher rate migration of LNAPL with high moisture content, and the rate of LNAPL migration decreases with presence of vibration and with lower moisture content. The migration time required to reach bottom of soil was longer for low moisture content with/without vibration as compared to the high moisture content with/without vibration.

Keywords Migration · Light non-aqueous phase liquid · Unsaturated laterite soil · Moisture content · With/without vibration · Digital image processing technique

---

M. F. A. Rashid · N. Alias (✉) · R. Sa'Ari
Department of Water and Environmental Engineering, School of Civil Engineering, Faculty of Engineering, Universiti Teknologi Malaysia, 81310 Johor Bahru, Johor, Malaysia
e-mail: noraliani@utm.my

K. Ahmad
Department of Geotechnics and Transportation, School of Civil Engineering, Faculty of Engineering, Universiti Teknologi Malaysia, 81310 Johor Bahru, Johor, Malaysia

M. Z. Ramli
Institute of Noise and Vibration, School of Civil Engineering, Faculty of Engineering, Universiti Teknologi Malaysia, 81310 Johor Bahru, Johor, Malaysia

H. Tatano and A. Collins (eds.), *Proceedings of the 4th Global Summit of Research Institutes for Disaster Risk Reduction*, Disaster and Risk Research: GADRI Book Series,
https://doi.org/10.1007/978-981-19-5566-2_16

16.1 Introduction

Increasing water demand significantly affects on quantity and quality of water available at local and global scales (IPCC 2016; Desa 2030; WWAP 2015). Groundwaters as one the water resource are also affected to leakage and spillage which contaminate the groundwater and are unsafe for consumption as well as industry usage. According to Harter (2003), human activities highly influence the undesirable changes in the quality of groundwater that ultimately leads to groundwater pollution. Pollution initiated by mankind has been a recurring problem ever since humans started to carry out animal rearing and crop farming once they had settled down into communities and villages (Maslia et al. 2016). Domenico and Schwartz (1998) stated that the characteristics of groundwater pollution consist of its loading history, the degree if localization, and the production of contaminant types produced. (Huling and Weaver 1991; The Environment Agency 2018) had stated that organic contaminants are able to exist in an aqueous phase, non-aqueous phase (as NAPL), or in the gaseous phase. Unfortunately, the impact of earthquake impact to the migration of pollutant into groundwater is not known. Therefore, this study will investigate the migration of light non-aqueous phase liquid in the groundwater for with and without vibration in different soil moisture contents.

16.2 Literature Review

Hydrocarbons are one of the hazardous chemicals that are immiscible in double-porosity medium. The flow rate and occurrence of hydrocarbon liquid migration is affected by the soil structure. Normally, soils have different soil structures and consist of non-homogenous characteristics. Double-porosity media displays two specific scales of porosity media and bimodal pore-size distribution focusing on intra- and inter-aggregate pores in aggregated soil such as in agricultural and compacted soil (Carminati et al. 2008; El-Zein et al. 2006; Li and Zhang 2009). According to Lewandowska et al. (2005), the characteristics of double-porosity soil could be re-created in an experiment on one-dimensional infiltration tests. Bagherieh et al. (2009) conducted consolidation and one-dimensional experiments using aggregated kaolin sample. Recently, more physical laboratory on double-porosity soil have started immerge from researchers such as (Alazaiza et al. 2017; Loke 2019; Abd Rashid et al. 2018; Kamaruddin et al. 2011; Loke et al. 2018; Tran Ngoc et al. 2014; Ngien 2012), and this has contributed to a deeper understanding of the soil characteristics in double-porosity soil.

A technique of digital image processing was used to observe and analyze the liquid migration in soil media. The usual methods used to monitor and record the migration of liquid require a lot of measuring apparatus, which involves the samples' original setup so as to solve the issue of using non-destructive and nonintrusive methods. Based on (Agaoglu et al. 2015), experiments involve non-invasive imaging techniques which used observation and characterization of systems for bigger precision. Furthermore, researchers such as (Loke 2019; Kamaruddin et al. 2011; Ngien 2012; Darnault et al. 2001, 1998) have conducted and analyzed the techniques of nonintrusive image for experiment of liquid migration. Results of the experiment indicated that the aggregated soil response approached on non-aggregated soil in compressibility and fluid retention when applied above a certain vertical stress.

The abovementioned researchers' experiments on the LNAPL migration have contributed to part of the knowledge with significant introduction in double-porosity soil. The characteristics of a soil depend on the physical properties of the soil such as mineralogy of soil, density, dry density or void ratio, grain size distribution, cohesion, particle shape, and internal friction angle. The researchers had stated that there had been problems in gathering reliable data on immiscible fluid movement behavior. Therefore, there is a need to study the migration of contaminants in laterite soil as an attempt to bridge the research gap to determine the rate of migration with different water contents in soil. This article will close the research gap by conducting experiments on LNAPL migration in laterite soil with the objectives to determine migration pattern on red-dyed oil in aggregated laterite soil using techniques of digital image processing with different moisture contents.

16.3 Methodology

The main objective of the study is to investigate the migration pattern of LNAPL through double-porosity soil with and without vibration impact for different soil moisture contents. Figure 16.1 displays the flowchart of experimental laboratory. The soil samples were compresed until 100 mm height as shown in Fig. 16.2a, b. All experiments were first set up at 23 °C room temperature as shown in Fig. 16.3. Every item in Fig. 16.3 is needed to be done with remained stationary throughout the test duration. Figure 16.2 shows the acrylic soil column which contained soil sample which was used to monitor and measure the hydrocarbon liquid migration that occurred inside the soil column with the aim of nonintrusion and non-destruction of the original soil sample setup. V-shaped mirror was set up, and the position of soil column of each sample was arranged for the red-dyed oil migration image acquisition as showed in Fig. 16.3.

The equipment for migration of hydrocarbon liquid with image acquisition system was a Nikon D90 DSLR digital camera with fixed sensor size of 23.6 × 15.8 mm and 3216 × 2136 pixels image format, 5.6 × 5.6 μm sizes for each pixel. The International Organization for Standardization (ISO) speed and aperture setup of digital camera during the experiments was set at 2500 and f/5.6, respectively, for all experiments. Samples 1 and 2 were first sheathed in white paper with a pre-drawn gridline (20 × 20 mm) onto the soil column as a control point on the reference image. Both experiments began by instantaneously pouring the 70 ml red-dyed toluene onto the center of the aggregated laterite surface in the soil column. The subsequent images were taken with specific time intervals to snap the red-dyed toluene migration. The

Fig. 16.1 Flowchart of experimental laboratory

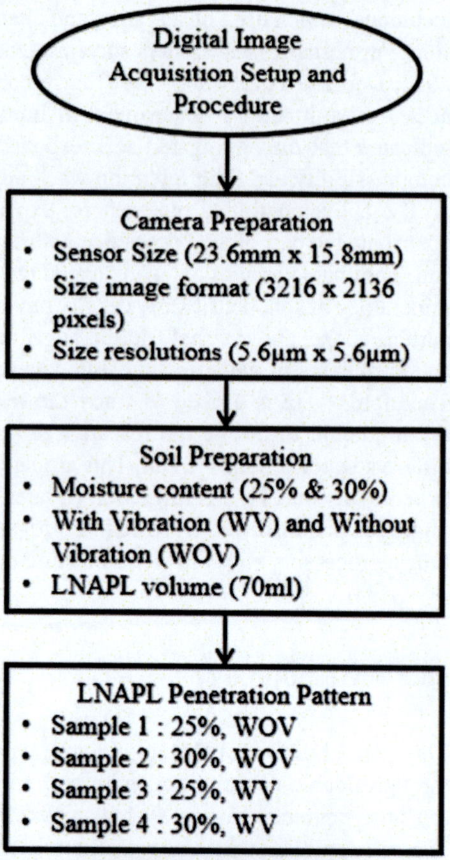

a: Soil column plan view b: 10 cm height of soil sample

Fig. 16.2 Preparation soil sample

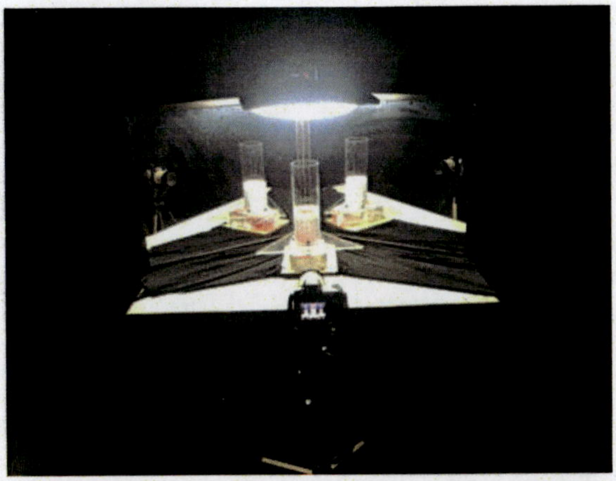

Fig. 16.3 Digital image experimental setup

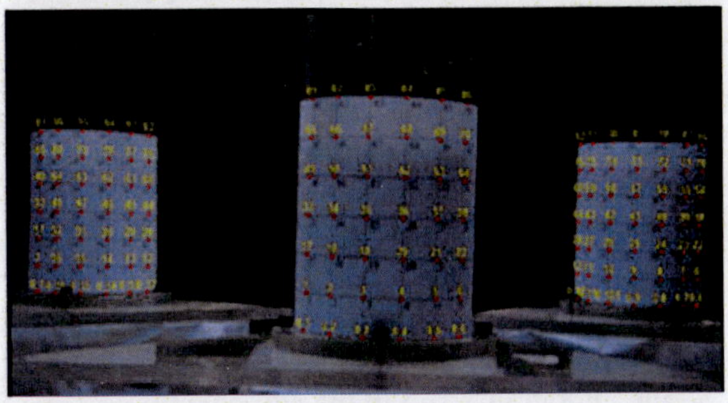

Fig. 16.4 Control points for each reference

images were captured and saved in the format of JPEG. Image processing using computer software specifically, Surfer Software and Matlab routine, the images in JPEG were then transferred to a computer from the digital camera. To extract the image coordinates of control points, the first step was digitized the control points on reference image. Area of control point's image refers to the boundary area of migration for sample 1 and sample 2 as shown in Fig. 16.4.

16.4 Results and Discussion

This section presents results and detailed discussion on classification of laterite soil properties and physical laboratory on LNAPL migration pattern. Employing DIPT to monitor LNAPL migration pattern and vibration impact leads to improvement of soil and decreases the migration rate of LNAPL through the double-porosity laterite soil for protection and restoration and provides essential and cost-effective support for planning and remediation of policies, the groundwater flow system regulations and engineering design.

Table 16.1 Laterite soil properties

Property	Value
Liquid limit	66%
Plastic limit	33%
Plasticity index	33%
Particle density	2.74 Mg/m3
USCS classification	Clay with high plasticity (CH)
Saturated permeability, $K_{average}$ (m/s)	4.62 × 10–7

Soil Characteristics Investigation

The laterite soil properties were tested for the purpose of obtaining the Atterberg Limits, particle size distribution, solid particle density, and saturated permeability. The properties of the laterite soil samples are display in Table 16.1.

Light Non-Aqueous Phase Liquid Migration Pattern

The acrylic soil column which contained different moisture content was used to monitor and measure the 70 ml volume of hydrocarbon liquid migration that occurred inside the soil column with the aim of nonintrusion and non-destruction of the original soil sample setup. The pattern of downward migration in the aggregated double-porosity laterite soil with and without vibration on 25% water content for Sample 1 and Sample 3, and 30% of water content for Sample 2 and Sample 4 can be seen as shown in Figs. 16.5, 16.6, 16.7 and 16.8.

The migration liquid in the soil column was one-dimesional; meanwhile, when the plot in the curve joint of the right and left boundaries merge apparently can form HSI plot clearly in two-dimesional. Red-dyed oil toluene was poured using a glass funnel instantaneously at center of the soil sample surface to ensure that the red-dyed oil toluene downward migrated moved in one direction as observed in Fig. 16.4. In these samples, hydrocarbon liquid would have migrated into the laterite soil sample before the whole soil sample surface was covered.

Selected HSI plots of hydrocarbon liquid migration recorded in Sample 1 at of 3, 180, and 430 s, Sample 2 at 3, 90 and 132 s, sample 3 at 3, 320, and 830 s and sample 4 at 3, 90, and 186 s were selected as can be seen in Figs. 16.5, 16.6, 16.7 and 16.8. Based on observation of the HSI intensity contour plot results in Sample 1, Sample 2, Sample 3, and Sample 4, it can be seen that the flow of migration pattern was consistently moving downward of x-axis horizontal line due to non-homogenous of double-porosity soil. All sample shows at 3 s after the instantaneous pouring, the red-dyed toluene migration reached a quartile of the height of soil sample.

At the beginning time, the migration rate was in between for all samples. However, the second image of HSI shows the migration rates were affected by moisture content and with/without vibration presence. It can be concluded that the results differentiated and identified on with and without vibration impact the soil Sample 1 and Sample 3 had the slowest migration compared to the Soil sample 2 and Sample 4 as a shorter time to red-dyed oil migrated to reach bottom of soil sample. Furthermore, time taken for red-dyed oil to reach at the bottom for soil sample with a vibration effect had the slowest migration compared to soil sample without vibration.

According to Fig. 16.9, it is observed that the migration rate of Sample 1 and Sample 3 is slower compared to Sample 2 and Sample 4. The migration rate of Sample 1 and Sample 3 is 1.25 and 0.75 mm/s, while Sample 2 and Sample 4 are 2.00 and 1.29 mm/s. This migration time was longer for Sample 1 and Sample 3 because of the immiscible hydrocarbon liquid interaction with presence of low

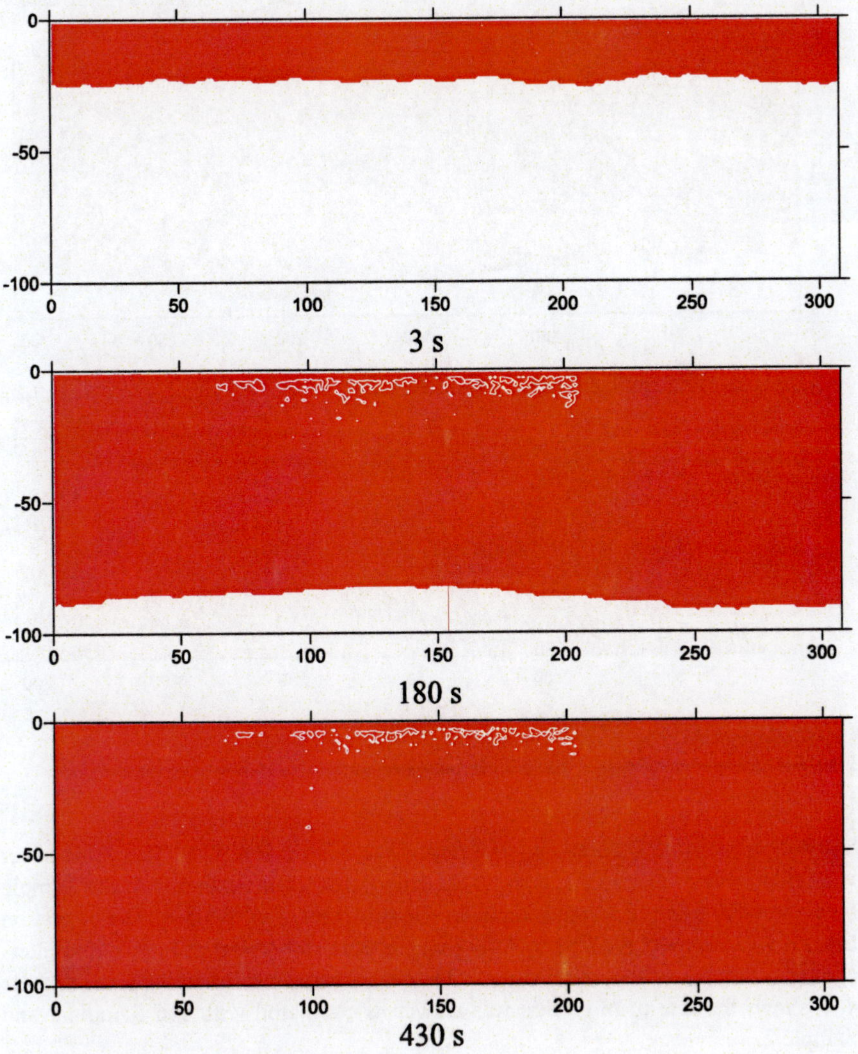

Fig. 16.5 Sample 1; 25% water content without vibration

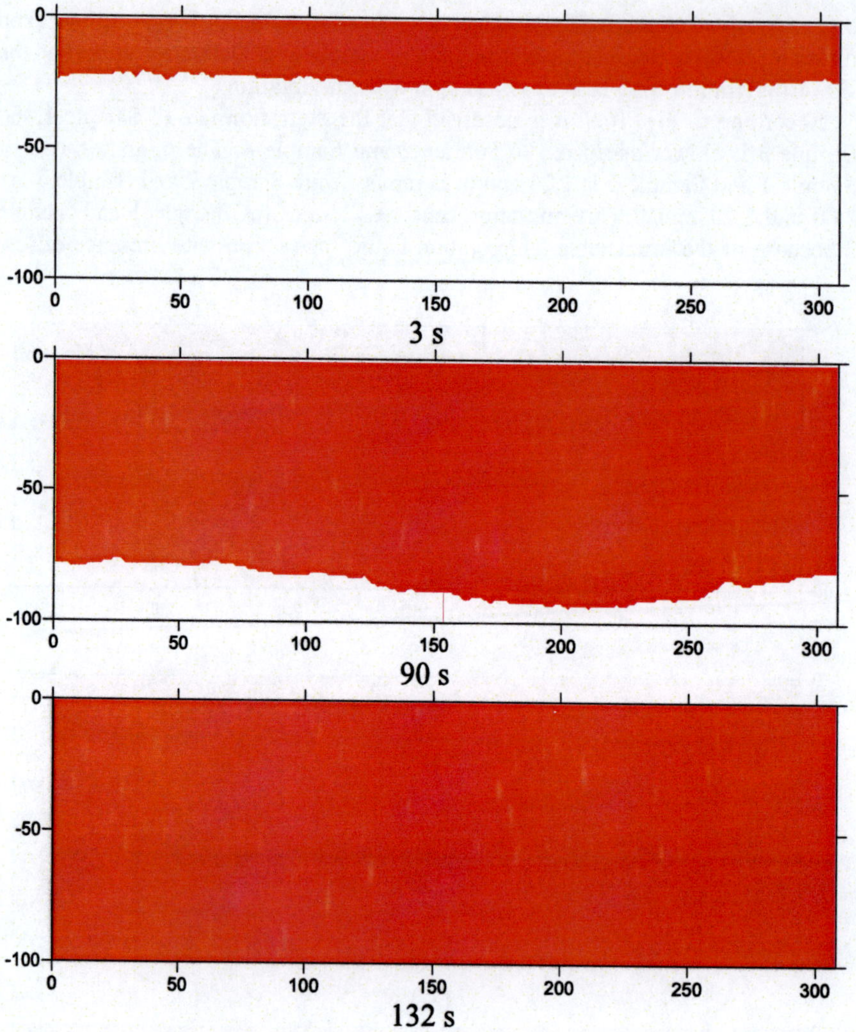

Fig. 16.6 Sample 2; 30% water content without vibration

moisture content 25% compared to Sample 2 and Sample 4 30%. The voids of air were filled with water, so it speeds up the rate of migration in Sample 2 and Sample 4 compared to Sample 1 and Sample 3. Vibration on soil also affects the migration time; it will compact the soil by filling up the pore in between od intra- and inter-aggregate that act as double-porosity soil as denser soil; on soil Sample 3 and Sample 4, it shows the rate of migration was slower as compared with soil Sample 1 and Sample 2.

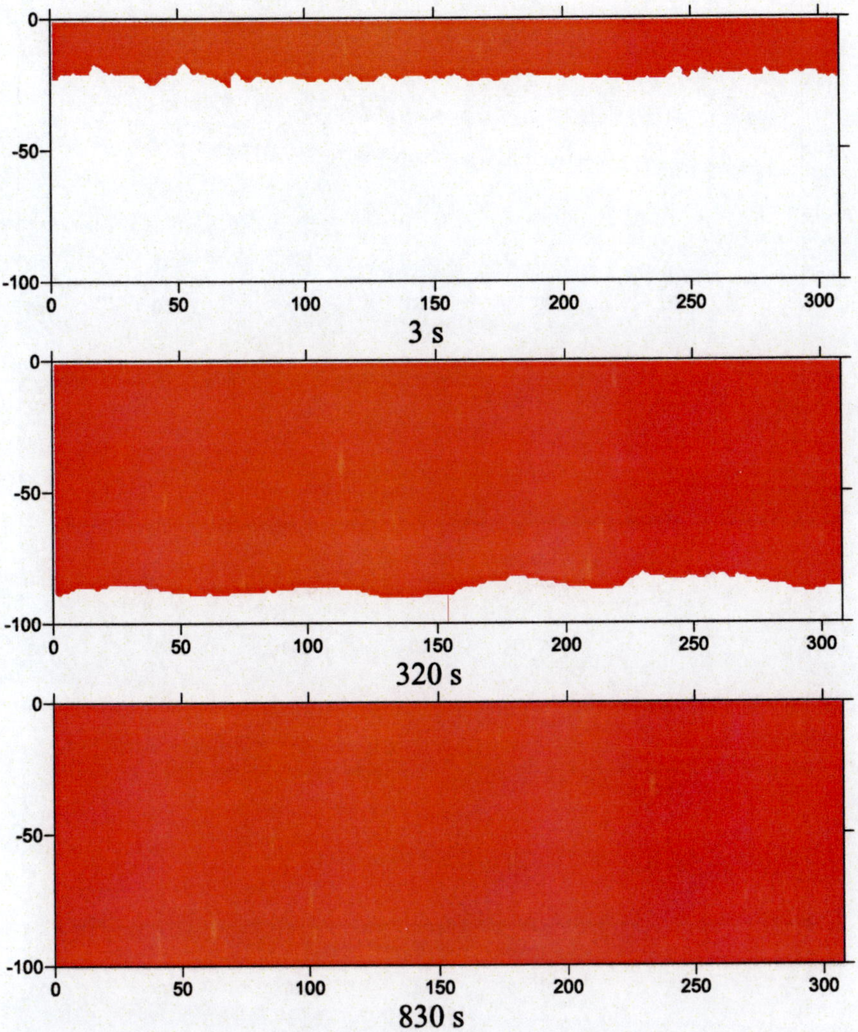

Fig. 16.7 Sample 3; 25% water content with vibration

16.5 Conclusion

Laboratory experiments on red-dyed oil migration through an aggregated laterite soil with 25 and 30% water content with and without vibration were conducted. The experiment study was deliberately designed to observe the red-dyed oil migration behavior and pattern in aggregated laterite soil placed in a soil column model. It applied digital processing technique using Surfer software version 10 and Matlab routine to analyze and extract HSI data obtained from the digital images. In conclusion, the observation of uniform downward migration experiment demonstrated that

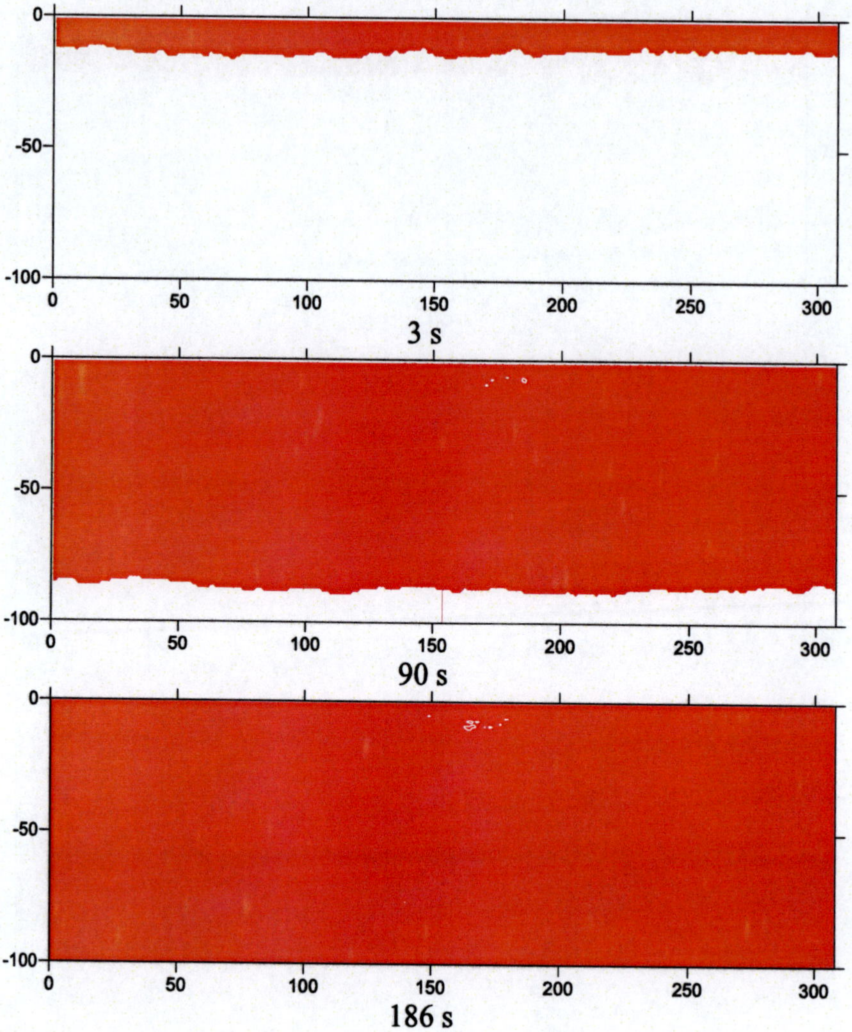

Fig. 16.8 Sample 4; 30% water content with vibration

the rate of dyed oil migration is faster in soil sample with 30% moisture content compared to 25% moisture content for without vibration impact. In contrast, the rate of dyed oil migration for vibrated samples was slower compared to without vibration samples. It was proven that presence of water content and soil structure of the soil sample has an influence on the migration behavior. It shows that the different moisture contents for double-porosity laterite soil will affect the migration speed rate, and with presence of vibration, it slower the migration is found to be valid. HSI intensity value contour plot of red-dyed oil migration behavior could explain details and practical information to facilitate current and future researchers on studying the

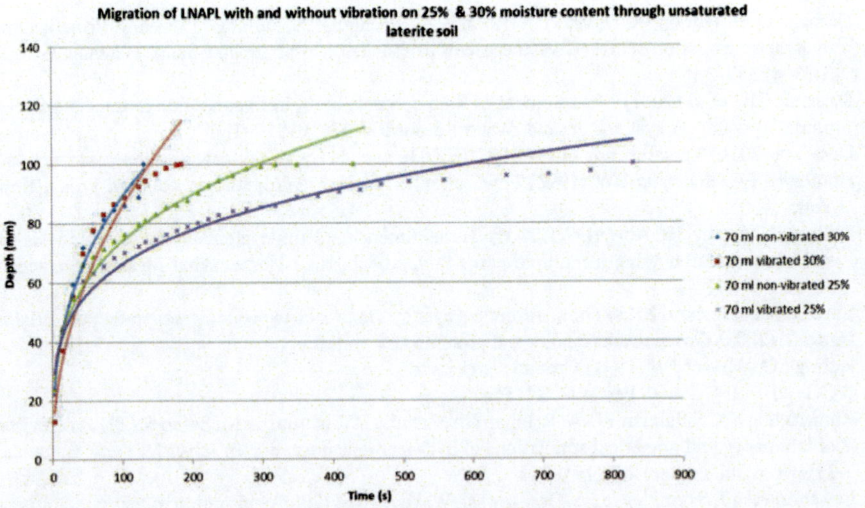

Fig. 16.9 LNAPL migration graph varies with different moisture content and with/without vibration

migration pattern of red-dyed oil using digital camera image camera close-range photogrammetry technique.

Acknowledgements This study was supported by the Research Management Center (RMC), Universiti Teknologi Malaysia under the Research University Grant—Tier 1 vote number Q.J130000.2522.20H53 and the Fundamental Research Grant under MOHE vote number R.J130000.7822.4F894. The authors would like to thank their respective Universities, Geotechnical Laboratory, Hydraulic Laboratory and Engineering Seismology and Earthquake Engineering Research Group (eSEER), School of Civil, Faculty of Engineering, Universiti Teknologi Malaysia for the cooperation given for this research. Highest appreciation is also expressed toward the Kings' Scholarship and UTMLead for supporting this research study and all whom were involved either directly or indirectly.

References

Abd Rashid MF et al (2018) Migration of light non-aqueous phase liquid in porous media. In: 4th International conference on WATER RESOURCES (ICWR 2018), no 8, pp 675–683

Agaoglu B, Copty NK, Scheytt T, Hinkelmann R (2015) Interphase mass transfer between fluids in subsurface formations: a review. Adv Water Resour 79:162–194

Alazaiza MYD, Ngien SK, Bob MM, Kamaruddin SA, Ishak WMF (2017) Assessment of the behaviour of soil structure in double-porosity kaolin media using light transmission visualization (LTV) method. Int J Geotech Eng 11(3):316–320

Bagherieh AR, Khalili N, Habibagahi G, Ghahramani A (2009) Drying response and effective stress in a double porosity aggregated soil. Eng Geol 105(1–2):44–50

Carminati A, Kaestner A, Lehmann P, Flühler H (2008) Unsaturated water flow across soil aggregate contacts. Adv Water Resour 31(9):1221–1232

Darnault CJG, Throop JA, Dicarlo DA, Rimmer A, Steenhuis TS, Parlange JY (1998) Visualization by light transmission of oil and water contents in transient two-phase flow fields. J Contam Hydrol 31(3–4):337–348

Darnault CJG et al (2001) Measurement of fluid contents by light transmission in transient three-phase oil-water-air systems in sand. Water Resour Res 37(7):1859–1868

Desa UN (2016) Transforming our world: The 2030 agenda for sustainable development

Domenico PA, Schwartz FW (1998) Physical and chemical hydrogeology, vol 506. Wiley, New York

El-Zein A, Carter JP, Airey DW (2006) Three-dimensional finite elements for the analysis of soil contamination using a multiple-porosity approach. Int J Numer Anal Methods Geomech 30(7):577–597

Environment Agency (2018) The environment agency's approach to groundwater protection. Bristol

Harter T (2003) Groundwater quality and groundwater pollution

Huling SG, Weaver JW (1991) Ground water issue

IPCC (2016) IPCC AR5 Part A WG II Final

Kamaruddin SA, Sulaiman WNA, Rahman NA, Zakaria MP, Mustaffar M, Sa'ari R (2011) A review of laboratory and numerical simulations of hydrocarbons migration in subsurface environments. J Environ Sci Technol 4(3):191–214

Lewandowska J, Szymkiewicz A, Gorczewska W, Vauclin M (2005) Infiltration in a double-porosity medium: experiments and comparison with a theoretical model. Water Resour Res 41(2):1–14

Li X, Zhang LM (2009) Characterization of dual-structure pore-size distribution of soil. Can Geotech J 46(2):129–141

Loke KF (2019) Migration of aqueous and non-aqueous phase liquids in deformable double-porosity subsurface under vibrated effect (Doctor Philosophy Thesis). Universiti Teknologi Malaysia

Loke KF, Abd Rahman N, Nazir R, Lewis RW (2018) Study of aqueous and non-aqueous phase liquid in fractured double-porosity soil using digital image processing. Geol Croat 71(2):55–63

Maslia ML et al (2016) Reconstructing historical VOC concentrations in drinking water for epidemiological studies at a US military base: summary of results. Water 8(10):449

Ngien SK (2012) Experimental and numerical analysis of non-aqueous phase liquids migration in double-porosity subsurface systems (Doctor of Philosophy Thesis). Universiti Teknologi Malaysia

Tran Ngoc TD, Lewandowska J, Bertin H (2014) Experimental evidence of the double-porosity effects in geomaterials. Acta Geophys 62(3):642–655

WWAP (United Nations World Water Assesment Programme) (2015) The United Nations, World Water Development Report 2015: water for a sustainable world. UNESCO, Paris

Part V
Closing Session—Resolution of the 4th Global Summit of GADRI

Chapter 17
Resolution of the 4th Global Summit of Research Institutes for Disaster Risk Reduction: Increasing the Effectiveness and Relevance of Our Institutes Held at the Disaster Prevention Research Institute (DPRI), Kyoto University, Kyoto, Japan from 13 to 15th March 2019

James Mori, Andrew Collins, Hirokazu Tatano, and Wilma James

The member institutes of Global Alliance of Disaster Research Institutes (GADRI) fully support the United Nations Office for Disaster Risk Reduction (UNISDR) Sendai Framework for Disaster Risk Reduction 2015–2030. Recognizing the urgency of meeting short-term goals, to support implementation of this plan, as described in the Science and Technology Road Map, GADRI advocates.

- Evidence-informed technical reporting for topics in interdisciplinary disaster risk reduction sciences, which currently include natural, engineering, social, and health sciences.
- Development of effective interfaces between technical information and policy making.
- Pro-active implementation of policies that will reduce human, economic, environmental, and cultural losses from disasters.

J. Mori · H. Tatano · W. James (✉)
Disaster Prevention Research Institute (DPRI), Kyoto University, Kyoto, Japan
e-mail: james.wilmatheonesta.6r@kyoto-u.ac.jp

J. Mori
URL: https://www.dpri.kyoto-u.ac.jp/en/

A. Collins
Disaster and Development Network, Northumbria University Newcastle, Newcastle-upon-Tyne, UK
URL: https://www.northumbria.ac.uk/about-us/academic-departments/geography-and-environmental-sciences/research/disaster-anddevelopment-network-ddn/

H. Tatano and A. Collins (eds.), *Proceedings of the 4th Global Summit of Research Institutes for Disaster Risk Reduction*, Disaster and Risk Research: GADRI Book Series,
https://doi.org/10.1007/978-981-19-5566-2_17

GADRI promotes changing roles of science and reflective practices of implementation that will contribute to the effectiveness of disaster risk reduction. This may include development of tools to support a new generation platform.[1]

GADRI will contribute to the[2] actions of the roadmap by providing a common voice and identifying priorities on important issues, such as;

- Contributing to the revision of hazard terminology.
- Contributing to understanding risks and vulnerabilities.
- Syntheses for all types of risks on a country or regional level, including examples of successes and contributions of local knowledge, to expand sponsor and user groups (e.g., DRRKC, Nations Synthesis Framework).[3]
- Effective education programs about disaster risk reduction for scientists, government and community officials, media, and the public.
- Improved research and practice on communication and interactions among and between technical experts, policy makers, implementation groups, and stakeholders.
- Development of new methodologies and technologies for data analyses, promotion of action data and encouragement of open data sharing, for research and risk reduction applications.

GADRI institutes will report their achievements at future Global Summits of Research Institutes for Disaster Risk Reduction.

References

GADRI web site. http://gadri.net/

Sendai Framework Agenda 2015–2030. https://www.undrr.org/publication/sendai-framework-disaster-risk-reduction-2015-2030

Nations Synthesis Framework. https://www.scj.go.jp/ja/member/iinkai/kagakubosai/pdf/kagakubosai-siryo2404-4a.pdf

[1] GADRI Web site. http://gadri.net/.

[2] Sendai Framework Agenda 2015–2030. https://www.undrr.org/publication/sendai-framework-disaster-risk-reduction-2015-2030.

[3] Nations Synthesis Framework. https://www.scj.go.jp/ja/member/iinkai/kagakubosai/pdf/kagakubosai-siryo2404-4a.pdf.

Part VI
Closing Ceremony

Chapter 18
Closing Ceremony

Wilma James

Greetings by Mr. Daisaku Kadokawa, Kyoto City Mayor, Kyoto, Japan at the Closing Ceremony (Fig. 1).

Mr. Daisaku Kadokawa opened his greetings by stating that he understands that the Global Alliance of Disaster Research Institutes (GADRI) is a research network that aims to contribute to the reduction of disaster risk and the improvement of disaster resilience from an academic point of view by collaborating with disaster research and disaster prevention research institutes around the world through sharing information, knowledge, experience, and ideas. It is admirable to note that to date the GADRI has grown to include approximately 150 disaster prevention expert institutes from 44 countries and regions around the world. The summit brought together these academic research institutes to discuss how research institutes in each country or in region, can contribute to reduce disaster risk and would have developed lively discussions on the agenda. Congratulations on the successful completion of the three-day conference.

It is interesting to note the participating institutions of GADRI and the Global Summit series continue to increase in terms of members and participants. This is a noteworthy highlight of the result of the efforts invested by everyone and especially Prof. Hirokazu Tatano, the Secretary-General, GADRI and the Professor at the Disaster Prevention Research Institutes (DPRI), Kyoto University. He expressed his deepest respect to Prof. Tatano and all related members for their efforts.

At this summit has introduced very important agendas to advance initiatives already in place, and these were preceded by lively discussions. We pray that the results of this summit will be put to good use in disaster prevention around the world.

Last year, the Kyoto City was hit by a series of natural disasters such as a large earthquake, heavy rain, and three typhoons, causing great damage throughout the

W. James (✉)
Kyoto University, Gokasho, Uji-shi, Kyoto 611-0011, Japan
e-mail: james.wilmatheonesta.6r@kyoto-u.ac.jp

H. Tatano and A. Collins (eds.), *Proceedings of the 4th Global Summit of Research Institutes for Disaster Risk Reduction*, Disaster and Risk Research: GADRI Book Series,
https://doi.org/10.1007/978-981-19-5566-2_18

Fig. 1 Mr. Daisaku Kadokawa, Kyoto City Mayor, Kyoto

city. The city is still doing its best to recover and reconstruct. During this time, as we have learned from various disasters, we have been steadily working on disaster prevention and mitigation with a sense of speed. It is a gift of Kyoto's regional and human power, as displayed by the voluntary disaster prevention associations, fire brigade, flood control teams, residents' associations, and social welfare council.

Kyoto has a history of more than a thousand years since the founding of Heian. It is a history that has refined and enhanced over the years with the ability to respond to and overcome all crises and that is, the "resilience" of the city. Kyoto City was selected as one of the Rockefeller Foundation's "100 Resilient Cities" three years ago. This month, Kyoto City will formulate the "Kyoto City Resilience Strategy," which will serve as a guideline for efforts to respond flexibly to all crises, including disasters, and to become a stronger and a more sustainable city in the future.

"Resilience" in line with the "SDGs", which covers 17 goals set by the United Nations, is based on the idea of "no one is left behind." Recently, in the advanced survey of "SDGs" conducted by Nihon Keizai Shimbun, Kyoto City was evaluated as No. 1 among 815 local governments nationwide. This is, of course, high praise for the result of the efforts by citizens and businesses.

By positioning the principles of "resilience" and "SDGs" as the philosophy of urban management of the city, Kyoto City will continue to work toward the realization of "resilient city" which will be a model city of the world.

Kyoto City will continue to contribute to disaster countermeasures around the world, aiming to become a world-class safe and secure city in collaboration with domestic and overseas specialized institutions, advanced cities, and businesses, including the Kyoto University, the Disaster Prevention Research Institute (DPRI),

Kyoto University and the Global Disaster Prevention Research Institutes (GADRI. We aspire to achieve that goal and we appreciate your understanding and cooperation.

We wish all the very best to the further development of the Global Disaster Prevention Research Institutes (GADRI), and wish you all good health and prosperity to all the participants.

Closing Remarks by Prof. Manabu Hashimoto

Distinguished guests, ladies and gentlemen,

It has been quite inspiring to see that all of you safely sailed through the ambitious and compact programme of the 4th Global Summit.

I would like to thank all of our keynote speakers and panelists as well as all of you active participants, for doing an outstanding job by sharing with us your valuable knowledge, insights, expertise and experiences. Without doubt and without your contributions, the 4th Global Summit could not have achieved its set goals. I learned that there are over 240 participants representing 107 institutions from 33 states.

Thank you also for our wrap-up team for excellently capturing the essence of the discussions and translating them to recommendations. Much work is ahead of us as we prepare to share the outcomes with the Global Platform for DRR in Geneva in May.

As we have learned through discussions and examples during the past three-days, there is a vast knowledge, expertise, and experience to draw from to improve disaster resilience and management. Although it may take a day or two to consolidate and absorb all that we have learned and shared, we hope to use this momentum to collectively contribute individually and as a team to achieving the goals set out in the SFDRR 2030 Agenda, 2016 SCIENCE AND TECHNOLOGY Roadmap and SDGs.

We also continue to be affected by climate change and disasters, and dealing with the consequences calls for our commitments to share our knowledge and experiences. As we resolutely move forward as a team, let us use this stimulus to improve disaster risk management on a regional and a global scale. We cannot prevent disasters but we can contribute to disaster risk reduction and resilience to disasters.

On behalf of all of us, once again, I would like to thank all of our distinguished keynote speakers for insightful speeches, the panelists, GADRI members of the Board of Directors and the Advisory Board, and all of you participants for your invaluable presence and contributions.

I would also like to thank all of our staff and especially those who have been directly involved in organizing this event.

Since the Global Summit is a biennial event, we look forward to seeing you all at EC-JRC, Italy in 2021.

Congratulations to all of you for successfully concluding the 4th Global Summit (Fig. 2).

Message by Prof. Kayo Inaba, Executive Vice-President, Gender Equality, International Affairs, and Public Relations Kyoto University

Good evening!

Fig. 2 Center—Prof. Kayo Inaba, and Prof. Manabu Hashimoto (center-right)

Distinguished guests, Ladies and Gentlemen!

First of all I would like to congratulate all of you for successfully concluding the 4th Global Summit of Research Institutes for Disaster Risk Reduction (4thGSRIDRR2019).

I was quite amazed with the rigorous programme and the outstanding array of keynote speakers.

As the Vice President for Gender Equality, International Affairs and Public Relations of the Kyoto University, I am truly amazed to see many female keynote speakers and participants too, contributing to this important event.

Quite impressive!

As for the programme, what a three-day marathon!

Congratulations to all of you for the successful closure of the 4th Global Summit.

All of you have done an exceptional job in sharing your valuable time, knowledge, insights and experiences on how to contribute further, to effectively build disaster risk reduction and resilience in our communities. I understand from Prof. Tatano that one of the goals of the 4th Global Summit was to present the outcomes of this conference at the Global Platform for DRR in Geneva in May 2019 and I thank you very much for your contributions.

I am personally convinced that if we continue to stay committed to work towards the goals set out in the 4th Global Summit resolution document and Sendai Framework Agenda for 2030, 2016 SCIENCE AND TECHNOLOGY Roadmap, and SDGs, we can, together, achieve DRR and disaster resilience in the world, and to ensure that our voices are being heard and DRR is integrated and implemented by various stakeholders, especially in governments and industries.

We hope to strengthen our approaches and share our experiences, and lessons learned with each other, to educate our communities, especially vulnerable communities on disaster risk reduction and risk management. As we all know, disasters continue to increase. We need stronger collaboration, like what we have seen during the past three days and resolutely move forward to ensure that they are placed high above in international agendas to support these goals.

Now with these words, let me, once again, express my sincere thanks to all of you for your valuable contributions and for making the 4th Global Summit a success.

Now enjoy this beautiful banquet.

****

The Calligraphy Performance and message shared by the Students of Todoh High School Calligraphy Club

Hello, everyone!

Thank you very much for giving us such a great opportunity.

We are the members of Todoh High School's calligraphy club.

Now, I will talk about our messages in this work.

At the upper left, we drew;

「京に彩づく文化の風 繋げよう 未来へ」

Kyoto is a very cultural city. We had like to learn about disaster risk reduction and preserve this peaceful society for future generations.

Actually, our school was hit and heavily damaged by a local heavy rain several years ago. We think we should tell the next generations about our experience.

In the middle;

「じしょうそくりつ」

This old Chinese word means that preparation is important in everything.

We wanted to show that it is important to prepare for disasters to reduce the damages.

And finally, in the words at the lower right, we express our ideas about disaster reduction, which means we need to correctly understand how to reduce disaster risks and find out a practical solution. And the most important thing is our strong bonds and cooperation.

We hope that fewer and fewer people will suffer from damages caused by natural disasters.

Thank you (Fig. 3).

.

Fig. 3 Calligraphy Performance by Todoh High School Calligraphy Club members

Final Art work of the calligraphy performance